An introduction to computational fluid dynamics
The finite volume method

An introduction to computational fluid dynamics
The finite volume method

H. K. VERSTEEG and W. MALALASEKERA

Harlow, England • London • New York • Boston • San Francisco • Toronto • Sydney • Singapore • Hong Kong
Tokyo • Seoul • Taipei • New Delhi • Cape Town • Madrid • Mexico City • Amsterdam • Munich • Paris • Milan

Pearson Education Limited
Edinburgh Gate
Harlow
Essex CM20 2JE
England

and Associated Companies throughout the world

Visit us on the World Wide Web at:
http://www.pearsoneduc.com

First published 1995

British Library Cataloguing in Publication Data
A catalogue entry for this title is available from the British Library.

ISBN 0-582-21884-5

Library of Congress Cataloguing-in-Publication Data
A catalog entry for this title is available from the Library of Congress.

Typeset by 21 in 10/12 Times

10 9 8
07 06 05 04 03

Printed in Malaysia, TCP

Contents

Preface

The use of computational fluid dynamics (CFD) to predict internal and external flows has risen dramatically in the past decade. In the 1980s the solution of fluid flow problems by means of CFD was the domain of the academic, postdoctoral or postgraduate researcher or the similarity trained specialist with many years of grounding in the area. The widespread availability of engineering workstations together with efficient solution algorithms and sophisticated pre- and post-processing facilities enable the use of commercial CFD codes by graduate engineers for research, development and design tasks in industry. The codes that are now on the market may be extremely powerful, but their operation still requires a high level of skill and understanding from the operator to obtain meaningful results in complex situations. The long learning curve, previously including apprenticeships of up to four years – more widely known as MPhil and PhD studies – meant that the users of the 1980s were, through their own experiences, very conscious of the limitations of CFD. However, the pressure on engineers in industry to come up with solutions to problems implies that there is not always the time available for the new type of user of the 1990s to learn about the pitfalls of CFD by osmosis and frequent failure.

It is the purpose of this book to fill a gap in the available literature for novice CFD users who, whilst developing CFD skills by using commercially available software, need a reader that provides the fundamentals of the fluid dynamics behind complex engineering flows and of the numerical solution algorithms on which the CFD codes are based. Although the material has been developed from first principles wherever possible, the book will be of greatest benefit to those who are familiar with the ideas of calculus, elementary vector and matrix algebra and basic numerical methods. Furthermore, we assume a knowledge of the conservation laws for mass, momentum and energy and an awareness of their application to fluid flow problems.

Although commercial CFD codes based on the finite element method have more recently entered the fray, the market is currently dominated by four codes, PHOENICS, FLUENT, FLOW3D and STAR-CD, that are all based on the finite volume method. This book intends to provide the theoretical background required for the effective use of this type of commercial code and covers the following subject areas:

Fluid dynamics

- Governing equations of viscous fluid flows
- Boundary conditions
- Introduction to the physics of turbulence
- Turbulence modelling in CFD

The finite volume method and its implementation in CFD codes

- Finite volume discretisation for the key transport phenomena in fluid flows: diffusion, convection and sources
- Discretisation procedures for unsteady phenomena
- Iterative solution processes (SIMPLE and its derivatives) to ensure correct coupling between all the flow variables
- Solution algorithms for systems of discretised equations (TDMA)
- Implementation of boundary conditions

The basic numerical techniques have been developed around a series of worked examples, which can be easily programmed on a PC. However, it is impossible to get to grips with the art of CFD without running a good quality code to explore the issues raised in this book in greater detail. As an illustration of the power of CFD we have presented a set of industrially relevant applications ranging from a benchmark simulation to very complex fire modelling. Throughout, one of the key messages is that CFD cannot be professed adequately without continued reference to experimental validation. The early ideas of the computational laboratory to supersede experimentation have fortunately gone out of fashion. Not all industrial companies have the high cost experimental infrastructure in place to support CFD activities, but the scientific literature contains a huge resource to the user of commercial codes. A vast and ever-increasing number of journals cover all aspects of CFD ranging from mathematically abstruse to applied work firmly rooted in industry. In addition to the necessary theoretical grounding the book, therefore, provides a set of connection points with up-to-date research literature giving the reader access to source material for code validation and further study.

After starting to teach CFD at senior undergraduate level we became acutely aware of the absence of a 'suitable' text pitched at 'the right level'. Undeniably, this book, which was developed from our course notes, was conceived with our own students as a target audience so, first and foremost, we hope that the book will be valuable as a learning and teaching resource to support CFD courses at undergraduate and postgraduate level. Nevertheless, with its intent to bridge the gap between introductory mathematics and fluid dynamics concepts, the academic CFD literature and applied industrial practice, we believe that this book will also be of use to professional engineers in industry, involved in R&D and design, who require a thorough but user-friendly reference guide to all the background knowledge needed to operate commercial CFD codes successfully.

We acknowledge Dr. S. Sivasegaram of Imperial College of Science Technology and Medicine, and Mr. R. K. Turton of Loughborough University for helpful comments on early drafts of this book. We are grateful to our wives, Helen and Anoma, for all the support and encouragement given to us during the compilation of this book.

March 1995
Loughborough

H. K. Versteeg
W. Malalasekera

Acknowledgements

The authors wish to acknowledge the following persons, organisations and publishers for permission to reproduce from their publications in this book. Professor H. Nagib for fig. 3.2; Professor S. Taneda and the Japan Society of Mechanical Engineers for fig. 3.7; Professor W. Fiszdon and the Polish Academy of Sciences for fig. 3.9; McGraw-Hill Inc. for fig. 3.13; The Combustion Institute for figs. 10.2 and 10.3; and Gordon and Breach Science Publishers for fig. 10.4. While every effort has been made to trace the owners of copyright, in some cases this has proved impossible and we would like to apologise to anyone whose rights we may have unwittingly infringed. All registered trade marks of CFD codes mentioned in this text are also acknowledged.

1

Introduction

1.1 What is CFD?

Computational Fluid Dynamics or CFD is the analysis of systems involving fluid flow, heat transfer and associated phenomena such as chemical reactions by means of computer-based simulation. The technique is very powerful and spans a wide range of industrial and non-industrial application areas. Some examples are:

- aerodynamics of aircraft and vehicles: lift and drag
- hydrodynamics of ships
- power plant: combustion in IC engines and gas turbines
- turbomachinery: flows inside rotating passages, diffusers etc.
- electrical and electronic engineering: cooling of equipment including micro-circuits
- chemical process engineering: mixing and separation, polymer moulding
- external and internal environment of buildings: wind loading and heating/ventilation
- marine engineering: loads on off-shore structures
- environmental engineering: distribution of pollutants and effluents
- hydrology and oceanography: flows in rivers, estuaries, oceans
- meteorology: weather prediction
- biomedical engineering: blood flows through arteries and veins

From the 1960s onwards the aerospace industry has integrated CFD techniques into the design, R&D and manufacture of aircraft and jet engines. More recently the methods have been applied to the design of internal combustion engines, combustion chambers of gas turbines and furnaces. Furthermore, motor vehicle manufacturers now routinely predict drag forces, under-bonnet air flows and the in-car environment with CFD. Increasingly CFD is becoming a vital component in the design of industrial products and processes.

The ultimate aim of developments in the CFD field is to provide a capability comparable to other CAE (Computer-Aided Engineering) tools such as stress analysis codes. The main reason why CFD has lagged behind is the tremendous complexity of the underlying behaviour, which precludes a description of fluid flows that is at the same time economical and sufficiently complete. The availability of affordable high performance computing hardware and the introduction of user-friendly interfaces have led to a recent upsurge of interest and CFD is poised to make an entry into the wider industrial community in the 1990s.

We estimate the minimum cost of suitable hardware to be between £5000 and £10000 (plus annual maintenance costs). The perpetual licence fee for commercial software typically ranges from £10000 to £50000 depending on the number of 'added extras' required. CFD software houses can usually arrange annual licences as an alternative. Clearly the investment costs of a CFD capability are not small, but the total expense is not normally as great as that of a high quality experimental facility. Moreover, there are several unique advantages of CFD over experiment-based approaches to fluid systems design:

- substantial reduction of lead times and costs of new designs
- ability to study systems where controlled experiments are difficult or impossible to perform (e.g. very large systems)
- ability to study systems under hazardous conditions at and beyond their normal performance limits (e.g. safety studies and accident scenarios)
- practically unlimited level of detail of results

The variable cost of an experiment, in terms of facility hire and/or man-hour costs, is proportional to the number of data points and the number of configurations tested. In contrast CFD codes can produce extremely large volumes of results at virtually no added expense and it is very cheap to perform parametric studies, for instance to optimise equipment performance.

We also note that, in addition to a substantial investment outlay, an organisation needs qualified people to run the codes and communicate their results and briefly consider the modelling skills required by CFD users. We complete this otherwise upbeat section by wondering whether the next constraint to the further spread of CFD amongst the industrial community could be a scarcity of suitably trained personnel instead of availability and/or cost of hardware and software.

1.2 How does a CFD code work?

CFD codes are structured around the numerical algorithms that can tackle fluid flow problems. In order to provide easy access to their solving power all commercial CFD packages include sophisticated user interfaces to input problem parameters and to examine the results. Hence all codes contain three main elements: (i) a pre-processor, (ii) a solver and (iii) a post-processor. We briefly examine the function of each of these elements within the context of a CFD code.

Pre-processor

Pre-processing consists of the input of a flow problem to a CFD program by means of an operator-friendly interface and the subsequent transformation of this input into a form suitable for use by the solver. The user activities at the pre-processing stage involve:

- Definition of the geometry of the region of interest: the computational *domain*.
- Grid generation–the sub-division of the domain into a number of smaller, non-overlapping sub-domains: a *grid* (or *mesh*) of *cells* (or *control volumes* or elements).
- Selection of the physical and chemical phenomena that need to be modelled.
- Definition of fluid properties.
- Specification of appropriate boundary conditions at cells which coincide with or touch the domain boundary.

The solution to a flow problem (velocity, pressure, temperature etc.) is defined at *nodes* inside each cell. The accuracy of a CFD solution is governed by the number of cells in the grid. In general, the larger the number of cells the better the solution accuracy. Both the accuracy of a solution and its cost in terms of necessary computer hardware and calculation time are dependent on the fineness of the grid. Optimal meshes are often non-uniform: finer in areas where large variations occur from point to point and coarser in regions with relatively little change. Efforts are under way to develop CFD codes with a (self-)adaptive meshing capability. Ultimately such programs will automatically refine the grid in areas of rapid variations. A substantial amount of basic development work still needs to be done before these techniques are robust enough to be incorporated into commercial CFD codes. At present it is still up to the skills of the CFD user to design a grid that is a suitable compromise between desired accuracy and solution cost.

Over 50% of the time spent in industry on a CFD project is devoted to the definition of the domain geometry and grid generation. In order to maximise productivity of CFD personnel all the major codes now include their own CAD-style interface and/or facilities to import data from proprietary surface modellers and mesh generators such as PATRAN and I-DEAS. Up-to-date pre-processors also give the user access to libraries of material properties for common fluids and a facility to invoke special physical and chemical process models (e.g. turbulence models, radiative heat transfer, combustion models) alongside the main fluid flow equations.

Solver

There are three distinct streams of numerical solution techniques: finite difference, finite element and spectral methods. In outline the numerical methods that form the basis of the solver perform the following steps:

- Approximation of the unknown flow variables by means of simple functions.
- Discretisation by substitution of the approximations into the governing flow equations and subsequent mathematical manipulations.
- Solution of the algebraic equations.

The main differences between the three separate streams are associated with the way in which the flow variables are approximated and with the discretisation processes. *Finite difference methods.* Finite difference methods describe the unknowns ϕ of the flow problem by means of point samples at the node points of a grid of co-ordinate lines. Truncated Taylor series expansions are often used to generate finite difference approximations of derivatives of ϕ in terms of point samples of ϕ at each grid point and its immediate neighbours. Those derivatives appearing in the governing equations are replaced by finite differences yielding an algebraic equation for the values of ϕ at each grid point. Smith (1985) gives a comprehensive account of all aspects of the finite difference method.

Finite Element Method. Finite element methods use simple piecewise functions (e.g. linear or quadratic) valid on elements to describe the local variations of unknown flow variables ϕ. The governing equation is precisely satisfied by the exact solution ϕ. If the piecewise approximating functions for ϕ are substituted into the equation it will not hold exactly and a residual is defined to measure the errors. Next the residuals (and hence the errors) are minimised in some sense by multiplying them by a set of weighting functions and integrating. As a result we obtain a set of algebraic equations for the unknown coefficients of the approximating functions. The theory of finite elements has been developed initially for structural stress analysis. A standard work for fluids applications is Zienkiewicz and Taylor (1991).

Spectral Methods. Spectral methods approximate the unknowns by means of truncated Fourier series or series of Chebyshev polynomials. Unlike the finite difference or finite element approach the approximations are not local but valid throughout the entire computational domain. Again we replace the unknowns in the governing equation by the truncated series. The constraint that leads to the algebraic equations for the coefficients of the Fourier or Chebyshev series is provided by a weighted residuals concept similar to the finite element method or by making the approximate function coincide with the exact solution at a number of grid points. Further information on this specialised method can be found in Gottlieb and Orszag (1977).

The finite volume method. The finite volume method was originally developed as a special finite difference formulation. This book shall be solely concerned with this most well-established and thoroughly validated general purpose CFD technique. It is central to four of the five main commercially available CFD codes: PHOENICS, FLUENT, FLOW3D and STAR-CD. The numerical algorithm consists of the following steps:

- Formal integration of the governing equations of fluid flow over all the (finite) control volumes of the solution domain.
- Discretisation involves the substitution of a variety of finite-difference-type approximations for the terms in the integrated equation representing flow processes such as convection, diffusion and sources. This converts the integral equations into a system of algebraic equations.
- Solution of the algebraic equations by an iterative method.

The first step, the control volume integration, distinguishes the finite volume method from all other CFD techniques. The resulting statements express the (exact) conservation of relevant properties for each finite size cell. This clear relationship between the numerical algorithm and the underlying physical conservation principle forms one of the main attractions of the finite volume method and makes its concepts much more simple to understand by engineers than finite element and spectral methods. The conservation of a general flow variable ϕ, for example a velocity component or enthalpy, within a finite control volume can be expressed as a balance between the various processes tending to increase or decrease it. In words we have:

$$\begin{bmatrix}\text{Rate of change}\\ \text{of } \phi \text{ in the the control}\\ \text{volume with}\\ \text{respect to time}\end{bmatrix} = \begin{bmatrix}\text{Net flux of}\\ \phi \text{ due to}\\ \text{convection into}\\ \text{the control volume}\end{bmatrix} + \begin{bmatrix}\text{Net flux of}\\ \phi \text{ due to}\\ \text{diffusion into the}\\ \text{control volume}\end{bmatrix} + \begin{bmatrix}\text{Net rate of creation}\\ \text{of } \phi \text{ inside the}\\ \text{control volume}\end{bmatrix}$$

CFD codes contain discretisation techniques suitable for the treatment of the key transport phenomena, convection (transport due to fluid flow) and diffusion (transport due to variations of ϕ from point to point) as well as for the source terms (associated with the creation or destruction of ϕ) and the rate of change with respect to time. The underlying physical phenomena are complex and non-linear so an iterative solution approach is required. The most popular solution procedures are the TDMA line-by-line solver of the algebraic equations and the SIMPLE algorithm to ensure correct linkage between pressure and velocity. Commercial codes may also give the user a selection of further, more recent, techniques such as Stone's algorithm and conjugate gradient methods.

Post-processor

As in pre-processing a huge amount of development work has recently taken place in the post-processing field. Owing to the increased popularity of engineering workstations, many of which have outstanding graphics capabilities, the leading CFD packages are now equipped with versatile data visualisation tools. These include:

- Domain geometry and grid display
- Vector plots
- Line and shaded contour plots
- 2D and 3D surface plots
- Particle tracking
- View manipulation (translation, rotation, scaling etc.)
- Colour postscript output

More recently these facilities may also include animation for dynamic result display and in addition to graphics all codes produce trusty alphanumeric output and have data export facilities for further manipulation external to the code. As in many other branches of CAE the graphics output capabilities of CFD codes have revolutionised the communication of ideas to the non-specialist.

1.3 Problem solving with CFD

In solving fluid flow problems we need to be aware that the underlying physics is complex and the results generated by a CFD code are at best as good as the physics (and chemistry) embedded in it and at worst as good as its operator. Elaborating on the latter issue first, the user of a code must have skills in a number of areas. Prior to setting up and running a CFD simulation there is a stage of identification and formulation of the flow problem in terms of the physical and chemical phenomena that need to be considered. Typical decisions that might be needed are whether to model a problem in two or three dimensions, to exclude the effects of ambient temperature or pressure variations on the density of an air flow, to choose to solve the turbulent flow equations or to neglect the effects of small air bubbles dissolved in tap water. To make the right choices requires good modelling skills, because in all but the simplest problems we need to make assumptions to reduce the complexity to a manageable level whilst preserving the salient features of the problem in hand. It is the appropriateness of the simplifications introduced at this stage that at least partly governs the quality of the information generated by CFD, so the user must continually stay aware of all the assumptions, clear-cut and tacit ones, that have been made.

A good understanding of the numerical solution algorithm is also crucial. Three mathematical concepts are useful in determining the success or otherwise of such algorithms: convergence, consistency and stability. **Convergence** is the property of a numerical method to produce a solution which approaches the exact solution as the grid spacing, control volume size or element size is reduced to zero. **Consistent** numerical schemes produce systems of algebraic equations which can be demonstrated to be equivalent to the original governing equation as the grid spacing tends to zero. **Stability** is associated with damping of errors as the numerical method proceeds. If a technique is not stable even roundoff errors in the initial data can cause wild oscillations or divergence.

Convergence is usually very difficult to establish theoretically and in practice we use Lax's equivalence theorem which states that for linear problems a necessary and sufficient condition for convergence is that the method is both consistent and stable. In CFD methods this theorem is of limited use since we shall see in Chapter 2 that the governing equations are non-linear. In such problems consistency and stability are necessary conditions for convergence, but not sufficient.

Our inability to prove conclusively that a numerical solution scheme is convergent is perhaps somewhat unsatisfying from a theoretical standpoint, but we need not be too concerned since the process of making the mesh spacing very close to zero is not feasible on computing machines with a finite representation of numbers (eight digits on Real*4). Roundoff errors would swamp the solution long before a grid spacing of zero is actually reached. Engineers need CFD codes that produce physically realistic results with good accuracy in simulations with finite (sometimes quite coarse) grids. Patankar (1980) has formulated rules which yield robust finite volume calculation schemes. These are discussed further in Chapter 5; here we highlight three crucial properties of robust methods: conservativeness, boundedness and transportiveness.

The finite volume approach guarantees local conservation of a fluid property ϕ for each control volume. Numerical schemes which possess the **conservativeness** property also ensure global conservation of the fluid property for the entire domain. This is clearly important physically and is achieved by means of consistent expressions for fluxes of ϕ through the cell faces of adjacent control volumes. The **boundedness** property is akin to stability and requires that in a linear problem without sources the solution is bounded by the maximum and minimum boundary values of the flow variable. Boundedness can be achieved by placing restrictions on the magnitude and signs of the coefficients of the algebraic equations. Although flow problems are non-linear it is important to study the boundedness of a finite volume scheme for closely related, but linear, problems.

Finally all flow processes contain effects due to convection and diffusion. In diffusive phenomena, such as heat conduction, a change of temperature at one location affects the temperature in more or less equal measure in all directions around it. Convective phenomena involve influencing exclusively in the flow direction so that a point only experiences effects due to changes at upstream locations. Finite volume schemes with the **transportiveness** property must account for the directionality of influencing in terms of the relative strength of diffusion to convection.

Conservativeness, boundedness and transportiveness are designed into all finite volume schemes and have been widely shown to lead to successful CFD simulations. Therefore, they are now commonly accepted as alternatives for the more mathematically rigorous concepts of convergence, consistency and stability. Good CFD often involves a delicate balancing act between solution accuracy and stability. The user needs a thorough appraisal of the extent to which conservativeness, boundedness and transportiveness requirements are satisfied by a code.

Performing the actual CFD computation itself requires operator skills of a different kind. Specification of the domain geometry and grid design are the main tasks at the input stage and subsequently the user needs to obtain a successful simulation result. The two aspects that characterise such a result are convergence of the iterative process and grid independence. The solution algorithm is iterative in nature and in a converged solution the so-called residuals – measures of the overall conservation of the flow properties – are very small. Progress towards a converged solution can be greatly assisted by careful selection of the settings of various relaxation factors and acceleration devices. There are no straightforward guidelines for making these choices since they are problem dependent. Optimisation of the solution speed requires considerable experience with the code itself, which can only be acquired by extensive use. There is no formal way of estimating the errors introduced by inadequate grid design for a general flow. Good initial grid design relies largely on an insight into the expected properties of the flow. A background in the fluid dynamics of the particular problem certainly helps and experience with gridding of similar problems is also invaluable. The only way to eliminate errors due to the coarseness of a grid is to perform a grid dependence study, which is a procedure of successive refinement of an initially coarse grid until certain key results do not change. Then the simulation is grid independent. A systematic search for grid-independent results forms an essential part of all high quality CFD studies.

Every numerical algorithm has its own characteristic error patterns. Well-known CFD euphemisms for the word error are terms such as numerical diffusion, false diffusion or even numerical flow. The likely error patterns can only be guessed on the basis of a thorough knowledge of the algorithms. At the end of a simulation the user must make a judgement whether the results are 'good enough'. It is impossible to assess the validity of the models of physics and chemistry embedded in a program as complex as a CFD code or the accuracy of its final results by any means other than comparison with experimental test work. Anyone wishing to use CFD in a serious way must realise that it is no substitute for experimentation, but a very powerful additional problem-solving tool. Validation of a CFD code requires highly detailed information concerning the boundary conditions of a problem and generates a large volume of results. To validate these in a meaningful way it is necessary to produce experimental data of similar scope. This may involve a programme of point flow velocity measurements with hot-wire or laser Doppler anemometry. However, if the environment is too hostile for such delicate laboratory equipment or if it is simply not available, static pressure and temperature measurements complemented by pitot-static tube traverses can also be useful to validate some aspects of a flow field.

Sometimes the facilities to perform experimental work may not (yet) exist in which case the CFD user must rely on (i) previous experience, (ii) comparisons with analytical solutions of similar but simpler flows and (iii) comparisons with high quality data from closely related problems reported in the literature. Excellent sources of the last type of information can be found in *Transactions of the ASME* (in particular the *Journal of Fluids Engineering, Journal of Engineering for Gas Turbines and Power* and *Journal of Heat Transfer*), *AIAA Journal, Journal of Fluid Mechanics* and *Proceedings of the IMechE.*

CFD computation involves the creation of a set of numbers that (hopefully) constitutes a realistic approximation of a real-life system. One of the advantages of CFD is that the user has an almost unlimited choice of the level of detail of the results, but in the prescient words of C. Hastings (1955), written in pre-IT days: 'The purpose of computing is **insight** not numbers'. The underlying message is rightly cautionary. We should make sure that the main outcome of any CFD exercise is improved understanding of the behaviour of a system, but since there are no cast iron guarantees with regard to the accuracy of a simulation we need to validate our results frequently and stringently.

It is clear that there are guidelines for good operating practice which can assist the user of a CFD code and repeated validation plays a key role as the final quality control mechanism. However, the main ingredients for success in CFD are experience and a thorough understanding of the physics of fluid flows and the fundamentals of the numerical algorithms. Without these it is very unlikely that the user gets the best out of a code. It is the intention of this book to provide all the necessary background material for a good understanding of the internal workings of a CFD code and its successful operation.

1.4 Scope of this book

This book seeks to present all the fundamental material needed for a good simulation of fluid flows by means of the finite volume method and is split into two parts. The first part, consisting of Chapters 2 and 3, is concerned with the fundamentals of fluid flows in three dimensions and turbulence. The treatment starts with the derivation of the governing partial differential equations of fluid flows in Cartesian co-ordinates. We stress the commonalties in the resulting conservation equations and arrive at the so-called transport equation which is the basic form for the development of the numerical algorithms that are to follow. Moreover, we look at the auxiliary conditions required to specify a well-posed problem from a general perspective and quote a set of recommended boundary conditions and a number of derived ones that are useful in CFD practice. Chapter 3 represents the development of the concepts of turbulence that are necessary for a full appreciation of the finer details of CFD in many engineering applications. We look at the physics of turbulence and the characteristics of some simple turbulent flows and at the consequences of the appearance of the random fluctuations on the flow equations. The resulting equations are not a closed or solvable set unless we introduce a turbulence model. We discuss the principal turbulence models that are used in industrial CFD, focusing our attention on the k–ε model which is very popular in general purpose flow computations. Some of the more recent developments that are likely to have a major impact on CFD in the near future are also reviewed.

Readers who are already familiar with the derivation of the three-dimensional flow equations can move on to section 2.4. without loss of continuity. Apart from the discussion of the k–ε turbulence model, to which we return later, the material in Chapters 2 and 3 is largely self-contained. This allows the use of this book by those wishing to concentrate principally on the numerical algorithms, but requiring an overview of the fluid dynamics and the mathematics behind it for occasional reference in the same text.

The second part of the book is devoted to the numerical algorithms of the finite volume method and covers the remaining Chapters 4 to 10. Discretisation schemes and solution procedures for steady flows are discussed in Chapters 4 to 7. Chapter 4 describes the basic approach and derives the central difference scheme for diffusion phenomena. In Chapter 5 we emphasise the key properties of discretisation schemes, conservativeness, boundedness and transportiveness, which are used as a basis for the further development of the upwind, hybrid and QUICK schemes for the discretisation of convective terms. The non-linear nature of the underlying flow phenomena and the linkage between pressure and velocity in variable density fluid flows requires special treatment which is the subject of Chapter 6. We introduce the SIMPLE algorithm and some of its more recent derivatives and also discuss the PISO algorithm. In Chapter 7 we describe the TDMA algorithm for the solution of the systems of algebraic equations that appear after the discretisation stage.

The theory behind all the numerical methods is developed around a set of worked examples which can be easily programmed on a PC. This presentation gives the opportunity for a detailed examination of all aspects of the discretisation schemes, which form the basic building blocks of practical CFD codes, including the characteristics of their solutions.

In Chapter 8 we assess the advantages and limitations of various schemes to deal with unsteady flows and Chapter 9 completes the development of the numerical algorithms by considering the practical implementation of the most common boundary conditions in the finite volume method.

The book is primarily aimed at supporting those who have access to a CFD package, so that the issues raised in the text can be explored in greater depth. Readers without access to a commercial CFD package can acquire the renowned TEAM CFD code free of charge from the public domain software bank HENSA. The solution procedures in this book are nevertheless sufficiently well documented for the interested reader to be able to start developing a CFD code from scratch.

In Chapter 10 we discuss ways in which advanced additional features such as models of combustion and buoyancy effects can be incorporated into a CFD code and evaluate the advantages of body-fitted co-ordinate systems. Finally we illustrate the application of the techniques developed in the previous chapters by means of a series of examples ranging from a benchmark test to the very complex subject of fire modelling. These clearly demonstrate the power of the finite volume method when used with appropriate backup of experimental validations.

2

Conservation Laws of Fluid Motion and Boundary Conditions

In this chapter we develop the mathematical basis for a comprehensive general purpose model of fluid flow and heat transfer from the basic principles of conservation of mass, momentum and energy. This leads to the governing equations of fluid flow and a discussion of the necessary auxiliary conditions – initial and boundary conditions. The main issues covered in this context are:

- Derivation of the system of partial differential equations (PDEs) that govern flows in Cartesian (x, y, z) co-ordinates
- Thermodynamic equations of state
- Newtonian model of viscous stresses leading to the Navier–Stokes equations
- Commonalities between the governing PDEs and the definition of the transport equation
- Integrated forms of the transport equation over a finite time interval and a finite control volume
- Classification of physical behaviours into three categories: elliptic, parabolic and hyperbolic
- Appropriate boundary conditions for each category
- Classification of fluid flows
- Auxiliary conditions for viscous fluid flows
- Problems with boundary condition specification in high Reynolds number and high Mach number flows

2.1 Governing equations of fluid flow and heat transfer

The governing equations of fluid flow represent mathematical statements of the **conservation laws of physics**.

- The mass of a fluid is conserved.
- The rate of change of momentum equals the sum of the forces on a fluid particle (Newton's second law).
- The rate of change of energy is equal to the sum of the rate of heat addition to and the rate of work done on a fluid particle (first law of thermodynamics).

The fluid will be regarded as a continuum. For the analysis of fluid flows at macroscopic length scales (say 1 μm and larger) the molecular structure of matter and molecular motions may be ignored. We describe the behaviour of the fluid in terms of macroscopic properties, such as velocity, pressure, density and temperature, and their space and time derivatives. These may be thought of as averages over suitably large numbers of molecules. A fluid particle or point in a fluid is then the smallest possible element of fluid whose macroscopic properties are not influenced by individual molecules.

We consider such a small element of fluid with sides δx, δy and δz (Figure 2.1).

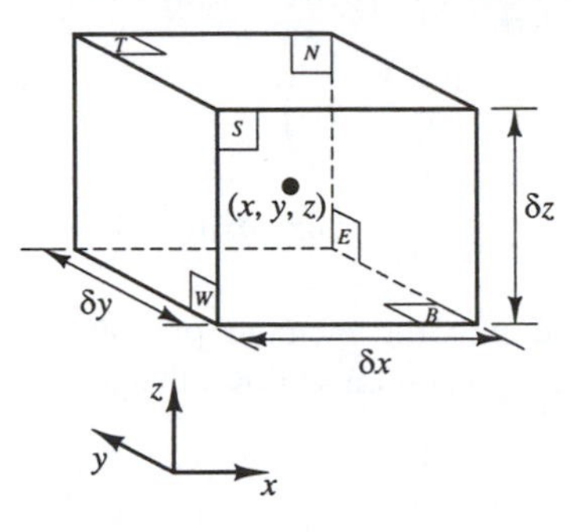

Fig. 2.1 Fluid element for conservation laws

The six faces are labelled *N, S, E, W, T, B* which stands for North, South, East, West, Top and Bottom. The positive directions along the co-ordinate axes are also given. The centre of the element is located at position (x, y, z). A systematic account of changes in the mass, momentum and energy of the fluid element due to fluid flow across its boundaries and, where appropriate, due to the action of sources inside the element, leads to the fluid flow equations.

All fluid properties are functions of space and time so we would strictly need to write $\rho(x, y, z, t)$, $p(x, y, z, t)$, $T(x, y, z, t)$ and $\mathbf{u}(x, y, z, t)$ for the density, pressure, temperature and the velocity vector respectively. To avoid unduly cumbersome notation we will not explicitly state the dependence on space co-ordinates and time. For instance, the density at the centre (x, y, z) of a fluid element at time t is denoted by ρ and the x-derivative of, say, pressure p at (x, y, z) and time t by $\partial p/\partial x$. This practice will also be followed for all other fluid properties.

The element under consideration is so small that fluid properties at the faces can be expressed accurately enough by means of the first two terms of a Taylor series expansion. So, for example, the pressure at the E and W faces, which are both at a distance of $1/2\delta x$ from the element centre, can be expressed as

$$p - \frac{\partial p}{\partial x}\frac{1}{2}\delta x \qquad \text{and} \qquad p + \frac{\partial p}{\partial x}\frac{1}{2}\delta x$$

2.1.1 Mass conservation in three dimensions

The first step in the derivation of the mass conservation equation is to write down a mass balance for the fluid element.

Rate of increase of mass in fluid element	=	Net rate of flow of mass into fluid element

The rate of increase of mass in the fluid element is

$$\frac{\partial}{\partial t}(\rho \delta x \delta y \delta z) = \frac{\partial \rho}{\partial t} \delta x \delta y \delta z \tag{2.1}$$

Next we need to account for the mass flow rate across a face of the element which is given by the product of density, area and the velocity component normal to the face. From Figure 2.2 it can be seen that the net rate of flow of mass into the element across its boundaries is given by

$$\begin{aligned} &\left(\rho u - \frac{\partial(\rho u)}{\partial x} \tfrac{1}{2}\delta x\right)\delta y \delta z - \left(\rho u + \frac{\partial(\rho u)}{\partial x} \tfrac{1}{2}\delta x\right)\delta y \delta z \\ &+ \left(\rho v - \frac{\partial(\rho v)}{\partial y} \tfrac{1}{2}\delta y\right)\delta x \delta z - \left(\rho v + \frac{\partial(\rho v)}{\partial y} \tfrac{1}{2}\delta y\right)\delta x \delta z \\ &+ \left(\rho w - \frac{\partial(\rho w)}{\partial z} \tfrac{1}{2}\delta z\right)\delta x \delta y - \left(\rho w + \frac{\partial(\rho w)}{\partial z} \tfrac{1}{2}\delta z\right)\delta x \delta y \end{aligned} \tag{2.2}$$

Flows which are directed into the element produce an increase of mass in the element and get a positive sign and those flows that are leaving the element are given a negative sign.

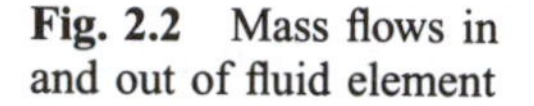

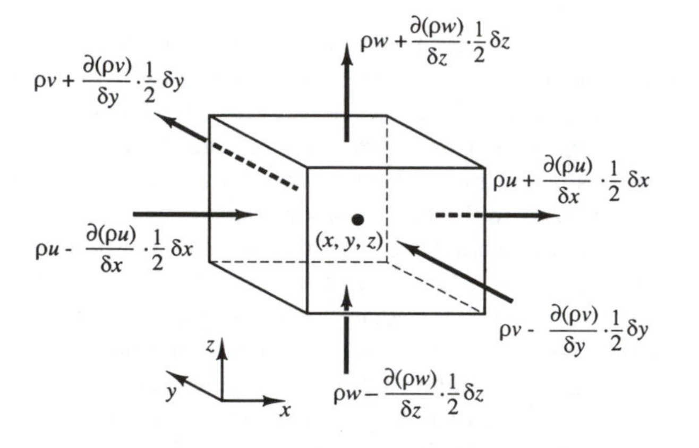

Fig. 2.2 Mass flows in and out of fluid element

The rate of increase of mass inside the element (2.1) is now equated to the net rate of flow of mass into the element across its faces (2.2). All terms of the resulting mass balance are arranged on the left hand side of the equals sign and the expression is divided by the element volume $\delta x \delta y \delta z$.

This yields

$$\frac{\partial \rho}{\partial t} + \frac{\partial(\rho u)}{\partial x} + \frac{\partial(\rho v)}{\partial y} + \frac{\partial(\rho w)}{\partial z} = 0 \tag{2.3}$$

or in more compact vector notation

$$\boxed{\frac{\partial \rho}{\partial t} + div(\rho \mathbf{u}) = 0} \tag{2.4}$$

Equation (2.4) is the **unsteady, three-dimensional mass conservation or continuity equation** at a point in a **compressible fluid**. The first term on the left hand side is the rate of change in time of the density (mass per unit volume). The second term describes the net flow of mass out of the element across its boundaries and is called the convective term.

For an **incompressible fluid** (i.e. a liquid) the density ρ is constant and equation (2.4) becomes

$$div\,\mathbf{u} = 0 \tag{2.5}$$

or in longhand notation

$$\frac{\partial u}{\partial x} + \frac{\partial v}{\partial y} + \frac{\partial w}{\partial z} = 0 \tag{2.6}$$

2.1.2 Rates of change following a fluid particle and for a fluid element

The momentum and energy conservation laws make statements regarding the changes of properties of a fluid particle. Each property of such a particle is a function of the position (x, y, z) of the particle and time t. Let the value of a property per unit mass be denoted by ϕ. The total or substantive derivative of ϕ with respect to time following a fluid particle, written as $D\phi/Dt$, is

$$\frac{D\phi}{Dt} = \frac{\partial \phi}{\partial t} + \frac{\partial \phi}{\partial x}\frac{dx}{dt} + \frac{\partial \phi}{\partial y}\frac{dy}{dt} + \frac{\partial \phi}{\partial z}\frac{dz}{dt}$$

A fluid particle follows the flow, so $dx/dt = u$, $dy/dt = v$ and $dz/dt = w$. Hence the substantive derivative of ϕ is given by

$$\frac{D\phi}{Dt} = \frac{\partial \phi}{\partial t} + u\frac{\partial \phi}{\partial x} + v\frac{\partial \phi}{\partial y} + w\frac{\partial \phi}{\partial z} = \frac{\partial \phi}{\partial t} + \mathbf{u}\,.\,grad\;\phi \tag{2.7}$$

$D\phi/Dt$ defines the rate of change of property ϕ per unit mass. As in the case of the mass conservation equation we are interested in developing equations for rates of change per unit volume. The rate of change of property ϕ per unit volume for a fluid particle is given by the product of $D\phi/Dt$ and density ρ, hence

$$\rho\frac{D\phi}{Dt} = \rho\left(\frac{\partial \phi}{\partial t} + \mathbf{u}\,.\,grad\;\phi\right) \tag{2.8}$$

The most useful forms of the conservation laws for fluid flow computation are concerned with changes of a flow property for a fluid element which is stationary in space. The relationship between the substantive derivative of ϕ, which follows a fluid particle, and rate of change of ϕ for a fluid element is now developed.

The mass conservation equation contains the mass per unit volume (i.e. the density ρ) as the conserved quantity. The sum of the rate of change of density and the convective term in the mass conservation equation (2.4) for a fluid element is

$$\frac{\partial \rho}{\partial t} + div(\rho\mathbf{u})$$

The generalisation of these terms for an arbitrary conserved property is

$$\frac{\partial(\rho\phi)}{\partial t} + div(\rho\phi\mathbf{u}) \tag{2.9}$$

Formula (2.9) expresses the rate of change of ϕ per unit volume plus the net flow of ϕ out of the fluid element per unit volume. It is now re-written to illustrate its relationship with the substantive derivative of ϕ:

$$\frac{\partial(\rho\phi)}{\partial t} + div(\rho\phi\mathbf{u}) = \rho\left[\frac{\partial\phi}{\partial t} + \mathbf{u}\cdot grad\ \phi\right] + \phi\left[\frac{\partial\rho}{\partial t} + div(\rho\mathbf{u})\right] = \rho\frac{D\phi}{Dt} \tag{2.10}$$

The term $\phi[\partial\rho/\partial t + div(\rho\mathbf{u})]$ is equal to zero by virtue of mass conservation (2.4). In words, relationship (2.10) states

Rate of increase of ϕ of fluid element	+	Net rate of flow of ϕ out of fluid element	=	Rate of increase of ϕ for a fluid particle

To construct the three components of the momentum equation and the energy equation the relevant entries for ϕ and their rates of change per unit volume as defined in (2.8) and (2.10) are given below:

x-momentum	u	$\rho\frac{Du}{Dt}$	$\frac{\partial(\rho u)}{\partial t} + div(\rho u\mathbf{u})$
y-momentum	v	$\rho\frac{Dv}{Dt}$	$\frac{\partial(\rho v)}{\partial t} + div(\rho v\mathbf{u})$
z-momentum	w	$\rho\frac{Dw}{Dt}$	$\frac{\partial(\rho w)}{\partial t} + div(\rho w\mathbf{u})$
Energy	E	$\rho\frac{DE}{Dt}$	$\frac{\partial(\rho E)}{\partial t} + div(\rho E\mathbf{u})$

Both the conservative (or divergence) form and non-conservative form of the rate of change can be used as alternatives to express the conservation of a physical quantity. The non-conservative forms are used in the derivations of momentum and energy equations for a fluid flow in sections 2.4 and 2.5 for brevity of notation and to emphasise that the conservation laws are fundamentally conceived as statements that apply to a particle of fluid. In the final section 2.8 we shall return to the conservative form which is used in finite volume CFD calculations.

2.1.3 Momentum equation in three dimensions

Newton's second law states that the rate of change of momentum of a fluid particle equals the sum of the forces on the particle.

Rate of increase of momentum of fluid particle	=	Sum of forces on fluid particle

The **rates of increase of x-, y- and z- momentum** per unit volume of a fluid particle are given by

$$\rho \frac{Du}{Dt} \qquad \rho \frac{Dv}{Dt} \qquad \rho \frac{Dw}{Dt} \tag{2.11}$$

We distinguish two types of **forces** on fluid particles:

- **surface forces**
  - pressure forces
  - viscous forces
- **body forces**
  - gravity force
  - centrifugal force
  - Coriolis force
  - electromagnetic force

It is common practice to highlight the contributions due to the surface forces as separate terms in the momentum equation and to include the effects of body forces as source terms.

The state of stress of a fluid element is defined in terms of the pressure and the nine viscous stress components shown in Figure 2.3. The pressure, a normal stress, is denoted by p. Viscous stresses are denoted by τ. The usual suffix notation τ_{ij} is applied to indicate the direction of the viscous stresses. The suffices i and j in τ_{ij} indicate that the stress component acts in the j-direction on a surface normal to the i-direction.

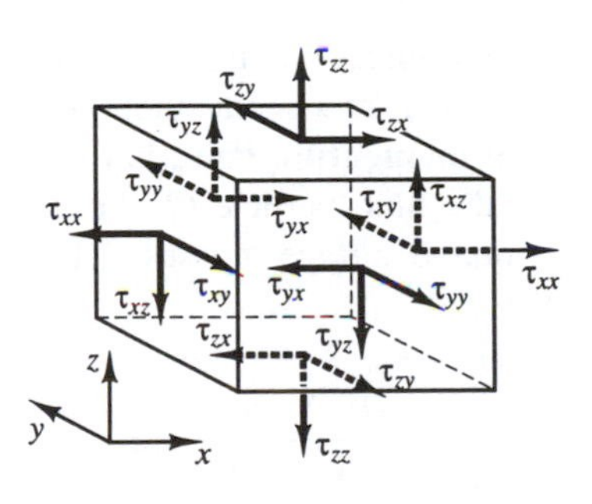

Fig. 2.3 Stress components on three faces of fluid element

First we consider the x-components of the forces due to pressure p and stress components τ_{xx}, τ_{yx} and τ_{zx} shown in Figure 2.4. The magnitude of a force resulting from a surface stress is the product of stress and area. Forces aligned with the direction of a co-ordinate axis get a positive sign and those in the opposite direction a negative sign. The net force in the x-direction is the sum of the force components acting in that direction on the fluid element.

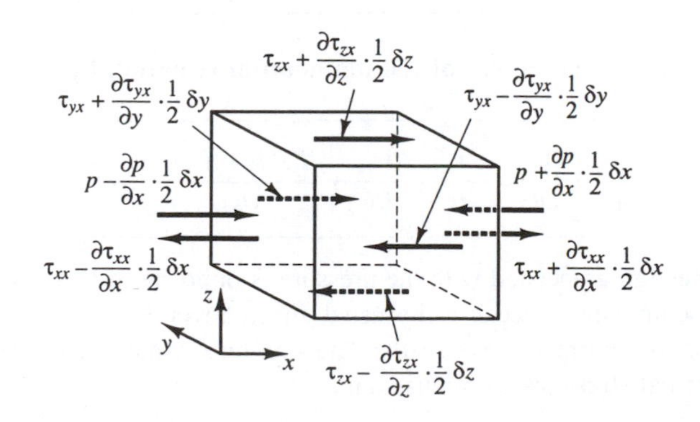

Fig. 2.4 Stress components in the x-direction

On the pair of faces (E, W) we have

$$\left[\left(p - \frac{\partial p}{\partial x}\frac{1}{2}\delta x\right) - \left(\tau_{xx} - \frac{\partial \tau_{xx}}{\partial x}\frac{1}{2}\delta x\right)\right]\delta y\delta z + \left[-\left(p + \frac{\partial p}{\partial x}\frac{1}{2}\delta x\right) + \left(\tau_{xx} + \frac{\partial \tau_{xx}}{\partial x}\frac{1}{2}\delta x\right)\right]\delta y\delta z = \left(-\frac{\partial p}{\partial x} + \frac{\partial \tau_{xx}}{\partial x}\right)\delta x\delta y\delta z \tag{2.12a}$$

The net force in the x-direction on the pair of faces (N, S) is

$$-\left(\tau_{yx} - \frac{\partial \tau_{yx}}{\partial y}\frac{1}{2}\delta y\right)\delta x\delta z + \left(\tau_{yx} + \frac{\partial \tau_{yx}}{\partial y}\frac{1}{2}\delta y\right)\delta x\delta z = \frac{\partial \tau_{yx}}{\partial y}\delta x\delta y\delta z \tag{2.12b}$$

Finally the net force in the x-direction on faces T and B is given by

$$-\left(\tau_{zx} - \frac{\partial \tau_{zx}}{\partial z}\frac{1}{2}\delta z\right)\delta x\delta y + \left(\tau_{zx} + \frac{\partial \tau_{zx}}{\partial z}\frac{1}{2}\delta z\right)\delta x\delta y = \frac{\partial \tau_{zx}}{\partial z}\delta x\delta y\delta z \tag{2.12c}$$

The total force per unit volume on the fluid due to these surface stresses is equal to the sum of (2.12a), (2.12b) and (2.12c) divided by the volume $\delta x\delta y\delta z$:

$$\frac{\partial(-p + \tau_{xx})}{\partial x} + \frac{\partial \tau_{yx}}{\partial y} + \frac{\partial \tau_{zx}}{\partial z} \tag{2.13}$$

Without considering the body forces in further detail their overall effect can be included by defining a source S_{Mx} of x-momentum per unit volume per unit time.

The ***x*-component of the momentum equation** is found by setting the rate of change of x-momentum of the fluid particle (2.11) equal to the total force in the x-direction on the element due to surface stresses (2.13) plus the rate of increase of x-momentum due to sources:

$$\boxed{\rho\frac{Du}{Dt} = \frac{\partial(-p + \tau_{xx})}{\partial x} + \frac{\partial \tau_{yx}}{\partial y} + \frac{\partial \tau_{zx}}{\partial z} + S_{Mx}} \tag{2.14a}$$

It is not too difficult to verify that the ***y*-component of the momentum equation** is given by

$$\boxed{\rho\frac{Dv}{Dt} = \frac{\partial \tau_{xy}}{\partial x} + \frac{\partial(-p + \tau_{yy})}{\partial y} + \frac{\partial \tau_{zy}}{\partial z} + S_{My}} \tag{2.14b}$$

and the ***z*-component of the momentum equation** by

$$\boxed{\rho\frac{Dw}{Dt} = \frac{\partial \tau_{xz}}{\partial x} + \frac{\partial \tau_{yz}}{\partial y} + \frac{\partial(-p + \tau_{zz})}{\partial z} + S_{Mz}} \tag{2.14c}$$

The sign associated with the pressure is opposite to that associated with the normal viscous stress, because the usual sign convention takes a tensile stress to be the positive normal stress so that the pressure, which is by definition a compressive normal stress, has a minus sign.

The effects of surface stresses are accounted for explicitly; the source terms S_{Mx}, S_{My} and S_{Mz} in (2.14a–c) include contributions due to body forces only. For example, the body force due to gravity would be modelled by $S_{Mx} = 0$, $S_{My} = 0$ and $S_{Mz} = -\rho g$.

2.1.4 Energy equation in three dimensions

The energy equation is derived from the **first law of thermodynamics** which states that the rate of change of energy of a fluid particle is equal to the rate of heat addition to the fluid particle plus the rate of work done on the particle.

Rate of increase of energy of fluid particle	=	Net rate of heat added to fluid particle	+	Net rate of work done on fluid particle

As before we will be deriving an equation for the **rate of increase of energy** of a fluid particle per unit volume which is given by

$$\rho \frac{DE}{Dt} \tag{2.15}$$

Work done by surface forces

The **rate of work done** on the fluid particle in the element by a **surface force** is equal to the product of the force and velocity component in the direction of the force. For example, the forces given by (2.12a–c) all act in the x-direction. The work done by these forces is given by

$$\begin{aligned}
&\left[\left(pu - \frac{\partial(pu)}{\partial x}\tfrac{1}{2}\delta x\right) - \left(\tau_{xx}u - \frac{\partial(\tau_{xx}u)}{\partial x}\tfrac{1}{2}\delta x\right)\right.\\
&\quad \left. - \left(pu + \frac{\partial(pu)}{\partial x}\tfrac{1}{2}\delta x\right) + \left(\tau_{xx}u + \frac{\partial(\tau_{xx}u)}{\partial x}\tfrac{1}{2}\delta x\right)\right]\delta y\delta z\\
&\quad + \left[-\left(\tau_{yx}u - \frac{\partial(\tau_{yx}u)}{\partial y}\tfrac{1}{2}\delta y\right) + \left(\tau_{yx}u + \frac{\partial(\tau_{yx}u)}{\partial y}\tfrac{1}{2}\delta y\right)\right]\delta x\delta z\\
&\quad + \left[-\left(\tau_{zx}u - \frac{\partial(\tau_{zx}u)}{\partial z}\tfrac{1}{2}\delta z\right) + \left(\tau_{zx}u + \frac{\partial(\tau_{zx}u)}{\partial z}\tfrac{1}{2}\delta z\right)\right]\delta x\delta y
\end{aligned}$$

The net rate of work done by these surface forces acting in the x-direction is given by

$$\left[\frac{\partial[u(-p+\tau_{xx})]}{\partial x} + \frac{\partial(u\tau_{yx})}{\partial y} + \frac{\partial(u\tau_{zx})}{\partial z}\right]\delta x\delta y\delta z \tag{2.16a}$$

Surface stress components in the y- and z-direction also do work on the fluid particle. A repetition of the above process gives the additional rates of work done on the fluid particle due to the work done by these surface forces:

$$\left[\frac{\partial(v\tau_{xy})}{\partial x} + \frac{\partial[v(-p+\tau_{yy})]}{\partial y} + \frac{\partial(v\tau_{zy})}{\partial z}\right]\delta x\delta y\delta z \tag{2.16b}$$

and

$$\left[\frac{\partial(w\tau_{xz})}{\partial x} + \frac{\partial(w\tau_{yz})}{\partial y} + \frac{\partial[w(-p+\tau_{zz})]}{\partial z}\right]\delta x\delta y\delta z \tag{2.16c}$$

The total rate of work done per unit volume on the fluid particle by all the surface forces is given by the sum of (2.16a–c) divided by the volume $\delta x \delta y \delta z$. The terms containing pressure can be collected together and written more compactly in vector form:

$$-\frac{\partial(up)}{\partial x} - \frac{\partial(vp)}{\partial y} - \frac{\partial(wp)}{\partial z} = -div(p\mathbf{u})$$

This yields the following **total rate of work done on the fluid particle by surface stresses**:

$$[-div(p\mathbf{u})] + \left[\frac{\partial(u\tau_{xx})}{\partial x} + \frac{\partial(u\tau_{yx})}{\partial y} + \frac{\partial(u\tau_{zx})}{\partial z} + \frac{\partial(v\tau_{xy})}{\partial x} + \frac{\partial(v\tau_{yy})}{\partial y} + \frac{\partial(v\tau_{zy})}{\partial z} + \frac{\partial(w\tau_{xz})}{\partial x} + \frac{\partial(w\tau_{yz})}{\delta y} + \frac{\partial(w\tau_{zz})}{\partial z}\right] \quad (2.17)$$

Energy flux due to heat conduction

The heat flux vector $\mathbf{q}$ has three components q_x, q_y and q_z (Figure 2.5).

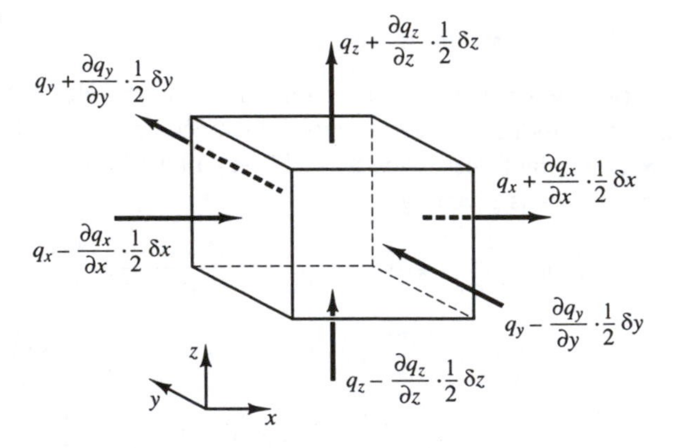

Fig. 2.5 Components of the heat flux vector

The **net rate of heat transfer to the fluid particle** due to heat flow in the x-direction is given by the difference between the rate of heat input across face W and the rate of heat loss across face E:

$$\left[\left(q_x - \frac{\partial q_x}{\partial x}\tfrac{1}{2}\delta x\right) - \left(q_x + \frac{\partial q_x}{\partial x}\tfrac{1}{2}\delta x\right)\right]\delta y \delta z = -\frac{\partial q_x}{\partial x}\delta x \delta y \delta z \quad (2.18a)$$

Similarly, the net rates of heat transfer to the fluid due to heat flows in the y- and z-direction are

$$-\frac{\partial q_y}{\partial y}\delta x \delta y \delta z \quad \text{and} \quad -\frac{\partial q_z}{\partial z}\delta x \delta y \delta z \quad (2.18b\text{–}c)$$

The total rate of heat added to the fluid particle per unit volume due to heat flow across its boundaries is the sum of (2.18a–c) divided by the volume $\delta x \delta y \delta z$

$$-\frac{\partial q_x}{\partial x} - \frac{\partial q_y}{\partial y} - \frac{\partial q_z}{\partial z} = -div\,\mathbf{q} \quad (2.19)$$

Fourier's law of heat conduction relates the heat flux to the local temperature gradient. So

$$q_x = -k \frac{\partial T}{\partial x} \qquad q_y = -k \frac{\partial T}{\partial y} \qquad q_z = -k \frac{\partial T}{\partial z}$$

This can be written in vector form as follows:

$$\mathbf{q} = -k \, grad \, T \tag{2.20}$$

Combining (2.19) and (2.20) yields the final form of the **rate of heat addition to the fluid particle due to heat conduction** across element boundaries:

$$-div \, \mathbf{q} = div(k \, grad \, T) \tag{2.21}$$

Energy equation

Thus far we have not defined the specific energy E of a fluid. Often the energy of a fluid is defined as the sum of internal (thermal) energy i, kinetic energy $\frac{1}{2}(u^2 + v^2 + w^2)$ and gravitational potential energy. This definition takes the view that the fluid element is storing gravitational potential energy. It is also possible to regard the gravitational force as a body force which does work on the fluid element as it moves through the gravity field.

Here we shall take the latter view and include the effects of potential energy changes as a source term. As before we define a source of energy S_E per unit volume per unit time. Conservation of energy of the fluid particle is ensured by equating the rate of change of energy of the fluid particle (2.15) to the sum of the net rate of work done on the fluid particle (2.17) and the net rate of heat addition to the fluid (2.21) and the rate of increase of energy due to sources. The **energy equation** is

$$\begin{aligned}\rho \frac{DE}{Dt} = &- div(p\mathbf{u}) + \left[\frac{\partial(u\tau_{xx})}{\partial x} + \frac{\partial(u\tau_{yx})}{\partial y} + \frac{\partial(u\tau_{zx})}{\partial z} + \frac{\partial(v\tau_{xy})}{\partial x}\right.\\ &\left. + \frac{\partial(v\tau_{yy})}{\partial y} + \frac{\partial(v\tau_{zy})}{\partial z} + \frac{\partial(w\tau_{xz})}{\partial x} + \frac{\partial(w\tau_{yz})}{\partial y} + \frac{\partial(w\tau_{zz})}{\partial z}\right]\\ &+ div(k \, grad \, T) + S_E\end{aligned} \tag{2.22}$$

In equation (2.22) we have $E = i + \frac{1}{2}(u^2 + v^2 + w^2)$.

Although (2.22) is a perfectly adequate energy equation it is common practice to extract the changes of the (mechanical) kinetic energy to obtain an equation for internal energy i or temperature T. The part of the energy equation attributable to the kinetic energy can be found by multiplying the x-momentum equation (2.14a) by velocity component u, the y-momentum equation (2.14b) by v and the z-momentum equation (2.14c) by w and adding the results together. It can be shown that this yields the following conservation equation for the kinetic energy:

$$\begin{aligned}\rho \frac{D\left[\frac{1}{2}(u^2 + v^2 + w^2)\right]}{Dt} = &- \mathbf{u} \, . \, grad \, p + u\left(\frac{\partial \tau_{xx}}{\partial x} + \frac{\partial \tau_{yx}}{\partial y} + \frac{\partial \tau_{zx}}{\partial z}\right)\\ &+ v\left(\frac{\partial \tau_{xy}}{\partial x} + \frac{\partial \tau_{yy}}{\partial y} + \frac{\partial \tau_{zy}}{\partial z}\right)\\ &+ w\left(\frac{\partial \tau_{xz}}{\partial x} + \frac{\partial \tau_{yz}}{\partial y} + \frac{\partial \tau_{zz}}{\partial z}\right) + \mathbf{u} \, . \, \mathbf{S}_M\end{aligned} \tag{2.23}$$

Subtracting (2.23) from (2.22) and defining a new source term as $S_i = S_E - \mathbf{u}.\mathbf{S}_M$ yields the internal energy equation

$$\rho \frac{Di}{Dt} = -p \; div \; \mathbf{u} + div(k \; grad \; T) + \tau_{xx}\frac{\partial u}{\partial x} + \tau_{yx}\frac{\partial u}{\partial y} + \tau_{zx}\frac{\partial u}{\partial z} + \tau_{xy}\frac{\partial v}{\partial x} + \tau_{yy}\frac{\partial v}{\partial y} + \tau_{zy}\frac{\partial v}{\partial z} + \tau_{xz}\frac{\partial w}{\partial x} + \tau_{yz}\frac{\partial w}{\partial y} + \tau_{zz}\frac{\partial w}{\partial z} + S_i \quad (2.24)$$

In the special case of an incompressible fluid we have $i = cT$, where c is the specific heat, and $div \; \mathbf{u} = 0$. This allows us to recast equation (2.24) into a temperature equation

$$\rho c \frac{DT}{Dt} = div(k \; grad \; T) + \tau_{xx}\frac{\partial u}{\partial x} + \tau_{yx}\frac{\partial u}{\partial y} + \tau_{zx}\frac{\partial u}{\partial z} + \tau_{xy}\frac{\partial v}{\partial x} + \tau_{yy}\frac{\partial v}{\partial y} + \tau_{zy}\frac{\partial v}{\partial z} + \tau_{xz}\frac{\partial w}{\partial x} + \tau_{yz}\frac{\partial w}{\partial y} + \tau_{zz}\frac{\partial w}{\partial z} + S_i \quad (2.25)$$

For compressible flows equation (2.22) is often re-arranged to give an equation for the **enthalpy**. The specific enthalpy h and the specific total enthalpy h_0 of a fluid are defined as

$$h = i + p/\rho \quad \text{and} \quad h_0 = h + \tfrac{1}{2}(u^2 + v^2 + w^2)$$

Combining these two definitions with the one for specific energy E we get

$$h_0 = i + p/\rho + \tfrac{1}{2}(u^2 + v^2 + w^2) = E + p/\rho \quad (2.26)$$

Substituting of (2.26) into equation (2.22) and some re-arrangement yields the **(total) enthalpy equation**

$$\frac{\partial(\rho h_0)}{\partial t} + div(\rho h_0 \mathbf{u}) = div(k \; grad \; T) + \frac{\partial p}{\partial t} + \left[\frac{\partial(u\tau_{xx})}{\partial x} + \frac{\partial(u\tau_{yx})}{\partial y} + \frac{\partial(u\tau_{zx})}{\partial z} + \frac{\partial(v\tau_{xy})}{\partial x} + \frac{\partial(v\tau_{yy})}{\partial y} + \frac{\partial(v\tau_{zy})}{\partial z} + \frac{\partial(w\tau_{xz})}{\partial x} + \frac{\partial(w\tau_{yz})}{\partial y} + \frac{\partial(w\tau_{zz})}{\partial z}\right] + S_h \quad (2.27)$$

It should be stressed that equations (2.24), (2.25) and (2.27) are **not** new (extra) conservation laws but merely alternative forms of the energy equation (2.22).

2.2 Equations of state

The motion of a fluid in three dimensions is described by a system of five partial differential equations: mass conservation (2.4), x-, y- and z-momentum equations (2.14a–c) and energy equation (2.22). Among the unknowns are four thermodynamic variables: ρ, p, i and T. In this brief discussion we point out the linkage between these four variables. Relationships between the thermodynamic variables can be obtained through the assumption of **thermodynamic equilibrium**. The fluid velocities may be large, but they are usually small enough that, even though the properties of a fluid particle change rapidly from place to place, the fluid can thermodynamically adjust itself to new conditions so quickly that the changes are effectively instantaneous. Thus the fluid always remains in thermodynamic equilibrium. The only exceptions are certain flows with strong shockwaves, but even some of those are often well enough approximated by equilibrium assumptions.

We can describe the state of a substance in thermodynamic equilibrium by means of just two state variables. **Equations of state** relate the other variables to the two state variables. If we use ρ and T as state variables we have state equations for pressure p and specific internal energy i:

$$p = p(\rho, T) \quad \text{and} \quad i = i(\rho, T) \tag{2.28}$$

For a **perfect gas** the following, well-known, equations of state are useful:

$$p = \rho RT \quad \text{and} \quad i = C_v T \tag{2.29}$$

The assumption of thermodynamic equilibrium eliminates all but the two thermodynamic state variables. In the flow of **compressible fluids** the equations of state provide the linkage between the energy equation on the one hand and mass conservation and momentum equations on the other. This linkage arises through the possibility of density variations as a result of pressure and temperature variations in the flow field.

Liquids and gases flowing at low speeds behave as **incompressible fluids**. Without density variations there is no linkage between the energy equation and the mass conservation and momentum equations. The flow field can often be solved by considering mass conservation and momentum equations only. The energy equation only needs to be solved alongside the others if the problem involves heat transfer.

2.3 Navier–Stokes equations for a Newtonian fluid

The governing equations contain as further unknowns the viscous stress components τ_{ij}. The most useful forms of the conservation equations for fluid flows are obtained by introducing a suitable model for the viscous stresses τ_{ij}. In many fluid flows the viscous stresses can be expressed as functions of the local deformation rate (or strain rate). In three-dimensional flows the local rate of deformation is composed of the linear deformation rate and the volumetric deformation rate.

All gases and many liquids are isotropic. Liquids which contain significant quantities of polymer molecules may exhibit anisotropic or directional viscous stress properties as a result of the alignment of the chain-like polymer molecules with the flow. Such fluids are beyond the scope of this introductory course and we shall continue the development by assuming that the fluids are isotropic.

The rate of linear deformation of a fluid element has nine components in three dimensions, six of which are independent in isotropic fluids (Schlichting, 1979). They are denoted by the symbol e_{ij}. The suffix system is identical to that for stress components (see section 2.4). There are three linear elongating deformation components:

$$e_{xx} = \frac{\partial u}{\partial x} \qquad e_{yy} = \frac{\partial v}{\partial y} \qquad e_{zz} = \frac{\partial w}{\partial z} \tag{2.30a}$$

There are also six shearing linear deformation components:

$$e_{xy} = e_{yx} = \tfrac{1}{2}\left(\frac{\partial u}{\partial y} + \frac{\partial v}{\partial x}\right) \qquad e_{xz} = e_{zx} = \tfrac{1}{2}\left(\frac{\partial u}{\partial z} + \frac{\partial w}{\partial x}\right)$$
$$e_{yz} = e_{zy} = \tfrac{1}{2}\left(\frac{\partial v}{\partial z} + \frac{\partial w}{\partial y}\right) \tag{2.30b}$$

The volumetric deformation is given by

$$\frac{\partial u}{\partial x} + \frac{\partial v}{\partial y} + \frac{\partial w}{\partial z} = div\ \mathbf{u} \tag{2.30c}$$

In a **Newtonian fluid the viscous stresses are proportional to the rates of deformation**. The three-dimensional form of Newton's law of viscosity for compressible flows involves two constants of proportionality: the (first) dynamic viscosity, μ, to relate stresses to linear deformations, and the second viscosity, λ, to relate stresses to the volumetric deformation. The nine viscous stress components, of which six are independent, are

$$\tau_{xx} = 2\mu \frac{\partial u}{\partial x} + \lambda\ div\ \mathbf{u} \quad \tau_{yy} = 2\mu \frac{\partial v}{\partial y} + \lambda\ div\ \mathbf{u} \quad \tau_{zz} = 2\mu \frac{\partial w}{\partial z} + \lambda\ div\ \mathbf{u}$$
$$\tau_{xy} = \tau_{yx} = \mu\left(\frac{\partial u}{\partial y} + \frac{\partial v}{\partial x}\right) \qquad \tau_{xz} = \tau_{zx} = \mu\left(\frac{\partial u}{\partial z} + \frac{\partial w}{\partial x}\right)$$
$$\tau_{yz} = \tau_{zy} = \mu\left(\frac{\partial v}{\partial z} + \frac{\partial w}{\partial y}\right) \tag{2.31}$$

Not much is known about the second viscosity λ, because its effect is small in practice. For gases a good working approximation can be obtained by taking the value $\lambda = -\frac{2}{3}\mu$ (Schlichting, 1979). Liquids are incompressible so the mass conservation equation is $div\ \mathbf{u} = 0$ and the viscous stresses are just twice the local rate of linear deformation times the dynamic viscosity.

Substitution of the above shear stresses (2.31) into (2.14a–c) yields the so-called Navier–Stokes equations named after the two 19th century scientists who derived them independently:

$$\rho \frac{Du}{Dt} = -\frac{\partial p}{\partial x} + \frac{\partial}{\partial x}\left[2\mu \frac{\partial u}{\partial x} + \lambda\ div\ \mathbf{u}\right] + \frac{\partial}{\partial y}\left[\mu\left(\frac{\partial u}{\partial y} + \frac{\partial v}{\partial x}\right)\right] + \frac{\partial}{\partial z}\left[\mu\left(\frac{\partial u}{\partial z} + \frac{\partial w}{\partial x}\right)\right] + S_{Mx} \tag{2.32a}$$

$$\rho \frac{Dv}{Dt} = -\frac{\partial p}{\partial y} + \frac{\partial}{\partial x}\left[\mu\left(\frac{\partial u}{\partial y} + \frac{\partial v}{\partial x}\right)\right] + \frac{\partial}{\partial y}\left[2\mu \frac{\partial v}{\partial y} + \lambda \, div \, \mathbf{u}\right] + \frac{\partial}{\partial z}\left[\mu\left(\frac{\partial v}{\partial z} + \frac{\partial w}{\partial y}\right)\right] + S_{My} \quad (2.32b)$$

$$\rho \frac{Dw}{Dt} = -\frac{\partial p}{\partial z} + \frac{\partial}{\partial x}\left[\mu\left(\frac{\partial u}{\partial z} + \frac{\partial w}{\partial x}\right)\right] + \frac{\partial}{\partial y}\left[\mu\left(\frac{\partial v}{\partial z} + \frac{\partial w}{\partial y}\right)\right] + \frac{\partial}{\partial z}\left[2\mu \frac{\partial w}{\partial z} + \lambda \, div \, \mathbf{u}\right] + S_{Mz} \quad (2.32c)$$

Often it is useful to re-arrange the viscous stress terms as follows:

$$\frac{\partial}{\partial x}\left[2\mu \frac{\partial u}{\partial x} + \lambda \, div \, \mathbf{u}\right] + \frac{\partial}{\partial y}\left[\mu\left(\frac{\partial u}{\partial y} + \frac{\partial v}{\partial x}\right)\right] + \frac{\partial}{\partial z}\left[\mu\left(\frac{\partial u}{\partial z} + \frac{\partial w}{\partial x}\right)\right]$$
$$= \frac{\partial}{\partial x}\left(\mu \frac{\partial u}{\partial x}\right) + \frac{\partial}{\partial y}\left(\mu \frac{\partial u}{\partial y}\right) + \frac{\partial}{\partial z}\left(\mu \frac{\partial u}{\partial z}\right)$$
$$+ \left[\frac{\partial}{\partial x}\left(\mu \frac{\partial u}{\partial x}\right) + \frac{\partial}{\partial y}\left(\mu \frac{\partial v}{\partial x}\right) + \frac{\partial}{\partial z}\left(\mu \frac{\partial w}{\partial x}\right)\right]$$
$$+ \frac{\partial}{\partial x}(\lambda \, div \, \mathbf{u}) = div(\mu \, grad \, u) + s_{Mx}$$

The viscous stresses in the y- and z-component equations can be re-cast in a similar manner. We clearly intend to simplify the momentum equations by ‘hiding’ the two smaller contributions to the viscous stress terms in the momentum source. Defining a new source by

$$S_M = S_M + s_M \quad (2.33)$$

the **Navier–Stokes equations** can be written in the most useful form for the development of the finite volume method:

$$\rho \frac{Du}{Dt} = -\frac{\partial p}{\partial x} + div(\mu \, grad \, u) + S_{Mx} \quad (2.34a)$$

$$\rho \frac{Dv}{Dt} = -\frac{\partial p}{\partial y} + div(\mu \, grad \, v) + S_{My} \quad (2.34b)$$

$$\rho \frac{Dw}{Dt} = -\frac{\partial p}{\partial z} + div(\mu \, grad \, w) + S_{Mz} \quad (2.34c)$$

If we use the Newtonian model for viscous stresses in the internal energy equation (2.24) we obtain after some re-arrangement

$$\rho \frac{Di}{Dt} = -p \, div \, \mathbf{u} + div(k \, grad \, T) + \Phi + S_i \quad (2.35)$$

All the effects due to viscous stresses in this internal energy equation are described by the dissipation function Φ which, after considerable algebra, can be shown to be equal to

$$\Phi = \mu\left\{2\left[\left(\frac{\partial u}{\partial x}\right)^2 + \left(\frac{\partial v}{\partial y}\right)^2 + \left(\frac{\partial w}{\partial z}\right)^2\right] + \left(\frac{\partial u}{\partial y} + \frac{\partial v}{\partial x}\right)^2 + \left(\frac{\partial u}{\partial z} + \frac{\partial w}{\partial x}\right)^2 + \left(\frac{\partial v}{\partial z} + \frac{\partial w}{\partial y}\right)^2\right\} + \lambda(div\,\mathbf{u})^2 \quad (2.36)$$

The dissipation function is non-negative since it only contains squared terms and represents a source of internal energy due to deformation work on the fluid particle. This work is extracted from the mechanical agency which causes the motion and is converted into internal energy or heat.

2.4 Conservative form of the governing equations of fluid flow

To summarize the findings thus far we quote in Table 2.1 the conservative or divergence form of the system of equations which governs the time-dependent three-dimensional fluid flow and heat transfer of a compressible Newtonian fluid.

Table 2.1 Governing equations of the flow of a compressible Newtonian fluid

Mass	$\frac{\partial \rho}{\partial t} + div(\rho\mathbf{u}) = 0$	(2.4)
x-momentum	$\frac{\partial(\rho u)}{\partial t} + div(\rho u\mathbf{u}) = -\frac{\partial p}{\partial x} + div(\mu\, grad\, u) + S_{Mx}$	(2.37a)
y-momentum	$\frac{\partial(\rho v)}{\partial t} + div(\rho v\mathbf{u}) = -\frac{\partial p}{\partial y} + div(\mu\, grad\, v) + S_{My}$	(2.37b)
z-momentum	$\frac{\partial(\rho w)}{\partial t} + div(\rho w\mathbf{u}) = -\frac{\partial p}{\partial z} + div(\mu\, grad\, w) + S_{Mz}$	(2.37c)
Internal energy	$\frac{\partial(\rho i)}{\partial t} + div(\rho i\mathbf{u}) = -p\, div\,\mathbf{u} + div(k\, grad\, T) + \Phi + S_i$	(2.38)
Equations of state	$p = p(\rho, T)$ and $i = i(\rho, T)$	(2.28)
	e.g. perfect gas	
	$p = \rho RT$ and $i = C_v T$	(2.29)

Momentum source S_M and dissipation function Φ are defined by (2.33) and (2.36) respectively.

It is interesting to note that the thermodynamic equilibrium assumption of section 2.2 has supplemented the five flow equations (PDEs) with two further algebraic equations. The further introduction of the Newtonian model, which expresses the viscous stresses in terms of gradients of velocity components has resulted in a system of seven equations with seven unknowns. With an equal number of equations and unknown functions this system is mathematically closed, i.e. it can be solved provided that suitable auxiliary conditions, initial and boundary conditions, are supplied.

2.5 Differential and integral forms of the general transport equations

It is clear from Table 2.1 that there are significant commonalities between the various equations. If we introduce a general variable ϕ the conservative form of all fluid flow equations, including equations for scalar quantities such as temperature and pollutant concentration etc., can usefully be written in the following form:

$$\frac{\partial(\rho\phi)}{\partial t} + div(\rho\phi\mathbf{u}) = div(\Gamma\ grad\ \phi) + S_\phi \tag{2.39}$$

In words

Rate of increase of ϕ of fluid element	+	Net rate of flow of ϕ out of fluid element	=	Rate of increase of ϕ due to diffusion	+	Rate of increase of ϕ due to sources

The equation (2.39) is the so-called transport equation for property ϕ. It clearly highlights the various transport processes: the **rate of change** term and the **convective** term on the left hand side and the **diffusive** term (Γ =diffusion coefficient) and the **source** term respectively on the right hand side. In order to bring out the common features we have, of course, had to hide the terms that are not shared between the equations in the source terms. Note that equations (2.39) can be made to work for the internal energy equation by changing i into T by means of an equation of state.

The equation (2.39) is used as the starting point for computational procedures in the finite volume method. By setting ϕ equal to 1, u, v, w and i (or T or h_0) and selecting appropriate values for the diffusion coefficient Γ and source terms we obtain special forms of Table 2.1 for each of the five partial differential equations for mass, momentum and energy conservation. The key step of the finite volume method, which is to be developed from Chapter 4 onwards, is the integration of (2.39) over a three-dimensional control volume CV yielding

$$\int_{CV} \frac{\partial(\rho\phi)}{\partial t} dV + \int_{CV} div(\rho\phi\mathbf{u})dV = \int_{CV} div(\Gamma\ grad\ \phi)dV + \int_{CV} S_\phi dV \tag{2.40}$$

The volume integrals in the second term on the left hand side, the convective term, and in the first term on the right hand side, the diffusive term, are re-written as integrals over the entire bounding surface of the control volume by using Gauss' divergence theorem. For a vector **a** this theorem states

$$\int_{CV} div\ \mathbf{a}dV = \int_{A} \mathbf{n}\,.\,\mathbf{a}dA \tag{2.41}$$

The physical interpretation of **n.a** is the component of vector **a** in the direction of the vector **n** normal to surface element dA. Thus the integral of the divergence of a vector **a** over a volume is equal to the component of **a** in the direction normal to the surface which bounds the volume summed (integrated) over the entire bounding surface A. Applying Gauss' divergence theorem, equation (2.40) can be written as follows:

$$\frac{\partial}{\partial t}\left(\int_{CV} \rho\phi dV\right) + \int_A \mathbf{n}\,.\,(\rho\phi\mathbf{u})dA = \int_A \mathbf{n}\,.\,(\Gamma\ grad\ \phi)dA + \int_{CV} S_\phi dV \tag{2.42}$$

The order of integration and differentiation has been changed in the first term on the left hand side of (2.42) to illustrate its physical meaning. This term signifies the **rate of change of the total amount of fluid property ϕ in the control volume**. The product $\mathbf{n}\,.\,(\rho\phi\mathbf{u})$ expresses the flux component of property ϕ due to fluid flow along the outward normal vector $\mathbf{n}$, so the second term on the left hand side of (2.42), the convective term, is therefore the **net rate of decrease of fluid property ϕ of the fluid element due to convection**.

A diffusive flux is positive in the direction of a negative gradient of the fluid property ϕ, i.e. along direction $-grad\ \phi$. For instance, heat is conducted in the direction of negative temperature gradients. Thus, the product $\mathbf{n}\,.\,(-\Gamma\ grad\ \phi)$ is the component of diffusion flux along the outward normal vector, and so out of the fluid element. Similarly, the product $\mathbf{n}\,.\,(\Gamma\ grad\ \phi)$, which is also equal to $\Gamma(-\mathbf{n}\,.\,(-grad\ \phi))$, can be interpreted as a positive diffusion flux in the direction of the inward normal vector $-\mathbf{n}$, i.e. into the fluid element. The first term on the right hand side of (2.42), the diffusive term, is thus associated with a flux into the element and represents the **net rate of increase of fluid property ϕ of the fluid element due to diffusion**. The final term on the right hand side of this equation gives the **rate of increase of property ϕ as a result of sources** inside the fluid element.

In words, relationship (2.42) for the fluid in the control volume can be expressed as follows:

Rate of increase of ϕ	+	Net rate of decrease of ϕ due to convection across the boundaries	=	Rate of increase of ϕ due to diffusion across the boundaries	+	Net rate of creation of ϕ

This discussion clarifies that integration of the partial differential equation generates a statement of the conservation of a fluid property for a finite size (macroscopic) control volume.

In steady state problems the rate of change term of (2.42) is equal to zero. This leads to the integrated form of the steady transport equation

$$\int_A \mathbf{n}\,.\,(\rho\phi\mathbf{u})dA = \int_A \mathbf{n}\,.\,(\Gamma\ grad\ \phi)dA + \int_{CV} S_\phi dV \tag{2.43}$$

In time-dependent problems it is also necessary to integrate with respect to time t over a small interval Δt from, say, t until $t + \Delta t$. This yields the most general integrated form of the transport equation:

$$\int_{\Delta t} \frac{\partial}{\partial t}\left(\int_{CV} (\rho\phi)\, dV\right) dt + \int_{\Delta t}\int_{A} \mathbf{n}\,.\,(\rho\phi\mathbf{u})dA\, dt = \int_{\Delta t}\int_{A} \mathbf{n}\,.\,(\Gamma_\phi\, grad\, \phi)dA\, dt + \int_{\Delta t}\int_{CV} S_\phi dV\, dt \tag{2.44}$$

2.6 Classification of physical behaviour

Now that we have derived the conservation equations of fluid flows the time has come to turn our attention to the issue of the initial and boundary conditions which are needed in conjunction with the equations to construct a well-posed mathematical model of a fluid flow. First we distinguish two principal categories of physical behaviour:

- Equilibrium problems
- Marching problems

Equilibrium problems

The problems in the first category are steady state situations, e.g. the steady state distribution of temperature in a rod of solid material, the equilibrium stress distribution of a solid object under a given applied load as well as many steady fluid flows. These and many other steady state problems are governed by **elliptic equations**. The prototype elliptic equation is Laplace's equation which describes the irrotational flow of an incompressible fluid and steady state conductive heat transfer. In two dimensions we have

$$\frac{\partial^2\phi}{\partial x^2} + \frac{\partial^2\phi}{\partial y^2} = 0 \tag{2.45}$$

A very simple example of an equilibrium problem is the steady state heat conduction (where $\phi = T$ in equation (2.45)) in an insulated rod of metal whose ends at $x = 0$ and $x = L$ are kept at constant, but different, temperatures T_0 and T_L (Figure 2.6).

This problem is one-dimensional and governed by the equation $kd^2T/dx^2 = 0$. Under the given boundary conditions the temperature distribution in the x-direction

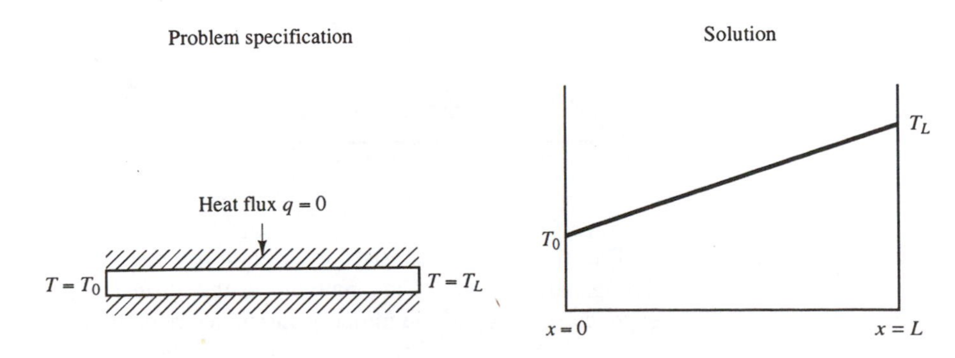

Fig. 2.6 Steady state temperature distribution of an insulated rod

will, of course, be a straight line. A unique solution to this and all elliptic problems can be obtained by specifying conditions on the dependent variable (here the temperature or its normal derivative, the heat flux) on all the boundaries of the solution domain. Problems requiring data over the entire boundary are called **boundary-value problems**.

An important feature of elliptic problems is that a disturbance in the interior of the solution, for example a change in temperature due to the sudden appearance of a small local heat source, changes the solution everywhere else. Disturbance signals travel in all directions through the interior solution. Consequently, the solutions to physical problems described by elliptic equations are always smooth even if the boundary conditions are discontinuous, which is a considerable advantage to the designer of numerical methods. To ensure that information propagates in all directions, the numerical techniques for elliptic problems must allow events at each point to be influenced by all its neighbours.

Marching problems

Transient heat transfer, all unsteady flows and wave phenomena are examples of problems in the second category, the marching or propagation problems. These problems are governed by **parabolic or hyperbolic equations**. However, not all marching problems are unsteady. We will see further on that certain steady flows are described by parabolic or hyperbolic equations. In these cases the flow direction acts as a time-like co-ordinate along which marching is possible.

Parabolic equations describe time-dependent problems which involve significant amounts of dissipation. Examples are unsteady viscous flows and unsteady heat conduction. The prototype parabolic equation is the diffusion equation

$$\frac{\partial \phi}{\partial t} = \alpha \frac{\partial^2 \phi}{\partial x^2} \tag{2.46}$$

The transient distribution of temperature (again $\phi = T$) in an insulated rod of metal whose ends at $x = 0$ and $x = L$ are kept at constant and equal temperature T_0 is governed by the diffusion equation. This problem arises when the rod cools down after an initially uniform source is switched off at time $t = 0$. The temperature distribution at the start is a parabola with a maximum at $x = L/2$ (Figure 2.7).

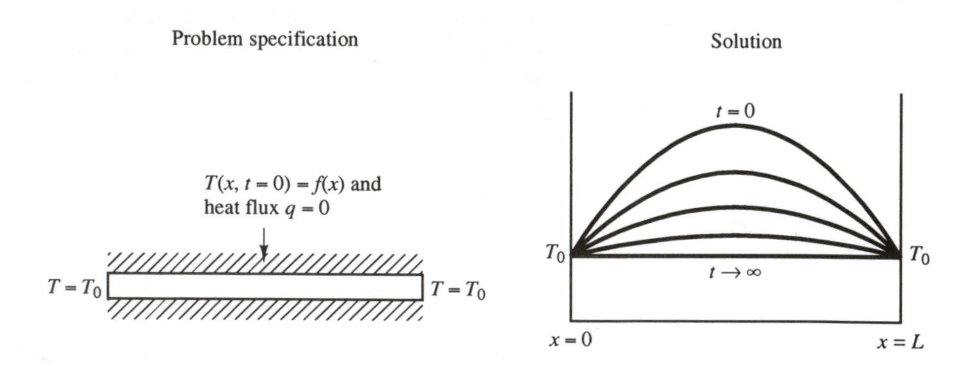

Fig. 2.7 Transient distribution of temperature in an insulated rod

The steady state consists of a uniform distribution of temperature $T = T_0$ throughout the rod. The solution of the diffusion equation (2.46) yields the exponential decay of the initial quadratic temperature distribution. Initial conditions are needed in the entire rod and conditions on all its boundaries are required for all times $t > 0$. This type of problem is termed an **initial–boundary-value problem**.

A disturbance at a point in the interior of the solution region (i.e. $0 < x < L$ and time $t_1 > 0$) can only influence events at later times $t > t_1$ (unless we allow time travel!). The solutions move forward in time and diffuse in space. The occurrence of dissipative effects ensures that the solutions are always smooth in the interior at times $t > 0$ even if the initial conditions contain discontinuities. The steady state is reached as time $t \rightarrow \infty$ and is elliptic. This change of character can be easily seen by setting $\partial\phi/\partial t = 0$ in equation (2.46). The governing equation is now equal to the one governing the steady temperature distribution in the rod.

Hyperbolic equations dominate the analysis of vibration problems. In general they appear in time-dependent processes with negligible amounts of dissipation. The prototype hyperbolic equation is the wave equation

$$\frac{\partial^2 \phi}{\partial t^2} = c^2 \frac{\partial^2 \phi}{\partial x^2} \tag{2.47}$$

The above form of the equation governs the transverse displacement ($\phi = y$) of a string under tension during small amplitude vibrations and also acoustic oscillations. The constant c is the wave speed. It is relatively straightforward to compute the time evolution of the fundamental mode of vibration of a string of length L using (2.47) .

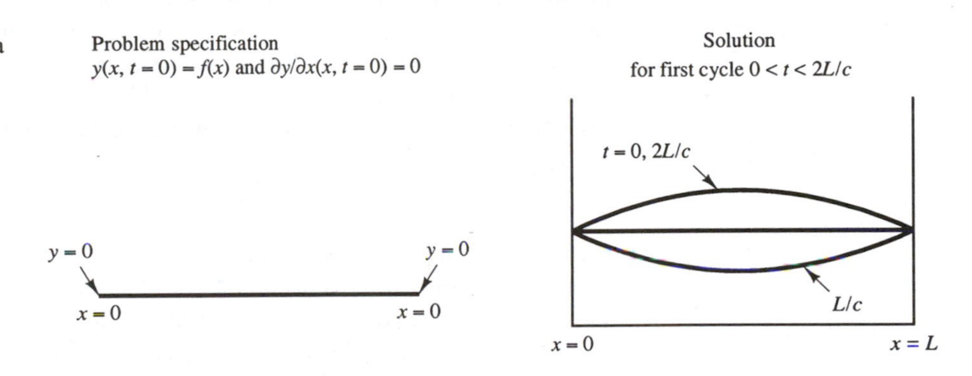

Fig. 2.8 Vibrations of a string under tension

Solutions to the wave equation (2.47) and other hyperbolic equations can be obtained by specifying two initial conditions on the displacement y of the string and one condition on all boundaries for times $t > 0$. Thus hyperbolic problems are also **initial-boundary-value problems**.

If the initial amplitude is given by a, the solution of this problem is

$$y(x, t) = a\cos\left(\frac{\pi c t}{L}\right) \sin\left(\frac{\pi x}{L}\right)$$

The solution shows that the vibration amplitude remains constant, which demonstrates the lack of damping in the system. This absence of damping has a further important consequence. Consider, for example, initial conditions corresponding to a near-triangular initial shape whose apex is a section of a circle with very small radius of curvature. This initial shape has a sharp discontinuity at the apex, but it can be represented by means of a Fourier series as a combination of sine waves. The governing equation is linear so each of the individual Fourier components (and also their sum) would persist in time without change of amplitude. The final result is that the discontinuity remains undiminished due to the absence of a dissipation mechanism to remove the kink in the slope.

Compressible fluid flows at speeds close to and above the speed of sound exhibit shockwaves and it turns out that the inviscid flow equations are hyperbolic at these speeds. The shockwave discontinuities are manifestations of the hyperbolic nature of such flows. Computational algorithms for hyperbolic problems are shaped by the need to allow for the possible existence of discontinuities in the interior of the solution.

It will be shown that disturbances at a point can only influence a limited region in space. The speed of disturbance propagation through an hyperbolic problem is finite and equal to the wave speed c. In contrast parabolic and elliptic models assume infinite propagation speeds.

2.7 The role of characteristics in hyperbolic equations

Hyperbolic equations have a special behaviour which is associated with the finite speed, namely the wave speed, at which information travels through the problem. This distinguishes hyperbolic equations from the two other types. To develop the ideas about the role of characteristic lines in hyperbolic problems we consider again a simple hyperbolic problem described by the wave equation (2.47). It can be shown (Open University, 1984) that a change of variables to $\zeta = x - ct$ and $\eta = x + ct$ transforms the wave equation into the following standard form:

$$\frac{\partial^2 \phi}{\partial \zeta \partial \eta} = 0 \tag{2.48}$$

The transformation requires repeated application of the chain rule for differentiation to express the derivatives of equation (2.47) in terms of the derivatives of the transform variables. The equation (2.48) can be solved very easily. The solution is, of course, $\phi(\zeta, \eta) = F_1(\zeta) + F_2(\eta)$, where F_1 and F_2 can be any function.

A return to the original variables yields the general solution of equation (2.47):

$$\phi(x,\ t) = F_1(x - ct) + F_2(x + ct) \tag{2.49}$$

The first component of the solution, function F_1, is constant if $x - ct$ is constant, and hence along lines of slope $dt/dx = 1/c$ in the x–t plane. The second component F_2 is constant if $x + ct$ is constant, so along lines of slope $dt/dx = -1/c$. The lines $x - ct =$ const and $x + ct =$ const are called the characteristics. Functions F_1 and F_2 represent the so-called **simple wave solutions** of the problem which are travelling waves with velocities $+c$ and $-c$ without change of shape or amplitude.

The particular forms of functions F_1 and F_2 can be obtained from the initial and boundary conditions of the problem. Let us consider a very long string $(-\infty < x < \infty)$ and let the following initial conditions hold:

$$\phi(x,\ 0) = f(x) \quad \text{and} \quad \partial\phi/\partial t(x,\ 0) = g(x) \tag{2.50}$$

Combining (2.49) and (2.50) we obtain

$$F_1(x) + F_2(x) = f(x) \quad \text{and} \quad -cF_1'(x) + cF_2'(x) = g(x) \tag{2.51}$$

It can be shown (Bland, 1988) that the particular solution of wave equation (2.47) with initial conditions (2.50) is given by

$$\phi(x, t) = \frac{1}{2}[f(x-ct) + f(x+ct)] + \frac{1}{2c}\int_{x-ct}^{x+ct} g(s)ds \tag{2.52}$$

Careful inspection of formula (2.52) shows that ϕ at point (x, t) in the solution domain depends only on the initial conditions in the interval $(x - ct, x + ct)$. It is particularly important to note that this implies that *the solution at (x,t) does not depend on initial conditions outside this interval.*

Figure 2.9 seeks to illustrate this point. The characteristics $x - ct =$ constant and $x + ct =$ constant through the point (x', t') intersect the x-axis at the points $(x' - ct', 0)$ and $(x' + ct', 0)$ respectively. The region in the $x - t$ plane enclosed by the x-axis and the two characteristics is termed the **domain of dependence**.

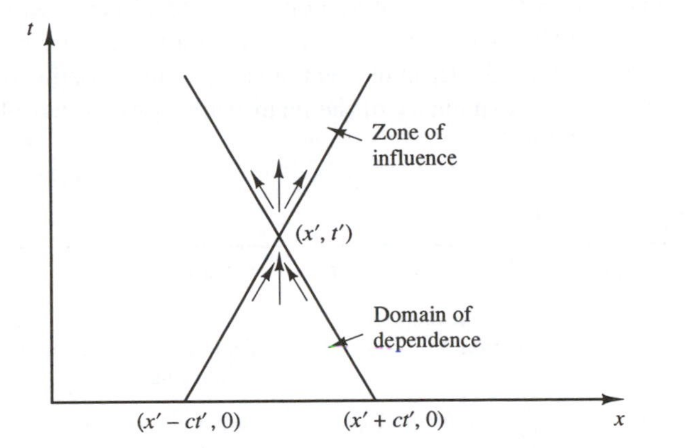

Fig. 2.9 Domain of dependence and zone of influence for an hyperbolic problem

In accordance with formula (2.52) the solution at (x', t') is influenced only by events inside the domain of dependence and not those outside. Physically this is caused by the limited propagation speed (equal to wave speed c) of mutual influences through the solution domain. Changes at the point (x', t') influence events at later times within the **zone of influence** shown in Figure 2.9, which is again bounded by the characteristics.

Figure 2.10a shows the situation for the vibrations of a string fixed at $x = 0$ and $x = L$. For points very close to the x-axis the domain of dependence is enclosed by two characteristics which originate at points on the x-axis. The characteristics through points such as P intersect the problem boundaries. The domain of

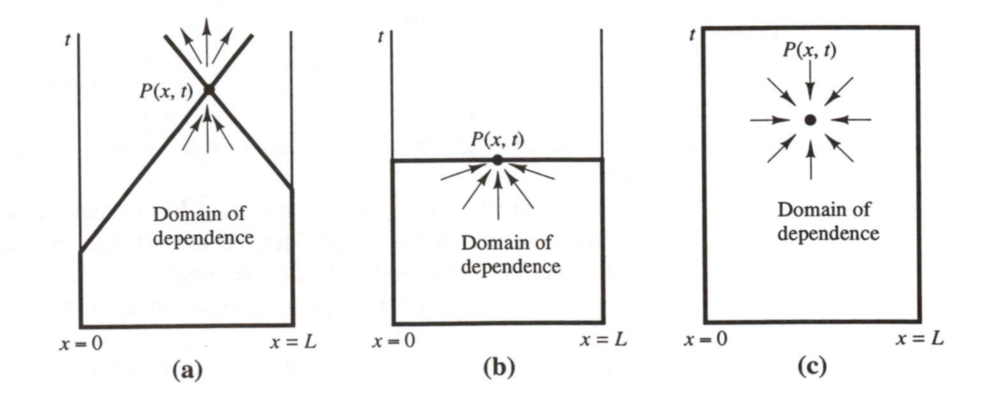

Fig. 2.10 Domains of dependence for (a) hyperbolic, (b) parabolic and (c) elliptic problem

dependence of P is bounded by these two characteristics and the lines $t = 0$, $x = 0$ and $x = L$.

The shape of the domains of dependence (see Figure 2.10b and c) in parabolic and elliptic problems is different because the speed of information travel is assumed to be infinite. The bold lines which demarcate the boundaries of each domain of dependence give the regions for which initial and/or boundary conditions are needed in order to be able to generate a solution at the point $P(x, t)$ in each case.

The way in which changes at one point affect events at other points depends on whether a physical problem represents a steady state or a transient phenomenon and whether the propagation speed of disturbances is finite or infinite. This has resulted in a classification of physical behaviours, and hence attendant PDEs, into elliptic, parabolic and hyperbolic problems. The distinguishing features of each of the categories was illustrated by considering three simple prototype second-order equations. In the following sections we shall discuss methods of classifying more complex PDEs and briefly state the limitations of the computational methods that will be developed later in this text in terms of the classification of the flow problems to be solved. A summary of the main features which have been identified so far is given in Table 2.2.

Table 2.2 Classification of physical behaviours

Problem type	*Equation type*	*Prototype equation*	*Conditions*	*Solution domain*	*Solution smoothness*
Equilibrium problems	Elliptic	$div\ grad\ \phi = 0$	Boundary conditions	Closed domain	Always smooth
Marching problems with dissipation	Parabolic	$\dfrac{\partial \phi}{\partial t} = \alpha\ div\ grad\ \phi$	Initial and boundary conditions	Open domain	Always smooth
Marching problems without dissipation	Hyperbolic	$\dfrac{\partial^2 \phi}{\partial t^2} = c^2\ div\ grad\ \phi$	Initial and boundary conditions	Open domain	May be discontinuous

2.8 Classification method for simple partial differential equations

A practical method of classifying PDEs is developed for a general second-order PDE in two co-ordinates x and y. Consider

$$a\frac{\partial^2 \phi}{\partial x^2} + b\frac{\partial^2 \phi}{\partial x \partial y} + c\frac{\partial^2 \phi}{\partial y^2} + d\frac{\partial \phi}{\partial x} + e\frac{\partial \phi}{\partial y} + f\phi + g = 0 \qquad (2.53)$$

At first we shall assume that the equation is linear and a, b, c, d, e, f and g are constants.

The classification of a PDE is governed by the behaviour of its highest order derivatives, so we need only consider the second-order derivatives. The class of a second-order PDE can be identified by searching for possible simple wave solutions. If they exist this indicates an hyperbolic equation. If not the equation is parabolic or elliptic.

Simple wave solutions occur if the characteristic equation (2.54) below has two

Table 2.3 Classification of linear second-order PDEs

$b^2 - 4ac$	Equation type	Characteristics
> 0	Hyperbolic	Two real characteristics
$= 0$	Parabolic	One real characteristic
< 0	Elliptic	No characteristics

real roots:

$$a\left(\frac{dy}{dx}\right)^2 - b\left(\frac{dy}{dx}\right) + c = 0 \tag{2.54}$$

The existence or otherwise of roots of the characteristic equation depends on the value of discriminant $(b^2 - 4ac)$. Table 2.3 outlines the three cases.

It is left as an exercise for the reader to verify the nature of the three prototype PDEs in section 2.6 by evaluating the discriminant.

The classification method by searching for the roots of the characteristic equation also applies if the coefficients a, b and c are functions of x and y or if the equation is non-linear. In the latter case a, b and c may be functions of dependent variable ϕ or its first derivatives. It is now possible that the equation type differs in various regions of the solution domain. As an example we consider the following equation:

$$y\frac{\partial^2 \phi}{\partial x^2} + \frac{\partial^2 \phi}{\partial y^2} = 0 \tag{2.55}$$

We look at the behaviour within the region $-1 < y < 1$. Hence $a = a(x, y) = y, b = 0$ and $c = 1$. The value of discriminant $(b^2 - 4ac)$ is equal to $-4y$. We need to distinguish three cases:

- If $y < 0 : b^2 - 4ac > 0$ so the equation is hyperbolic.
- If $y = 0 : b^2 - 4ac = 0$ so the equation is parabolic.
- If $y > 0 : b^2 - 4ac < 0$ hence the equation is elliptic.

Equation (2.55) is of mixed type. The equation is locally hyperbolic, parabolic or elliptic depending on the value of y. For the non-linear case similar remarks apply. The classification of the PDE depends on the local values of a, b and c.

Second-order PDEs in N independent variables $(x_1, x_2, ..., x_N)$ can be classified by re-writing them first in the following form with $A_{jk} = A_{kj}$:

$$\sum_{j=1}^{N}\sum_{k=1}^{N} A_{jk}\frac{\partial^2 \phi}{\partial x_j \partial x_k} + H = 0 \tag{2.56}$$

Fletcher (1991) explains that the equation can be classified on the basis of the eigenvalues of a matrix with entries A_{jk}. Hence we need to find values for λ for which

$$\det\left[A_{jk} - \lambda I\right] = 0 \tag{2.57}$$

The classification rules are:

- if any eigenvalue $\lambda = 0$, the equation is parabolic
- if all eigenvalues $\lambda \neq 0$ and they are all of the same sign, the equation is elliptic
- if all eigenvalues $\lambda \neq 0$ and all but one are of the same sign, the equation is hyperbolic

In the cases of the Laplace equation, the diffusion equation and the wave equation it is simple to verify that this method yields the same results as the solution of characteristic equation (2.54).

2.9 Classification of fluid flow equations

Systems of first-order PDEs with more than two independent variables are similarly cast in matrix form. Their classification involves finding eigenvalues of the resulting matrix. Systems of second-order PDEs or mixtures of first- and second-order PDEs can also be classified with this method. The first stage of the method involves the introduction of auxiliary variables which express each second-order equation in first-order equations. Care must be taken to select the auxiliary variables in such a way that the matrix which appears is non-singular.

The Navier–Stokes equation and its reduced forms can be classified using such a matrix approach. The details are beyond the scope of this short introduction to the subject. We quote the main results in Table 2.4 and refer the interested reader to Fletcher (1991) for a full discussion.

Table 2.4 Classification of the main categories of fluid flow

	Steady flow	*Unsteady flow*
Viscous flow	Elliptic	Parabolic
Inviscid flow	M < 1 elliptic M > 1 hyperbolic	Hyperbolic
Thin shear layers	Parabolic	Parabolic

The classifications in Table 2.4 are the 'formal' classifications of the flow equations. In practice many fluid flows behave in a complex way. The steady Navier–Stokes equations and the energy (or enthalpy) equations are formally elliptic and the unsteady equations are parabolic.

The mathematical classification of inviscid flow equations is different from the Navier–Stokes and energy equations due to the complete absence of the (viscous) higher order terms. The classification of the resulting equation set depends on the extent to which fluid compressibility plays a role and hence on the magnitude of the Mach number M. The elliptic nature of inviscid flows at Mach numbers below 1 originates from the action of pressure. If $M < 1$ the pressure can propagate disturbances at the speed of sound which is greater than the flow speed. But if $M > 1$ the fluid velocity is greater than the propagation speed of disturbances and the pressure is unable to influence events in the upstream direction. Limitations to the zone of influence are a key feature of hyperbolic phenomena, so the supersonic inviscid flow equations are hyperbolic. Below we shall see a simple example which demonstrates this behaviour.

In thin shear layer flows all velocity derivatives in the flow (x- and z-)direction are much smaller than those in the cross-stream (y-)direction. Boundary layers, jets, mixing layers and wakes as well as fully developed duct flows fall within this category. In these conditions the governing equations contain only one (second-order) diffusion term and are therefore classified as parabolic.

As an illustration of the complexities which may arise in inviscid flows we analyse the potential equation which governs steady, isentropic, inviscid, compressible flow past a slender body (Shapiro, 1953) with a free stream Mach number M_∞:

$$\left(1 - M_\infty^2\right)\frac{\partial^2 \phi}{\partial x^2} + \frac{\partial^2 \phi}{\partial y^2} = 0 \tag{2.58}$$

Taking $x_1 = x$ and $x_2 = y$ in equation (2.56) we have matrix elements $A_{11} = 1 - M_\infty^2$, $A_{12} = A_{21} = 0$ and $A_{22} = 1$. To classify the equation we need to solve

$$\det\begin{vmatrix} (1 - M_\infty^2) - \lambda & 0 \\ 0 & 1 - \lambda \end{vmatrix} = 0$$

The two solutions are $\lambda_1 = 1$ and $\lambda_2 = 1 - M_\infty^2$. If the free stream Mach number is smaller than 1 (subsonic flow) both eigenvalues are greater than zero and the flow is elliptic. If the Mach number is greater than 1 (supersonic flow) the second eigenvalue is negative and the flow is hyperbolic. The reader is left to demonstrate that these results are identical to those obtained by considering the discriminant of characteristic equation (2.54).

It is interesting to note that we have discovered an instance of hyperbolic behaviour in a steady flow where both independent variables are space co-ordinates. The flow direction behaves in a time-like manner in hyperbolic inviscid flows and also in the parabolic thin shear layers. These problems are of the marching type and flows can be computed by marching in the time-like direction of increasing x.

The above example shows the dependence of the classification of compressible flows on the parameter M_∞. The general equations of inviscid compressible flow (the Euler equations) exhibit similar behaviour, but the classification parameter is now the local Mach number M. This complicates matters greatly when flows around and above $M = 1$ are to be computed. Such flows may contain shock discontinuities and regions of subsonic (elliptic) flow and supersonic (hyperbolic) flow, whose exact locations are not known a priori. Figure 2.11 gives a sketch of the flow around an aerofoil at a Mach number somewhat greater than 1.

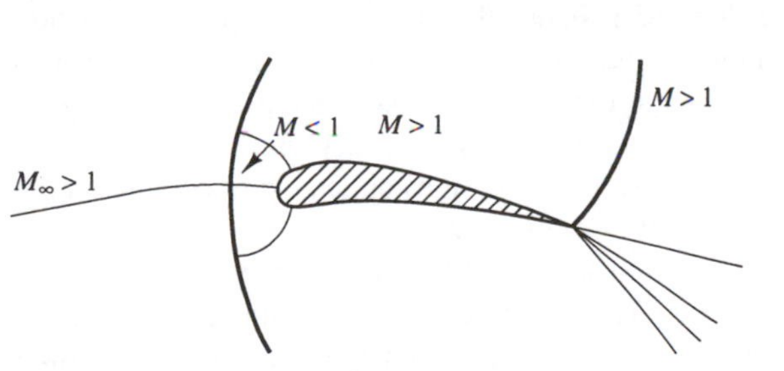

Fig. 2.11 Sketch of flow around an aerofoil at supersonic Mach free stream speed

2.10 Auxiliary conditions for viscous fluid flow equations

The complicated mixture of elliptic, parabolic and hyperbolic behaviours has implications for the way in which boundary conditions enter into a flow problem, in particular at locations where flows are bounded by fluid boundaries. Unfortunately few theoretical results regarding the range of permissible boundary conditions are available for compressible flows. CFD practice is guided here by physical arguments and the success of its simulations. The **boundary conditions for a compressible viscous flow** are given in Table 2.5.

In the table suffices n and t indicate directions normal (outward) and tangential to the boundary respectively and F is the given surface stress.

It is unnecessary to specify outlet or solid wall boundary conditions for the density because of the special character of the continuity equation which describes

Table 2.5 Boundary conditions for compressible viscous flow

Initial conditions for unsteady flows:	
• Everywhere in the solution region ρ, $\mathbf{u}$ and T must be given at time $t = 0$	
Boundary conditions for unsteady and steady flows:	
• On solid walls	$\mathbf{u} = \mathbf{u}_w$ (no-slip condition) $T = T_w$ (fixed temperature) or $k\partial T/\partial n = -q_w$ (fixed heat flux)
• On fluid boundaries	inlet: ρ, $\mathbf{u}$ and T must be known as a function of position outlet: $-p + \mu \partial u_n/\partial n = F_n$ and $\mu \partial u_t/\partial n = F_t$ (stress continuity)

the changes of density experienced by a fluid particle along its path for a known velocity field. At the inlet the density needs to be known. Everywhere else the density emerges as part of the solution and no boundary values need to be specified. For an **incompressible viscous flow** there are no conditions on the density, but all the other above conditions apply without modification.

Commonly outflow boundaries are positioned at locations where the flow is approximately unidirectional and where surface stresses take known values. For high Reynolds number flows far from solid objects in an external flow or in the fully developed flow out of a duct there is no change in any of the velocity components in the direction across the boundary and $F_n = -p$ and $F_t = 0$. This gives the outflow condition which is almost universally used in the finite volume method:

$$\text{specified pressure, } \partial u_n/\partial n = 0 \quad \text{and} \quad \partial T/\partial n = 0$$

Gresho (1991) reviews the intricacies of open boundary conditions in incompressible flow and states that there are some 'theoretical concerns' regarding open boundary conditions which use $\partial u_n/\partial n = 0$; however, its success in CFD practice leaves him to recommend it as the simplest and cheapest form when compared with theoretically more satisfying selections.

Figure 2.12 illustrates the application of boundary conditions for a typical internal and external viscous flow.

General purpose CFD codes also often include inlet and outlet pressure boundary conditions. The pressures are set at fixed values and sources and sinks of mass placed on the boundaries to carry the correct mass flow into and out of the solution zone across the constant pressure boundaries. Furthermore, symmetric and cyclic boundary conditions are supplied to take advantage of special geometrical features of the solution region:

- Symmetry boundary condition: $\partial \phi/\partial n = 0$
- Cyclic boundary condition: $\phi_1 = \phi_2$

Figure 2.13 shows typical boundary geometries for which symmetry and cyclic boundary conditions (bc) may be useful.

2.11 Problems in transonic and supersonic compressible flows

Difficulties arise when calculating flows at speeds near to and above the speed of sound. At these speeds the Reynolds number is usually very high and the viscous regions in the flow are usually very thin. The flow in a large part of the solution

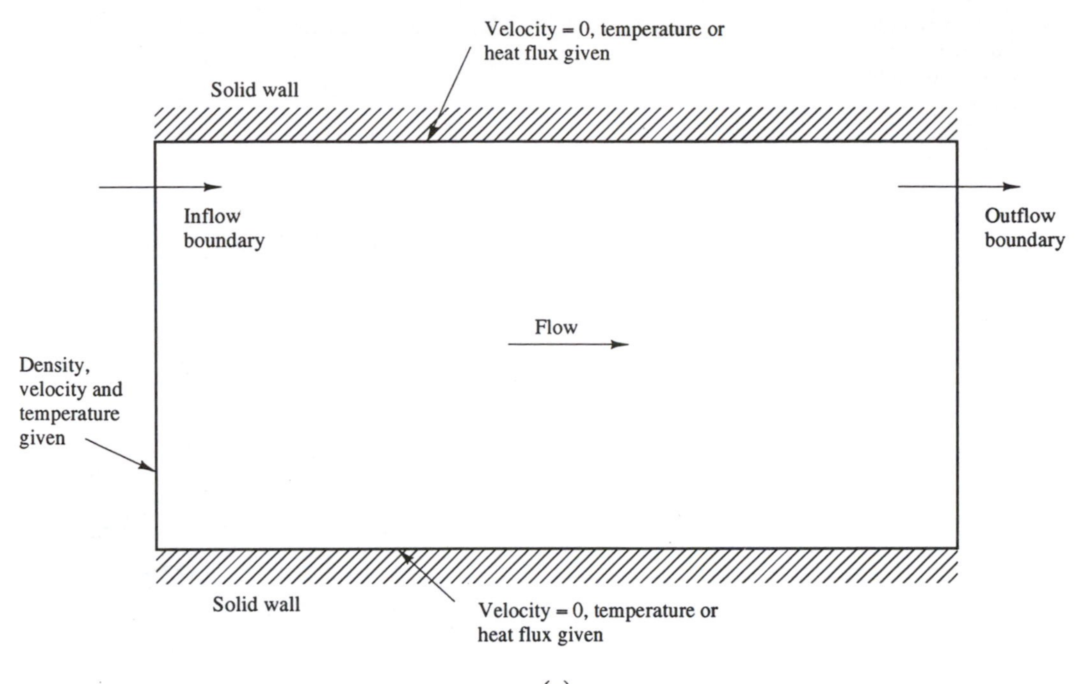

(a)

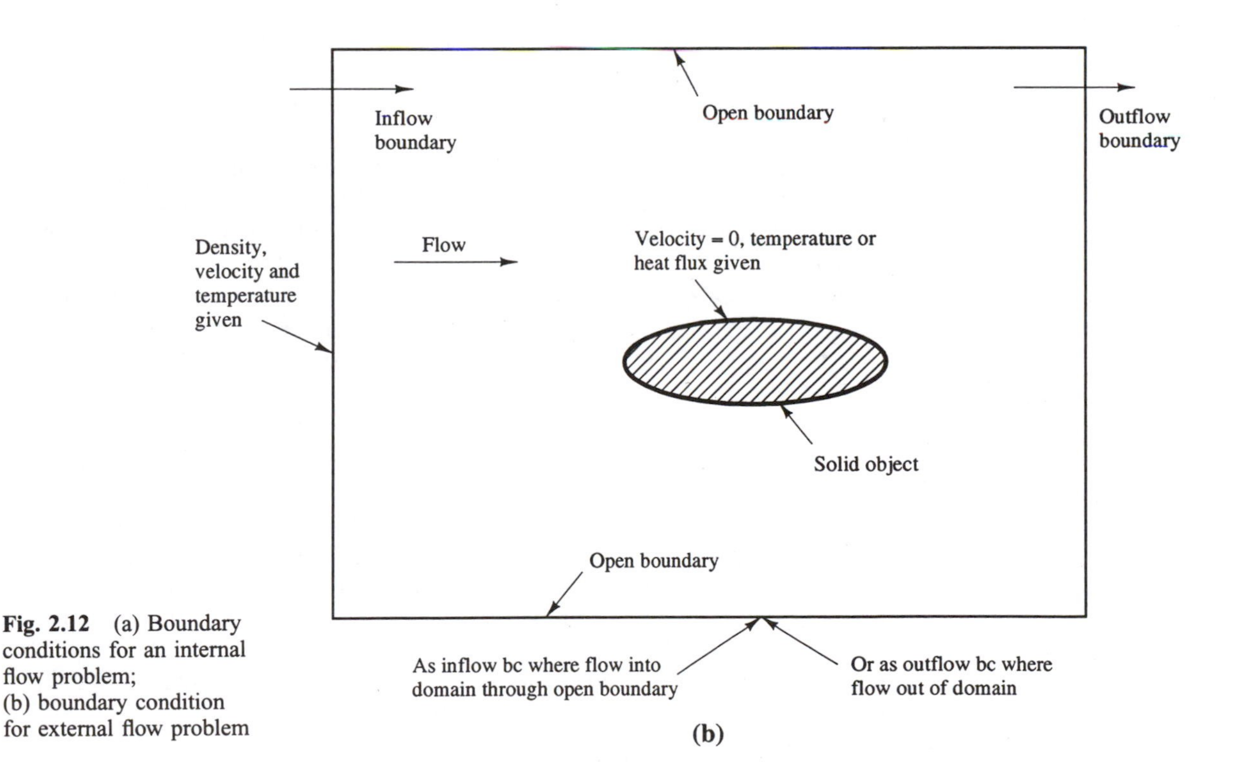

(b)

Fig. 2.12 (a) Boundary conditions for an internal flow problem; (b) boundary condition for external flow problem

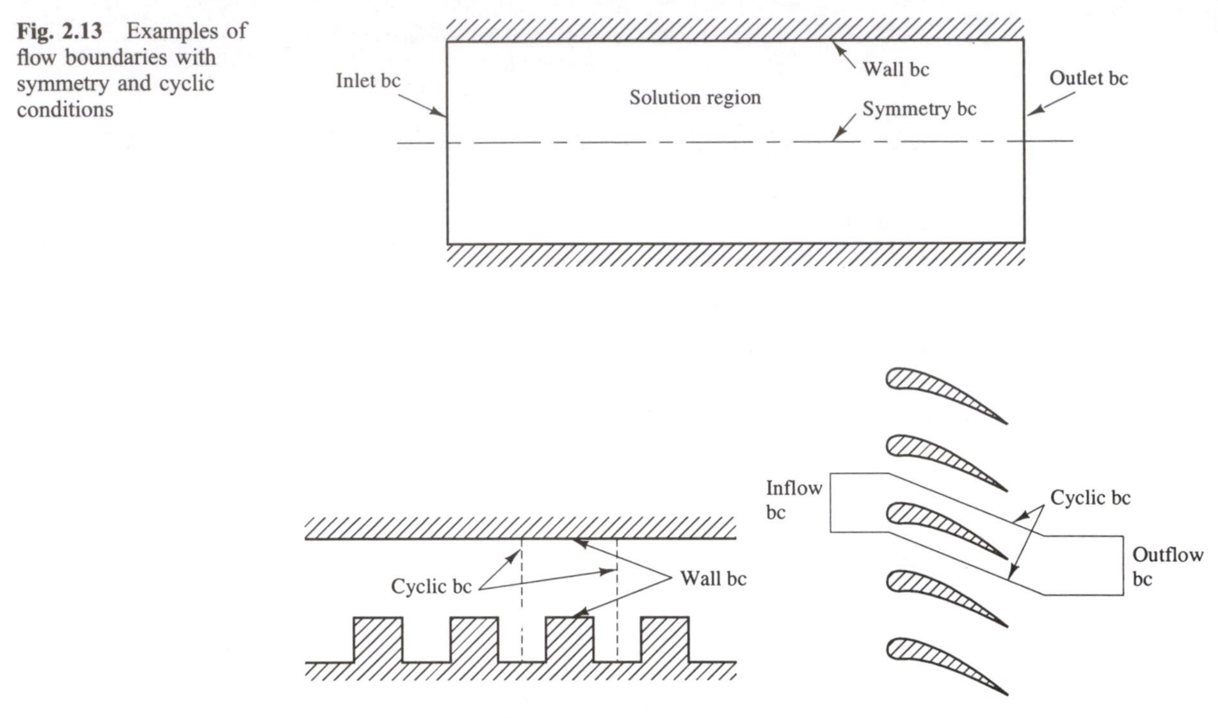

Fig. 2.13 Examples of flow boundaries with symmetry and cyclic conditions

region behaves effectively inviscid. This gives rise to problems in external flows, because the part of the flow where the boundary conditions are applied behaves in an inviscid way which differs from the (viscous) region of flow on which the overall classification is based.

The standard SIMPLE pressure correction algorithm for finite volume calculations (see Chapter 6.4) needs to be modified. The transient version of the algorithm needs to be adopted to make use of the favourable character of parabolic/hyperbolic procedures. To cope with the appearance of shockwaves in the solution interior and with reflections from the domain boundaries artificial damping needs to be introduced. It is further necessary to ensure that the limited domain of dependence of effectively inviscid (hyperbolic) flows at Mach numbers greater than 1 is adequately modelled. Issa and Lockwood (1977) and McGuirk and Page (1990) give lucid papers which identify the main issues relevant to the finite volume method.

Open (far field) boundary conditions give the most serious problems for the designer of general purpose CFD codes. Subsonic inviscid compressible flow equations require fewer inlet conditions (normally only ρ and $\mathbf{u}$ are specified) than viscous flow equations and only one outlet condition (typically specified pressure). Supersonic inviscid compressible flows require the same number of inlet boundary conditions as viscous flows, but do not admit any outflow boundary conditions because the flow is hyperbolic.

Without knowing a great deal about the flow before solving a problem it is very difficult to specify the precise number and nature of the allowable boundary conditions on any fluid/fluid boundary in the far field. Issa and Lockwood's work (1977) reports the solution of a shock/boundary layer interaction problem where part of the far field boundary conditions are obtained from an inviscid solution performed prior to the viscous solution. The usual (viscous) outlet condition $\partial(\rho u_n)/\partial n$ is applied on the remainder of the far field boundary.

Fletcher (1991) notes that under-specification of boundary conditions normally leads to failure to obtain a unique solution. Over-specification, however, gives rise to flow solutions with severe and unphysical 'boundary layers' close to the boundary where the condition is applied.

If the location of the outlet or far field boundaries is chosen far enough away from the region of interest within the solution domain it is possible to get physically meaningful results. Most careful solutions test the sensitivity of the interior solution to the positioning of outflow and far field boundaries. If results do not change in the interior the boundary conditions are 'transparent' and the results are acceptable.

These complexities make it very difficult for general purpose finite volume CFD codes to cope with general subsonic, transonic and/or supersonic viscous flows. Although all commercially available codes claim to be able to make computations in all flow regimes they perform most effectively at Mach numbers well below 1 as a consequence of all the problems outlined above.

2.12 Summary

We have derived the complete set of governing equations of fluid flow from basic conservation principles. The thermodynamic equilibrium assumption and the Newtonian model of viscous stresses were enlisted to close the system mathematically. Since no particular assumptions were made with regard to the viscosity it is straightforward to accommodate a variable viscosity which is dependent on local conditions. This facilitates the inclusion of fluids with temperature-dependent viscosity and those with non-Newtonian characteristics within the framework of equations.

We have identified a common differential form for all the flow equations, the so-called transport equation and developed integrated forms which are central to the finite volume CFD method: for steady state processes

$$\boxed{\int_A \mathbf{n}\,.\,(\rho\phi\mathbf{u})dA = \int_A \mathbf{n}\,.\,(\Gamma\ grad\ \phi)dA + \int_{CV} S_\phi\, dV} \tag{2.43}$$

and for time-dependent processes

$$\boxed{\begin{aligned}\int_{\Delta t} \frac{\partial}{\partial t}\left(\int_{CV} (\rho\phi)dV\right)dt + \int_{\Delta t}\int_A \mathbf{n}\,.\,(\rho\phi\mathbf{u})dA\,dt \\ = \int_{\Delta t}\int_A \mathbf{n}\,.\,(\Gamma_\phi\ grad\ \phi)dA\,dt + \int_{\Delta t}\int_{CV} S_\phi\, dV\,dt\end{aligned}} \tag{2.44}$$

The auxiliary conditions – initial and boundary conditions – needed to solve a fluid flow problem were also discussed. It emerged that there are three types of distinct physical behaviour, elliptic, parabolic and hyperbolic, and the governing fluid flow equations were formally classified. Problems with this formal classification were identified as resulting from: (i) boundary-layer-type behaviour in flows at high

Reynolds numbers and (ii) compressibility effects at Mach numbers around and above 1. These lead to severe difficulties in the specification of boundary conditions for completely general purpose CFD procedures working at any Reynolds number and Mach number.

Experience with the finite volume method has yielded a set of auxiliary conditions that give physically realistic flow solutions in many industrially relevant problems. The most complete problem specification includes, in addition to the initial values of all flow variables, the following boundary conditions:

- complete specification of the distribution of all variables ϕ (except pressure) at all **inlets** to the flow domain of interest
- specification of pressure at one location inside the flow domain
- set gradient of all variables ϕ to zero in the flow direction at suitably positioned **outlets**
- specification of all variables ϕ (except pressure and density) or their normal gradients at solid walls

3

Turbulence and its Modelling

All flows encountered in engineering practice, both simple ones such as two-dimensional jets, wakes, pipe flows and flat plate boundary layers and more complicated three-dimensional ones, become unstable above a certain Reynolds number (UL/ν where U and L are characteristic velocity and length scales of the mean flow and ν is the kinematic viscosity). At low Reynolds numbers flows are laminar. At higher Reynolds numbers flows are observed to become turbulent. A chaotic and random state of motion develops in which the velocity and pressure change continuously with time within substantial regions of flow.

Flows in the laminar regime are completely described by the equations developed in Chapter 2. In simple cases the continuity and Navier–Stokes equations can be solved analytically (Schlichting, 1979). More complex flows can be tackled numerically with CFD techniques such as the finite volume method without additional approximations.

Many, if not most, flows of engineering significance are turbulent so the turbulent flow regime is not just of theoretical interest. Fluid engineers need access to viable tools capable of representing the effects of turbulence. This chapter gives a brief introduction to the physics of turbulence and to its modelling in CFD.

In sections 3.1 and 3.2, the nature of turbulent flows and the physics of the transition from laminar flow to turbulence are examined. Next, in section 3.3, the consequences of the appearance of the fluctuations associated with turbulence on the time-averaged Navier–Stokes equations are analysed and in section 3.4 the characteristics of some simple two-dimensional turbulent flows are described. The velocity fluctuations give rise to additional stresses on the fluid, the so-called Reynolds stresses. An engineering approach to the modelling of these extra stress terms and its implementation will be discussed in section 3.5.

3.1 What is turbulence?

Here we take a brief look at the main characteristics of turbulent flows. The Reynolds number of a flow gives a measure of the relative importance of inertia forces (associated with convective effects) and viscous forces. In experiments on fluid systems it is observed that at values below the so-called critical Reynolds number Re_{crit} the flow is smooth and adjacent layers of fluid slide past each other in an orderly fashion. If the applied boundary conditions do not change with time the flow is steady. This regime is called **laminar** flow.

At values of the Reynolds number above Re_{crit} a complicated series of events takes place which eventually leads to a radical change of the flow character. In the final state the flow behaviour is random and chaotic. The motion becomes intrinsically unsteady even with constant imposed boundary conditions. The velocity and all other flow properties vary in a random and chaotic way. This regime is called **turbulent** flow. A typical point velocity measurement might exhibit the form shown in Figure 3.1.

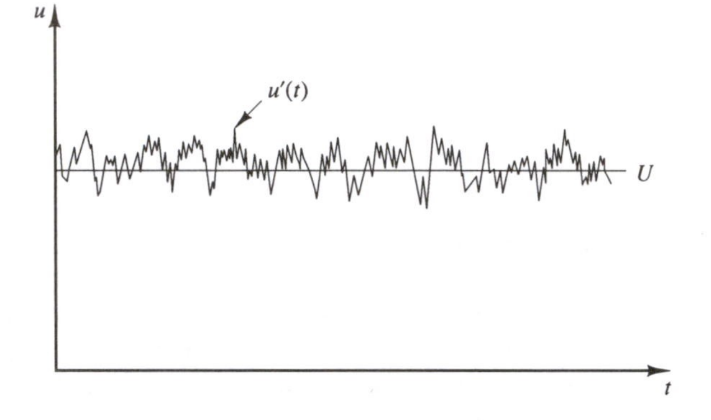

Fig. 3.1 Typical point velocity measurement in turbulent flow

The **random** nature of a turbulent flow precludes computations based on a complete description of the motion of all the fluid particles. Instead the velocity in Figure 3.1 can be decomposed into a steady mean value U with a fluctuating component $u'(t)$ superimposed on it: $u(t) = U + u'(t)$. In general, it is most attractive to characterise a turbulent flow by the mean values of flow properties (U, V, W, P etc.) and the statistical properties of their fluctuations (u', v', w', p' etc.).

Even in flows where the mean velocities and pressures vary in only one or two space dimensions, turbulent fluctuations always have a **three-dimensional** spatial character. Furthermore, visualisations of turbulent flows reveal rotational flow structures, so-called turbulent eddies, with a **wide range of length scales**. Figure 3.2, which depicts a cross-sectional view of a turbulent boundary layer on a flat plate, shows eddies whose length scale is comparable to that of the flow boundaries as well as eddies of intermediate and small size.

Particles of fluid which are initially separated by a long distance can be brought close together by the eddying motions in turbulent flows. As a consequence, heat, mass and momentum are very effectively exchanged. For example, a streak of dye which is introduced at a point in a turbulent flow will rapidly break up and be dispersed right across the flow. Such **effective mixing** gives rise to high values of diffusion coefficients for mass, momentum and heat.

The largest turbulent eddies interact with and extract energy from the mean flow by a process called vortex stretching. The presence of mean velocity gradients in sheared flows distorts the rotational turbulent eddies. Suitably aligned eddies are stretched because one end is forced to move faster than the other.

Fig. 3.2 Visualisation of a turbulent boundary layer

The characteristic velocity ϑ and characteristic length ℓ of the larger eddies are of the same order as the velocity scale U and length scale L of the mean flow. Hence a 'large eddy' Reynolds number $(= \vartheta\ell/\nu)$ formed by combining these eddy scales with the kinematic viscosity will be large in all turbulent flows (since UL/ν is also large) so these large eddies are dominated by inertia effects and viscous effects are negligible.

The large eddies are therefore effectively inviscid and angular momentum is conserved during vortex stretching. This causes the rotation rate to increase and the radius of their cross-sections to decrease. Thus the process creates motions at smaller transverse length scales and also at smaller time scales. The stretching work done by the mean flow on the large eddies provides the energy which maintains the turbulence.

Smaller eddies are themselves stretched strongly by somewhat larger eddies and more weakly by the mean flow. In this way the kinetic energy is handed down from large eddies to progressively smaller and smaller eddies in what is termed the **energy cascade**. All the fluctuating properties of a turbulent flow contain energy across a wide range of frequencies or wavenumbers $(= 2\pi f/U$ where f = frequency). This is demonstrated in Figure 3.3 which gives the energy spectrum of turbulence downstream of a grid.

The smallest scale of motion which can occur in a turbulent flow is dictated by viscosity. The Reynolds number of the smallest eddies based on their characteristic velocity υ and characteristic length $\eta (= \upsilon\eta/\nu)$ is equal to 1. At these scales (lengths on the order of 0.1 to 0.01 mm and frequencies around 10 kHz in typical turbulent engineering flows) viscous effects become important. Work is performed against the action of viscous stresses, so that the energy associated with the eddy motions is dissipated and converted into thermal internal energy. This dissipation results in **increased energy losses** associated with turbulent flows.

The structure of the **largest eddies** is highly **anisotropic** (directional) and flow dependent due to their strong interaction with the mean flow. The diffusive action of viscosity tends to smear out directionality at small scales. At high mean flow Reynolds numbers the **smallest eddies** in a turbulent flow are, therefore, **isotropic** (non-directional).

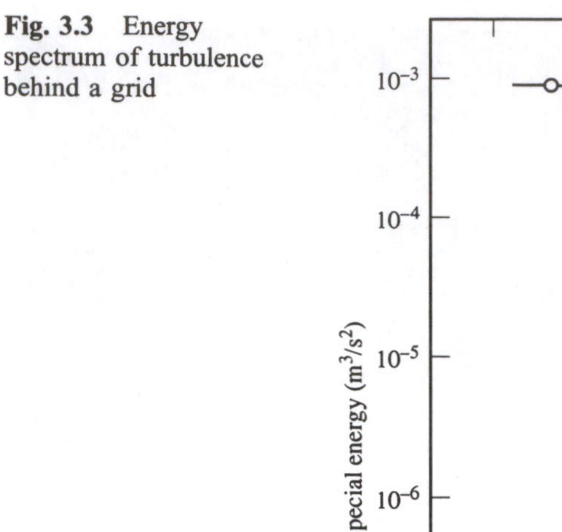

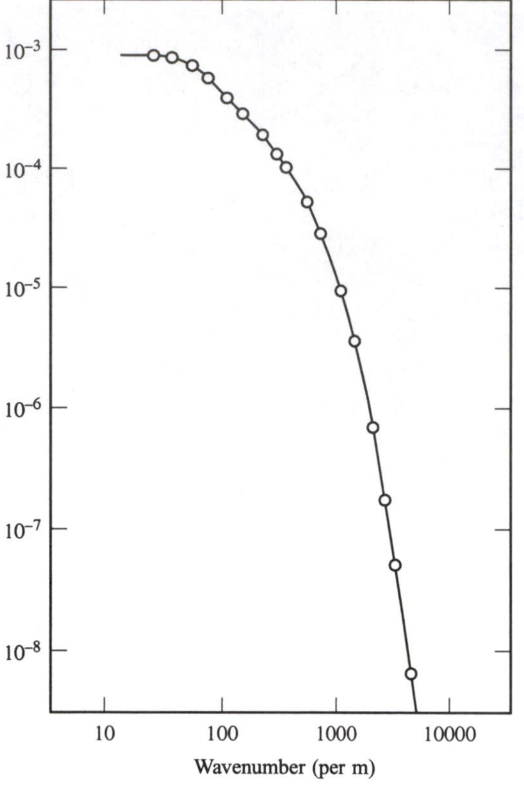

Fig. 3.3 Energy spectrum of turbulence behind a grid

3.2 Transition from laminar to turbulent flow

The initial cause of the **transition** to turbulence can be explained by considering the stability of laminar flows to small disturbances. A sizeable body of theoretical work is devoted to the analysis of the inception of transition: **hydrodynamic instability**. In many relevant instances the transition to turbulence is associated with sheared flows. Linear hydrodynamic stability theory seeks to identify conditions which give rise to the amplification of disturbances. Of particular interest in an engineering context is the prediction of the values of the Reynolds numbers $Re_{x,crit}(= Ux_{crit}/\nu)$ at which disturbances are amplified and $Re_{x,tr}(= Ux_{tr}/\nu)$ at which transition to fully turbulent flow takes place.

A mathematical discussion of the theory is beyond the scope of this brief introduction. White (1991) gives a useful overview of theory and experiments. The subject matter is fairly complex but its confirmation has led to a series of experiments which reveal an insight into the physical processes causing the transition from laminar to turbulent flow. Most of our knowledge stems from work on two-dimensional incompressible flows. All such flows are sensitive to two-dimensional disturbances with a relatively long wavelength, several times the transverse distance over which velocity changes take place (e.g. six times the thickness of a flat plate boundary layer).

Hydrodynamic stability of laminar flows

Two fundamentally different instability mechanisms operate, which are associated with the shape of the two-dimensional laminar velocity profile of the base flow. Flows with a velocity distribution which contains a point of inflexion as shown in Figure 3.4a are always unstable with respect to infinitesimal disturbances if the Reynolds number is large enough. This instability was first identified by making an inviscid assumption in the equations describing the evolution of the disturbances. Subsequent refinement of the theory by inclusion of the effect of viscosity changed its results very little, so this type of instability is known as **inviscid instability**. Velocity profiles of the type shown in Figure 3.4a are associated with jet flows, mixing layers and wakes and also with boundary layers over flat plates under the influence of an adverse pressure gradient ($\partial p/\partial x > 0$). The role of viscosity is to dampen out fluctuations and stabilise the flow at low Reynolds numbers.

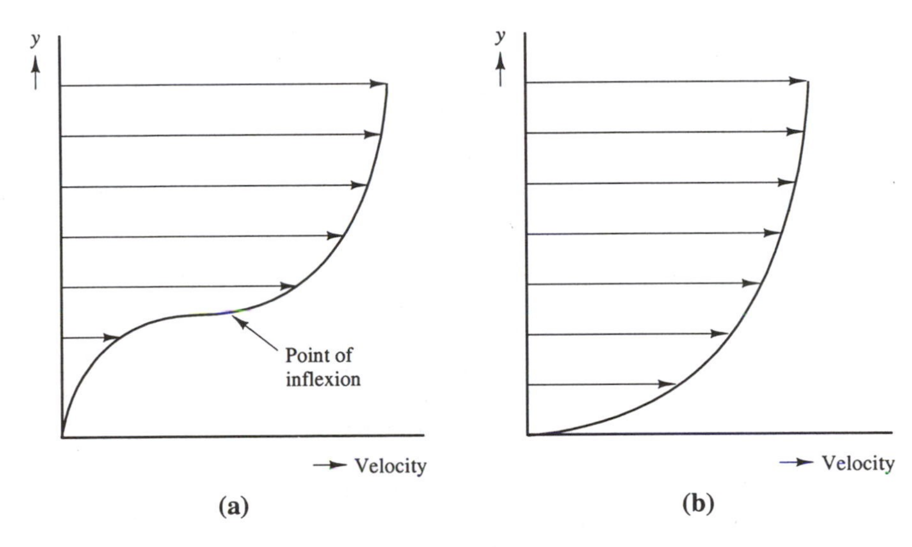

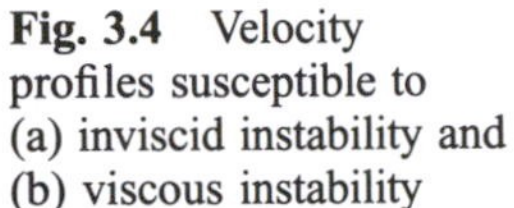
Fig. 3.4 Velocity profiles susceptible to (a) inviscid instability and (b) viscous instability

Flows with laminar velocity distributions without a point of inflexion such as the profile shown in Figure 3.4b are susceptible to **viscous instability**. The approximate inviscid theory predicts unconditional stability for these velocity profiles, which are invariably associated with flows near solid walls such as pipe, channel and boundary layer flows without adverse pressure gradients ($\partial p/\partial x \leq 0$). Viscous effects play a more complex role providing damping at low and high Reynolds numbers, but contributing to the destabilisation of the flows at intermediate Reynolds numbers.

Transition to turbulence

The point where instability first occurs is always upstream of the point of transition to fully turbulent flow. The distance between the point of instability where the Reynolds number equals $Re_{x,crit}$ and the point of transition $Re_{x,tr}$ depends on the degree of amplification of the unstable disturbances. The point of instability and the onset of the transition process can be predicted with the linear theory of hydrodynamic instability. There is, however, no comprehensive theory regarding the path leading from initial instability to fully turbulent flows. Below we describe the main, experimentally observed, characteristics of three simple flows: jets, flat plate boundary layers and pipe flows.

Jet flow: an example of a flow with a point of inflexion. Flows which possess one or more points of inflexion amplify long wavelength disturbances at all Reynolds numbers typically above about 10. The transition process is explained by considering the sketch of a jet flow (Figure 3.5).

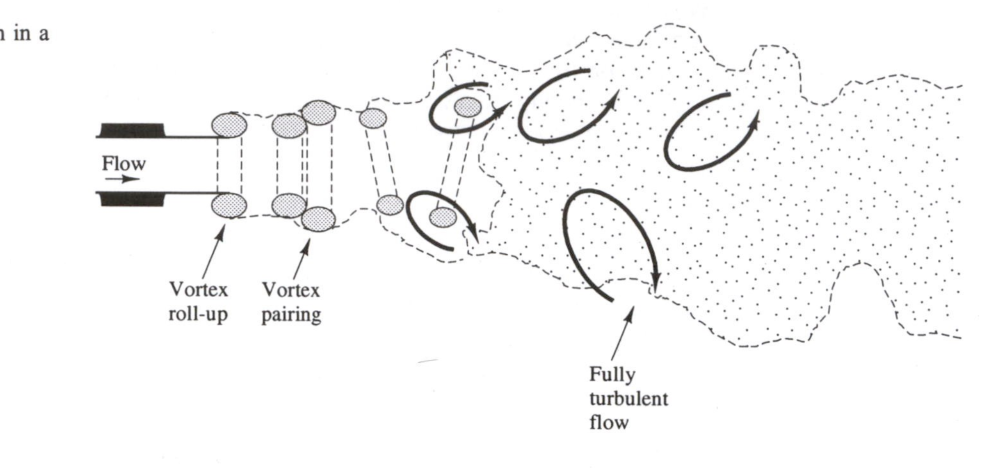

Fig. 3.5 Transition in a jet flow

After the flow emerges from the orifice the laminar exit flow produces the rolling up of a vortex fairly close to the orifice. Subsequent amplification involves the formation of a single vortex of greater strength through the pairing of vortices. A short distance further downstream three-dimensional disturbances cause the vortices to become heavily distorted and less distinct. The flow breaks down generating a large number of small scale eddies and undergoes rapid transition to the fully turbulent regime. Mixing layers and wakes behind bluff bodies exhibit a similar sequence of events leading to transition and turbulent flow.

Boundary layer on a flat plate: an example of a flow without a point of inflexion. In flows with a velocity distribution without a point of inflexion viscous instability theory predicts that there is a finite region of Reynolds numbers around $Re_\delta = 1000$ (δ = boundary layer thickness) where infinitesimal disturbances are amplified. The developing flow over a flat plate is such a flow and the transition process has been extensively researched for this case.

The precise sequence of events is sensitive to the level of disturbance of the incoming flow. However, if the flow system creates sufficiently smooth conditions the instability of a boundary layer flow to relatively long wavelength disturbances can be clearly detected. A sketch of the processes leading to transition and fully turbulent flow is given in Figure 3.6.

If the incoming flow is laminar numerous experiments confirm the predictions of the theory that initial linear instability occurs around $Re_{x,crit} = 91000$. The unstable two-dimensional disturbances are called Tollmien–Schlichting (T–S) waves. These disturbances are amplified in the flow direction.

The subsequent development depends on the amplitude of the waves at maximum (linear) amplification. Since amplification takes place over a limited range of Reynolds numbers, it is possible that the amplified waves are attenuated further downstream and that the flow remains laminar. If the amplitude is large enough a

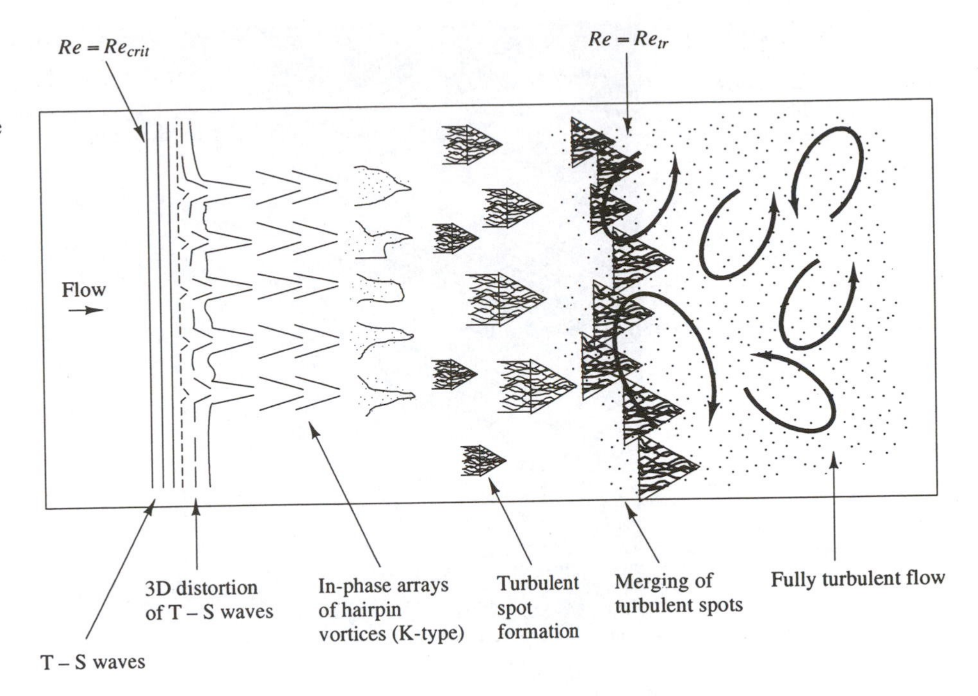

Fig. 3.6 Plan view sketch of transition processes in boundary layer flow over a flat plate

secondary, non-linear, instability mechanism causes the Tollmien–Schlichting waves to become three dimensional and finally evolve into hairpin Λ-vortices. In the most common mechanism of transition, the so-called K-type transition, the hairpin vortices are aligned.

Above the hairpin vortices a high shear region is induced which subsequently intensifies, elongates and rolls up. Further stages of the transition process involve a cascading breakdown of the high shear layer into smaller units with the frequency spectra of measurable flow parameters approaching randomness. Regions of intense and highly localised changes occur at random times and locations near the solid wall. Triangular turbulent spots burst from these locations. These turbulent spots are carried along with the flow and grow by spreading sideways which causes increasing amounts of laminar fluid to take part in the turbulent motion.

Transition of a natural flat plate boundary layer involves the formation of turbulent spots at active sites and the subsequent merging of different turbulent spots convected downstream by the flow. This takes place at Reynolds numbers $Re_{x,tr} \approx 10^6$. Figure 3.7 is a plan view snapshot of a flat plate boundary layer that illustrates this process.

Pipe flow transition. The transition in a pipe flow represents an example of a special category of flows without an inflexion point. The viscous theory of hydrodynamic stability predicts that these flows are unconditionally stable to infinitesimal disturbances at all Reynolds numbers. In practice, transition to turbulence takes place between $Re(= UD/\nu)$ 2000 and 10^5. Various details are still unclear, which illustrates the limitations of current stability theories.

The cause of the apparent failure of the theory is almost certainly the role played by distortions of the inlet velocity profile and the finite amplitude disturbances due to entry effects. Experiments show that in pipe flows, as in flat plate boundary layers, turbulent spots appear in the near wall region. These grow, merge and finally subsequently fill the pipe cross-section to form turbulent slugs. In industrial pipe

Fig. 3.7 Merging of turbulent spots and transition to turbulence in a natural flat plate boundary layer

flows the intermittent formation of turbulent slugs takes place at Reynolds numbers around 2000 giving rise to alternate turbulent and laminar regions along the length of the pipe. At Reynolds numbers above 2300 the turbulent slugs link up and the entire pipe is filled with turbulent flow.

Final comments

It is clear from the above descriptions of transition in jets, flat plate boundary layers and pipe flows that there are a number of common features in the transition processes: (i) the amplification of initially small disturbances, (ii) the development of areas with concentrated rotational structures, (iii) the formation of intense small scale motions and finally (iv) the growth and merging of these areas of small scale motions into fully turbulent flows.

The transition to turbulence is strongly affected by factors such as pressure gradient, disturbance levels, wall roughness and heat transfer. The discussions only apply to subsonic incompressible flows. The appearance of significant compressibility effects in flows at Mach numbers above about 0.7 greatly complicates the stability theory.

It should be noted that although a great deal has been learnt from simple flows there is no comprehensive theory of transition. Recent advances in supercomputer technology have made it possible to simulate the events leading up to transition, including turbulent spot formation, by solving the complete, time-dependent Navier–Stokes equations at modest Reynolds numbers for a number of very simple geometries. Kleiser and Zang (1991) give a review of the state of the art which highlights very favourable agreement between experiments and (extremely expensive) computations.

For engineering purposes the major case where the transition process influences a sizeable fraction of the flow is that of external wall boundary layer flows at intermediate Reynolds numbers. This occurs in certain turbomachines, helicopter rotors and some low speed aircraft wings. Cebeci (1989) presents an engineering calculation method based on a combination of inviscid far field and boundary layer computations in conjunction with a linear stability analysis to identify the critical and transition Reynolds numbers. Transition is deemed to have occurred at the point where an (arbitrary) amplification factor $e^9 (\approx 8000)$ of initial disturbances is found. The procedure, which includes a mixing length model (see section 3.5.1) for the fully turbulent part of the boundary layer, has proved very effective for aerofoil calculations, but requires a substantial amount of empirical input and therefore lacks generality.

Commercially available general purpose CFD procedures often ignore transition entirely and classify flows as either laminar or fully turbulent. The transition region often comprises only a very small fraction of the size of the flow domain and in those cases it is assumed that the errors made by neglecting its detailed structure are only small.

3.3 Effect of turbulence on time-averaged Navier–Stokes equations

The crucial difference between visualisations of laminar and turbulent flows is the appearance of eddying motions of a wide range of length scales in turbulent flows. A typical flow domain of 0.1 by 0.1 m with a high Reynolds number turbulent flow might contain eddies down to 10 to 100 μm size. We would need computing meshes of 10^9 up to 10^{12} points to be able to describe processes at all length scales. The fastest events take place with a frequency on the order of 10 kHz so we would need to discretise time into steps of about 100 μs. Speziale (1991) states that the direct simulation of a turbulent pipe flow at a Reynolds number of 500000 requires a computer which is 10 million times faster than a current generation CRAY supercomputer.

With present day computing power it has only recently started to become possible to track the dynamics of eddies in very simple flows at transitional Reynolds number (see section 3.2). The computing requirements for the direct solution of the time-dependent Navier–Stokes equations of fully turbulent flows at high Reynolds numbers are truly phenomenal and must await major developments in computer hardware.

Meanwhile, engineers need computational procedures which can supply adequate information about the turbulent processes, but which avoid the need to predict the effects of each and every eddy in the flow. Fortunately this category of CFD users is almost always satisfied with information about the time-averaged properties of the flow (e.g. mean velocities, mean pressures and mean stresses etc.). In this section we examine the effects of the appearance of turbulent fluctuations on the mean flow properties.

Reynolds equations

First we define the mean Φ of a flow property φ as follows:

$$\Phi = \frac{1}{\Delta t} \int_0^{\Delta t} \varphi(t)\, dt \tag{3.1}$$

In theory we should take the limit of time interval Δt approaching infinity, but Δt is large enough if it exceeds the time scales of the slowest variations (due to the largest eddies) of property φ. This definition of the mean of a flow property is adequate for steady mean flows. In time-dependent flows the mean of a property at time t is taken to be the average of the instantaneous values of the property over a large number of repeated identical experiments: the so-called 'ensemble average'.

The flow property φ is time dependent and can be thought of as the sum of a steady mean component Φ and a time-varying fluctuating component φ' with zero mean value; hence $\varphi(t) = \Phi + \varphi'(t)$. From now on we shall not write down the time dependence of φ and φ' explicitly, so we write $\varphi = \Phi + \varphi'$. The time average of the fluctuations φ' is, by definition, zero:

$$\overline{\varphi'} = \frac{1}{\Delta t}\int_0^{\Delta t} \varphi'(t)dt \equiv 0 \tag{3.2}$$

Information regarding the fluctuating part of the flow can, for example, be obtained from the root-mean-square (rms) of the fluctuations:

$$\varphi_{rms} = \sqrt{\overline{(\varphi')^2}} = \left[\frac{1}{\Delta t}\int_0^{\Delta t} (\varphi')^2\, dt\right]^{1/2} \tag{3.3}$$

The rms values of the velocity components are of particular importance since they can be easily measured with a velocity probe sensitive to the turbulent fluctuations (e.g. a hot-wire anemometer) and simple electrical circuitry. The kinetic energy k (per unit mass) associated with the turbulence is defined as

$$k = \frac{1}{2}\left(\overline{u'^2} + \overline{v'^2} + \overline{w'^2}\right) \tag{3.4}$$

The turbulence intensity T_i is linked to the kinetic energy and a reference mean flow velocity U_{ref} as follows:

$$T_i = \frac{\left(\frac{2}{3}k\right)^{1/2}}{U_{ref}} \tag{3.5}$$

Before deriving the mean flow equations for a turbulent flow we summarise the following rules which govern the time averages of fluctuating properties $\varphi = \Phi + \varphi'$ and $\psi = \Psi + \psi'$ and their combinations, derivatives and integrals:

$$\overline{\varphi'} = \overline{\psi'} = 0;\quad \overline{\Phi} = \Phi;\quad \overline{\frac{\partial \varphi}{\partial s}} = \frac{\partial \Phi}{\partial s};\quad \overline{\int \varphi ds} = \int \Phi ds$$

$$\overline{\varphi + \psi} = \Phi + \Psi;\ \overline{\varphi\psi} = \Phi\Psi + \overline{\varphi'\psi'};\ \overline{\varphi\Psi} = \Phi\Psi;\ \overline{\varphi'\Psi} = 0 \tag{3.6}$$

These relationships can be easily verified by application of (3.1) and (3.2) noting that the time-averaging operation is itself an integration and that therefore the order of time averaging and a further integration or differentiation can be swapped.

Since *div* and *grad* are both differentiations the above rules can be extended to a fluctuating vector quantity $\mathbf{a} = \mathbf{A} + \mathbf{a}'$ and its combinations with a fluctuating scalar $\varphi = \Phi + \varphi'$:

$$\overline{div\,\mathbf{a}} = div\,\mathbf{A};\quad \overline{div(\varphi\mathbf{a})} = div\overline{(\varphi\mathbf{a})} = div(\Phi\mathbf{A}) + div\left(\overline{\varphi'\mathbf{a}'}\right);$$

$$\overline{div\ grad\ \varphi} = div\ grad\ \Phi \tag{3.7}$$

To illustrate the influence of turbulent fluctuations on the mean flow we consider the instantaneous continuity and Navier–Stokes equations for an incompressible flow with constant viscosity. This considerably simplifies the algebra involved without detracting from the main messages. As usual we take Cartesian co-ordinates so that the velocity vector $\mathbf{u}$ has x-component u, y-component v and z-component w:

$$div\, \mathbf{u} = 0 \tag{3.8}$$

$$\frac{\partial u}{\partial t} + div(u\mathbf{u}) = -\frac{1}{\rho}\frac{\partial p}{\partial x} + \nu\, div\, grad\, u \tag{3.9a}$$

$$\frac{\partial v}{\partial t} + div(v\mathbf{u}) = -\frac{1}{\rho}\frac{\partial p}{\partial y} + \nu\, div\, grad\, v \tag{3.9b}$$

$$\frac{\partial w}{\partial t} + div(w\mathbf{u}) = -\frac{1}{\rho}\frac{\partial p}{\partial z} + \nu\, div\, grad\, w \tag{3.9c}$$

To investigate the effects of fluctuations we replace in equations (3.8) and (3.9a–c) the flow variables $\mathbf{u}$ (hence also u, v and w) and p by the sum of a mean and fluctuating component. Thus

$$\mathbf{u} = \mathbf{U} + \mathbf{u}';\, u = U + u';\; v = V + v';\; w = W + w';\; p = P + p'$$

Then the time average is taken applying the rules stated in (3.7). Considering the continuity equation (3.8) first we note that $\overline{div \mathbf{u}} = div\, \mathbf{U}$. This yields the continuity equation for the mean flow

$$div\, \mathbf{U} = 0 \tag{3.10}$$

A similar process is now carried out on the x-momentum equation (3.9a). The time averages of the individual terms in this equation can be written as follows:

$$\overline{\frac{\partial u}{\partial t}} = \frac{\partial U}{\partial t}; \qquad \overline{div(u\mathbf{u})} = div(U\mathbf{U}) + div\left(\overline{u'\mathbf{u}'}\right)$$

$$\overline{-\frac{1}{\rho}\frac{\partial p}{\partial x}} = -\frac{1}{\rho}\frac{\partial P}{\partial x}; \qquad \overline{\nu\, div\, grad\, u} = \nu\, div\, grad\, U$$

Substitution of these results gives the time-average x-momentum equation

$$\frac{\partial U}{\partial t} + div(U\mathbf{U}) + div\left(\overline{u'\mathbf{u}'}\right) = -\frac{1}{\rho}\frac{\partial P}{\partial y} + \nu\, div\, grad\, U \tag{3.11a}$$

(I) (II) (III) (IV) (V)

Repetition of this process on equations (3.9b) and (3.9c) yields the time-average y- and z-momentum equations

$$\frac{\partial V}{\partial t} + div(V\mathbf{U}) + div\left(\overline{v'\mathbf{u}'}\right) = -\frac{1}{\rho}\frac{\partial P}{\partial y} + \nu\, div\, grad\, V \tag{3.11b}$$

(I) (II) (III) (IV) (V)

$$\frac{\partial W}{\partial t} + div(W\mathbf{U}) + div\left(\overline{w'\mathbf{u}'}\right) = -\frac{1}{\rho}\frac{\partial P}{\partial z} + \nu\, div\, grad\, W \tag{3.11c}$$

(I) (II) (III) (IV) (V)

It is important to note that the terms (I), (II), (IV) and (V) in (3.11a–c) also appear in the instantaneous equations (3.9a–c), but the process of time averaging has introduced new terms (III) in the resulting time-average momentum equations. The terms involve products of fluctuating velocities and constitute convective momentum transfer due to the velocity fluctuations. It is customary to place these terms on the right hand side of the equations (3.11a–c) to reflect their role as additional turbulent stresses on the mean velocity components U, V and W:

$$\frac{\partial U}{\partial t} + div(U\mathbf{U}) = -\frac{1}{\rho}\frac{\partial P}{\partial x} + \nu\, div\, grad\, U + \left[-\frac{\partial \overline{u'^2}}{\partial x} - \frac{\partial \overline{u'v'}}{\partial y} - \frac{\partial \overline{u'w'}}{\partial z}\right] \tag{3.12a}$$

$$\frac{\partial V}{\partial t} + div(V\mathbf{U}) = -\frac{1}{\rho}\frac{\partial P}{\partial y} + \nu\, div\, grad\, V + \left[-\frac{\partial \overline{u'v'}}{\partial x} - \frac{\partial \overline{v'^2}}{\partial y} - \frac{\partial \overline{v'w'}}{\partial z}\right] \tag{3.12b}$$

$$\frac{\partial W}{\partial t} + div(W\mathbf{U}) = -\frac{1}{\rho}\frac{\partial P}{\partial z} + \nu\, div\, grad\, W + \left[-\frac{\partial \overline{u'w'}}{\partial x} - \frac{\partial \overline{v'w'}}{\partial y} - \frac{\partial \overline{w'^2}}{\partial z}\right] \tag{3.12c}$$

The extra stress terms have been written out in longhand to clarify their structure. They result from six additional stresses, three normal stresses and three shear stresses:

$$\tau_{xx} = -\rho\overline{u'^2} \qquad \tau_{yy} = -\rho\overline{v'^2} \qquad \tau_{zz} = -\rho\overline{w'^2}$$
$$\tau_{xy} = \tau_{yx} = -\rho\overline{u'v'} \qquad \tau_{xz} = \tau_{zx} = -\rho\overline{u'w'} \qquad \tau_{yz} = \tau_{zy} = -\rho\overline{v'w'} \tag{3.13}$$

These extra turbulent stresses are termed the **Reynolds stresses**. In turbulent flows the normal stresses $-\rho\overline{u'^2}$, $-\rho\overline{v'^2}$ and $-\rho\overline{w'^2}$ are always non-zero because they contain squared velocity fluctuations. The shear stresses $-\rho\overline{u'v'}$, $-\rho\overline{u'w'}$, and $-\rho\overline{v'w'}$ are associated with correlations between different velocity components. If, for instance, u' and v' were statistically independent fluctuations the time average of their product $\overline{u'v'}$ would be zero. However, the turbulent shear stresses are also non-zero and usually very large compared to the viscous stresses in a turbulent flow. The equation set (3.12a–c) is called the **Reynolds equations**.

Similar extra turbulent transport terms arise when we derive a transport equation for an arbitrary scalar quantity. The **time average transport equation for scalar** φ is

$$\frac{\partial \Phi}{\partial t} + div(\Phi\mathbf{U}) = div(\Gamma_\Phi^*\, grad\, \Phi) + \left[-\frac{\partial \overline{u'\varphi'}}{\partial x} - \frac{\partial \overline{v'\varphi'}}{\partial y} - \frac{\partial \overline{w'\varphi'}}{\partial z}\right] + S_\Phi \tag{3.14}$$

So far we have assumed that the fluid density is constant, but in practical flows the mean density may vary and the instantaneous density always exhibits turbulent fluctuations. Bradshaw *et al* (1981) state that small density fluctuations do not appear to affect the flow significantly. If rms velocity fluctuations are on the order of 5% of the mean speed they show that density fluctuations are unimportant up to Mach numbers 3 to 5. In free turbulent flows we shall see in section 3.4 that velocity fluctuations can easily reach values around 20% of the mean velocity. In such circumstances density fluctuations start to affect the turbulence around Mach numbers of 1. To summarise the results of the current section we quote, without proof, in Table 3.1 the density-weighted averaged (or Favre-averaged, see Anderson *et al*, 1984) form of the mean flow equations for compressible turbulent flows where the effects of density fluctuations are neglible but the mean density variations are not. This form is widely used in commercial CFD packages. The symbol ρ stands for the mean density.

Table 3.1 Turbulent flow equations for compressible flows

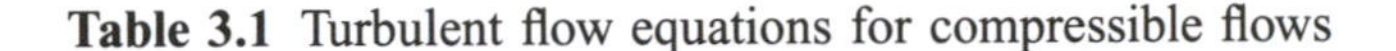

Continuity

$$\frac{\partial \rho}{\partial t} + div(\rho \mathbf{U}) = 0 \tag{3.15}$$

Reynolds equations

$$\frac{\partial(\rho U)}{\partial t} + div(\rho U \mathbf{U}) = -\frac{\partial P}{\partial x} + div(\mu \ grad \ U) + \left[-\frac{\partial\left(\rho \overline{u'^2}\right)}{\partial x} - \frac{\partial(\rho \overline{u'v'})}{\partial y} - \frac{\partial(\rho \overline{u'w'})}{\partial z} \right] + S_{Mx} \tag{3.16a}$$

$$\frac{\partial(\rho V)}{\partial t} + div(\rho V \mathbf{U}) = -\frac{\partial P}{\partial y} + div(\mu \ grad \ V) + \left[-\frac{\partial(\rho \overline{u'v'})}{\partial x} - \frac{\partial\left(\rho \overline{v'^2}\right)}{\partial y} - \frac{\partial(\rho \overline{v'w'})}{\partial z} \right] + S_{My} \tag{3.16b}$$

$$\frac{\partial(\rho W)}{\partial t} + div(\rho W \mathbf{U}) = -\frac{\partial P}{\partial z} + div(\mu \ grad \ W) + \left[-\frac{\partial(\rho \overline{u'w'})}{\partial x} - \frac{\partial(\rho \overline{v'w'})}{\partial y} - \frac{\partial\left(\rho \overline{w'^2}\right)}{\partial z} \right] + S_{Mz} \tag{3.16c}$$

Scalar transport equation

$$\frac{\partial(\rho \Phi)}{\partial t} + div(\rho \Phi \mathbf{U}) = div(\Gamma_\Phi \ grad \ \Phi) + \left[-\frac{\partial(\rho \overline{u'\varphi'})}{\partial x} - \frac{\partial(\rho \overline{v'\varphi'})}{\partial y} - \frac{\partial(\rho \overline{w'\varphi'})}{\partial z} \right] + S_\Phi \tag{3.17}$$

Closure problem – the need for turbulence modelling

The instantaneous continuity and Navier–Stokes equations (3.8) and (3.9a–c) form a closed set of four equations with four unknowns u, v, w and p. In the introduction to this section it was demonstrated that these equations could not be solved directly in the foreseeable future.

Engineers are content to focus their attention on certain mean quantities. However, in performing the time-averaging operation on the momentum equations we throw away all details concerning the state of the flow contained in the instantaneous fluctuations. As a result we obtain six additional unknowns, the Reynolds stresses, in the time averaged momentum equations. Similarly, time average scalar transport equations show extra terms containing $\overline{u'\varphi'}$, $\overline{v'\varphi'}$ and $\overline{w'\varphi'}$. The complexity of turbulence usually precludes simple formulae for the extra stresses and turbulent scalar transport terms. It is the main task of turbulence modelling to develop computational procedures of sufficient accuracy and generality for engineers to predict the Reynolds stresses and the scalar transport terms.

3.4 Characteristics of simple turbulent flows

Most of the theory of turbulent flow and its modelling was initially developed by careful examination of the turbulence structure of thin shear layers. In such flows large velocity changes are concentrated in thin regions. Expressed more formally, the rates of change of flow variables in the (x-)direction of the flow are negligible compared to the rates of change in the cross-stream (y-)direction ($\partial/\partial x \ll \partial/\partial y$). Furthermore, the cross-stream width δ of the region over which changes take place is always small compared to any length scale L in the flow direction ($\delta/L \ll 1$). In the context of this brief introduction we review the overall characteristics of some simple two-dimensional incompressible turbulent flows with constant imposed pressure. The following flows will be considered here:

Free turbulent flows

- mixing layer
- jet
- wake

Boundary layers near solid walls

- flat plate boundary layer
- pipe flow

Given an engineer's recognised interest in mean quantities we review data for the mean velocity distribution $U = U(y)$ and the pertinent Reynolds stresses $-\rho\overline{u'^2}, -\rho\overline{v'^2}, -\rho\overline{w'^2}$ and $-\rho\overline{u'v'}$. Local values of the above-mentioned quantities can be measured very effectively by means of hot-wire anemometry (Comte-Bellot, 1976). More recently laser doppler anemometers have been widely used for mean flow and turbulence measurements (Buchhave *et al*, 1979).

3.4.1 Free turbulent flows

Among the simplest flows of significant engineering importance are those in the category of free turbulent flows: mixing layers, jets and wakes. A mixing layer forms at the interface of two regions: one with fast and the other with slow moving fluid. In a jet a region of high speed flow is completely surrounded by stationary fluid. A wake is formed behind an object in a flow, so here a slow-moving region is surrounded by fast-moving fluid. Figure 3.8 gives a sketch of the development of the mean velocity distribution in the streamwise direction for these free turbulent flows.

It is clear that velocity changes across an initially thin layer are important in all three flows: transition to turbulence occurs after a very short distance in the flow direction from the point where the different streams initially meet, the turbulence causes vigorous mixing of adjacent fluid layers and rapid widening of the region across which the velocity changes take place.

Figure 3.9 shows a visualisation of a jet flow. It is immediately clear that the turbulent part of the flow contains a wide range of length scales. Large eddies with a

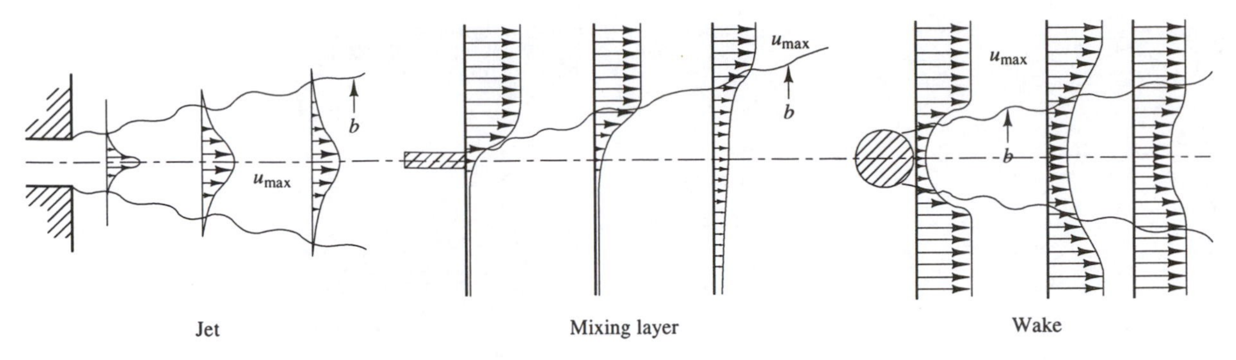

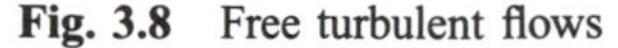
Fig. 3.8 Free turbulent flows

size comparable to the width across the flow are occurring alongside eddies of very small size.

The visualisation correctly suggests that the flow inside the jet region is fully turbulent, but the flow in the outer region far away from the jet is smooth and largely unaffected by the turbulence. The position of the edge of the turbulent zone is determined by the (time-dependent) passage of individual large eddies. Close to the edge these will occasionally penetrate into the surrounding region. During the resulting bursts of turbulent activity in the outer region – called **intermittency** – fluid from the surroundings is drawn into the turbulent zone. This process is termed **entrainment** and is the main cause of the spreading of turbulent flows (including wall boundary layers) in the flow direction.

Initially fast moving jet fluid will lose momentum to speed up the stationary surrounding fluid. Owing to the entrainment of the surrounding fluid the velocity gradients decrease in magnitude in the flow direction. This causes the decrease of the mean speed of the jet at its centreline. Similarly the difference between the speed of the wake fluid and its fast-moving surroundings will decrease in the flow direction. In mixing layers the width of the layer containing the velocity change continues to increase in the flow direction but the overall velocity difference between the two outer regions is unaltered.

Experimental observations of many such turbulent flows show that after a certain distance their structure becomes independent of the exact nature of the flow source. Only the local environment appears to control the turbulence in the flow. The appropriate length scale is the cross-stream layer width (or half width) b. We find that if y is the distance in the cross-stream direction

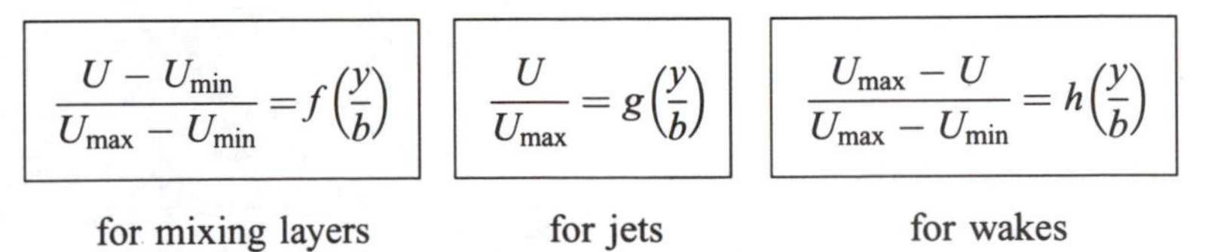

$$\frac{U - U_{\text{min}}}{U_{\text{max}} - U_{\text{min}}} = f\left(\frac{y}{b}\right) \qquad \frac{U}{U_{\text{max}}} = g\left(\frac{y}{b}\right) \qquad \frac{U_{\text{max}} - U}{U_{\text{max}} - U_{\text{min}}} = h\left(\frac{y}{b}\right)$$

for mixing layers for jets for wakes

In these formulae U_{max} and U_{min} represent the maximum and minimum mean velocity at a distance x downstream of the source (see Figure 3.8). Hence, if these local mean velocity scales are chosen and x is large enough, the functions f, g and h are independent of distance x in the flow direction. Such flows are called self-preserving.

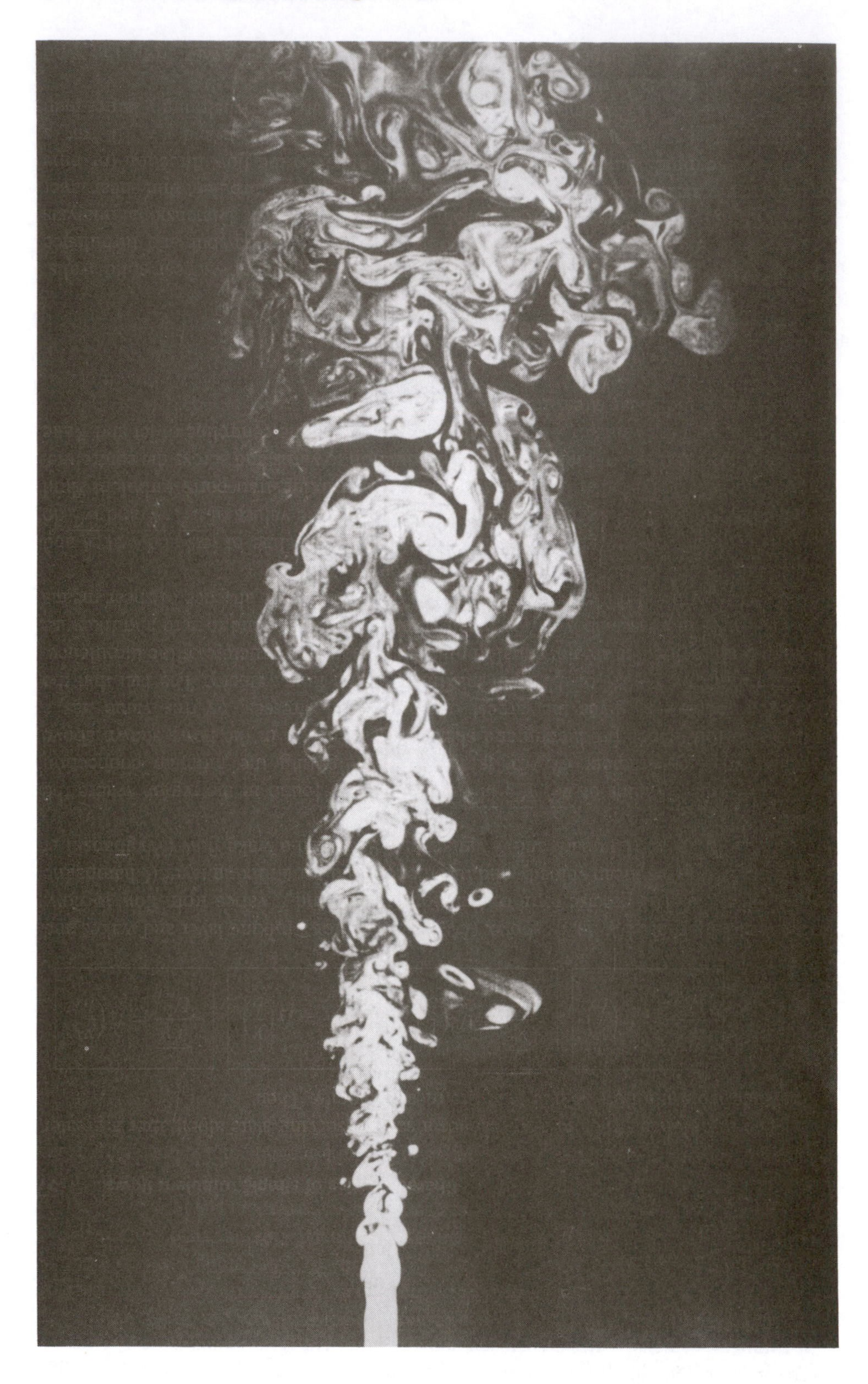

Fig. 3.9 Visualisation of a jet flow

The turbulence structure also reaches a self-preserving state albeit after a greater distance from the flow source than the mean velocity. Then

$$\frac{\overline{u'^2}}{U_{ref}^2} = f_1\left(\frac{y}{b}\right) \quad \frac{\overline{v'^2}}{U_{ref}^2} = f_2\left(\frac{y}{b}\right) \quad \frac{\overline{w'^2}}{U_{ref}^2} = f_3\left(\frac{y}{b}\right) \quad \frac{\overline{u'v'}}{U_{ref}^2} = f_4\left(\frac{y}{b}\right)$$

The velocity scale U_{ref} is, as above, $(U_{max} - U_{min})$ for a mixing layer and wakes and U_{max} for jets. The precise form of functions f, g, h and f_i varies from flow to flow. Figure 3.10 gives mean velocity and turbulence data for a mixing layer (Champagne *et al*, 1976), a jet (Gutmark and Wygnanski, 1976) and a wake flow (Wygnanski *et al*, 1986).

The largest values of $\overline{u'^2}$, $\overline{v'^2}$, $\overline{w'^2}$ and $-\overline{u'v'}$ are found in the region where the mean velocity gradient $\partial U/\partial y$ is largest highlighting the intimate connection between turbulence production and sheared mean flows. In the flows shown above the component u' gives the largest of the normal stresses; its rms value has a maximum of 15–40% of the local maximum mean flow velocity. The fact that the fluctuating velocities are not equal implies an anisotropic structure of the turbulence.

As $|y/b|$ increases above 1 the mean velocity gradients tend to zero and likewise the values of the turbulence properties drop off to zero. The absence of shear means that turbulence cannot be sustained in this region.

The mean velocity gradient is also zero at the centreline of jets and wakes and hence no turbulence is produced here. Nevertheless, the values of $\overline{u'^2}$, $\overline{v'^2}$ and $\overline{w'^2}$ do not decrease very much because vigorous eddy mixing transports turbulent fluid from nearby regions of high turbulence production towards and across the centreline. By symmetry the value of $-\overline{u'v'}$ has to become zero at the centreline of jet and wake flows since the shear stress must change sign here.

3.4.2 Flat plate boundary layer and pipe flow

Next we will examine the characteristics of two turbulent flows near solid walls. Owing to the presence of the solid boundary the flow behaviour and turbulence structure are considerably different from free turbulent flows. Dimensional analysis has greatly assisted in correlating the experimental data. In turbulent thin shear layer flows a Reynolds number based on a length scale L in the flow direction (or pipe radius), Re_L, is always very large (e.g. $U = 1$ m/s, $L = 0.1$ m and $\nu = 10^{-6}\text{m}^2/\text{s}$ gives $Re_L = 10^5$). This implies that the inertia forces are overwhelmingly larger than the viscous forces at these scales.

If we form a Reynolds number based on a distance y away from the wall ($Re_y = Uy/\nu$) we see that if the value of y is on the order of L the above argument holds. Inertia forces dominate in the flow far away from the wall. As y is decreased to zero, however, a Reynolds number based on y will also decrease to zero. Just before y reaches zero there will be a range of values of y for which Re_y is on the order of 1. At this distance from the wall and closer the viscous forces will be equal in order of magnitude to the inertia forces or larger. To sum up, in flows along solid boundaries there is usually a substantial region of inertia-dominated flow far away from the wall and a thin layer within which viscous effects are important.

Close to the wall the flow is influenced by viscous effects and does not depend on free stream parameters. The mean flow velocity only depends on the distance y from

Fig. 3.10 Mean velocity distributions and turbulence properties for (a) two-dimensional mixing layer, (b) planar turbulent jet and (c) wake behind a solid strip

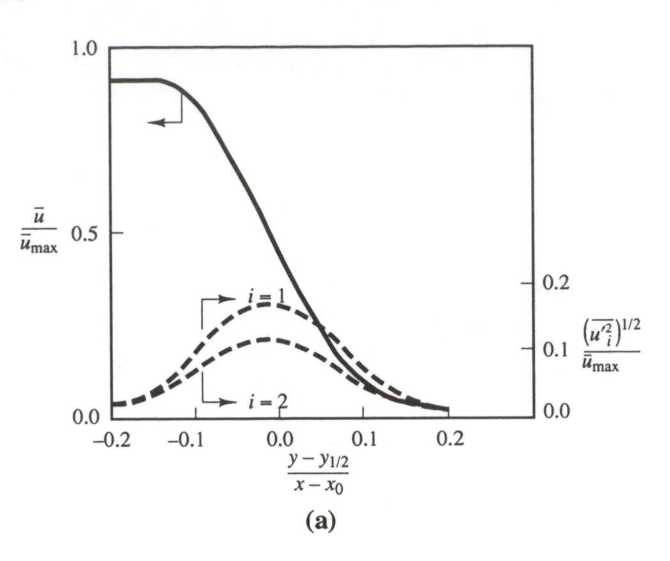

(a)

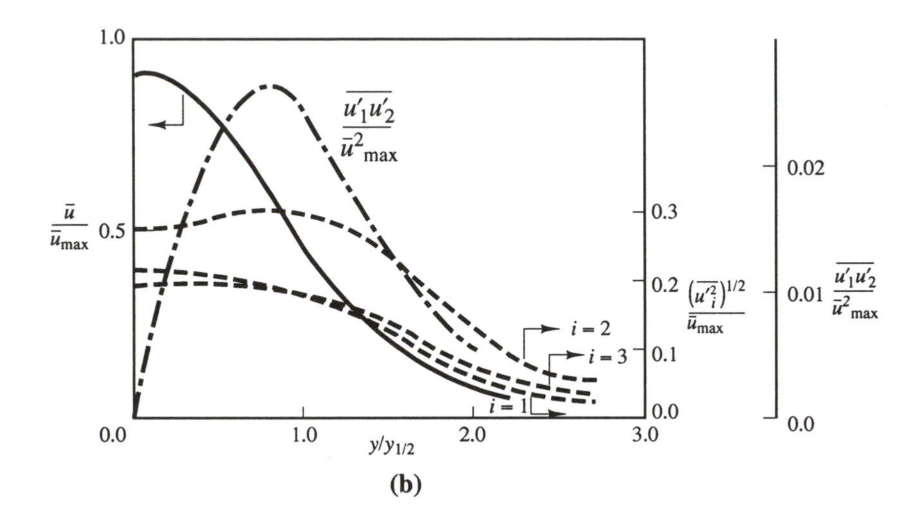

(b)

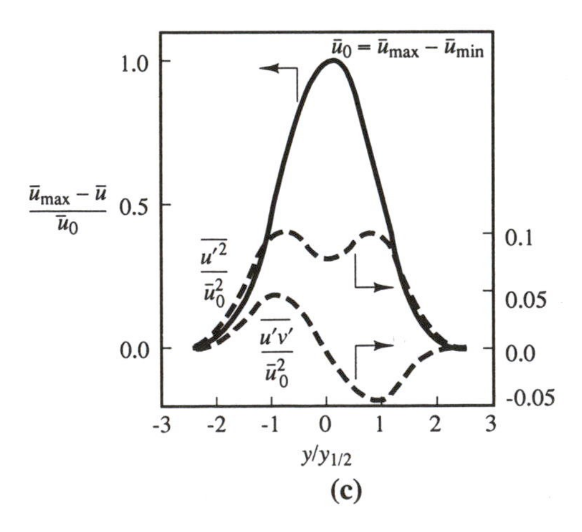

(c)

the wall, fluid density ρ and viscosity μ and the wall shear stress τ_w. So

$$U = f(y,\ \rho,\ \mu,\ \tau_w)$$

Dimensional analysis shows that

$$u^+ = \frac{U}{u_\tau} = f\left(\frac{\rho u_\tau y}{\mu}\right) = f(y^+) \tag{3.18}$$

Formula (3.18) is called the **law of the wall** and contains the definitions of two important dimensionless groups u^+ and y^+. Note that the appropriate velocity scale is $u_\tau = (\tau_w/\rho)^{\frac{1}{2}}$, the so-called friction velocity.

Far away from the wall we expect the velocity at a point to be influenced by the retarding effect of the wall through the value of the wall shear stress, but not by the viscosity itself. The length scale appropriate to this region is the boundary layer thickness δ. In this region we have

$$U = g(y,\ \delta,\ \rho,\ \tau_w)$$

Dimensional analysis yields

$$u^+ = \frac{U}{u_\tau} = g\left(\frac{y}{\delta}\right)$$

The most useful form emerges if we view the wall shear stress as the cause of a velocity deficit $U_{max} - U$ which decreases the closer we get to the edge of the boundary layer or the pipe centreline. Thus

$$\frac{U_{max} - U}{u_\tau} = g\left(\frac{y}{\delta}\right) \tag{3.19}$$

This formula is called the **velocity-defect law**.

Linear sub-layer – the fluid layer in contact with a smooth wall

At the solid surface the fluid is stationary. Turbulent eddying motions must also stop very close to the wall. In the absence of turbulent (Reynolds) shear stress effects the fluid closest to the wall is dominated by viscous shear. This layer is in practice extremely thin ($y^+ < 5$) and we may assume that the shear stress is approximately constant and equal to the wall shear stress τ_w throughout the layer. Thus

$$\tau(y) = \mu \frac{\partial U}{\partial y} \cong \tau_w$$

After integration with respect to y and application of boundary condition $U = 0$ if $y = 0$ we obtain a linear relationship between the mean velocity and the distance from the wall:

$$U = \frac{\tau_w y}{\mu}$$

After some simple algebra and making use of the definitions of u^+ and y^+ this leads to

$$u^+ = y^+ \tag{3.20}$$

Because of the linear relationship between velocity and distance from the wall the fluid layer adjacent to the wall is often known as the **linear sub-layer**.

Log-law layer – the turbulent region close to a smooth wall

Outside the viscous sublayer ($30 < y^+ < 500$) a region exists where viscous and turbulent effects are both important. The shear stress τ varies slowly with distance from the wall and within this inner region it is assumed to be constant and equal to the wall shear stress. One further assumption regarding the length scale of turbulence (mixing length $\ell_m = \kappa y$, see section 3.5.1 and Schlichting, 1979) allows us to derive a dimensionally correct form of the functional relationship between u^+ and y^+

$$u^+ = \frac{1}{\kappa}\ln y^+ + B = \frac{1}{\kappa}\ln(Ey^+) \tag{3.21}$$

Numerical values for the constants are found from measurements. We find $\kappa = 0.4$ and $B = 5.5$ (or $E = 9.8$) for smooth walls; wall roughness causes a decrease in the value of B. The values of κ and B are universal constants valid for all turbulent flows past smooth walls at high Reynolds number. Because of the logarithmic relationship between u^+ and y^+ formula (3.21) is often called the **log-law** and the layer where y^+ takes values between 30 and 500 the **log-law layer**.

Outer layer – the inertia-dominated region far from the wall

Experimental measurements show that the log-law is valid in the region $0.02 < y/\delta < 0.2$. For larger values of y the velocity-defect law (3.19) provides the correct form. In the overlap region the log-law and velocity-defect law have to become equal. Tennekes and Lumley (1972) show that a matched overlap is obtained by assuming the following logarithmic form:

$$\frac{U_{\max} - U}{u_\tau} = \frac{1}{\kappa}\ln\left(\frac{y}{\delta}\right) + A \tag{3.22}$$

where A is a constant. This velocity-defect law is often called the **law of the wake**.

Figure 3.11 from Schlichting (1979) shows the close agreement between theoretical equations (3.20) and (3.21) in their respective areas of validity and experimental data.

The turbulent boundary layer adjacent to a solid surface is composed of two regions:

- The inner region: 10 to 20% of the total thickness of the wall layer; the shear stress is (almost) constant and equal to the wall shear stress τ_w. Within this region there are three zones; in order of increasing distance from the wall we have:
  - the linear sub-layer: viscous stresses dominate the flow adjacent to the surface
  - the buffer layer: viscous and turbulent stresses are of similar magnitude
  - the log-law layer: turbulent (Reynolds) stresses dominate.
- The outer region or law-of-the-wake layer: inertia-dominated core flow far from wall; free from direct viscous effects.

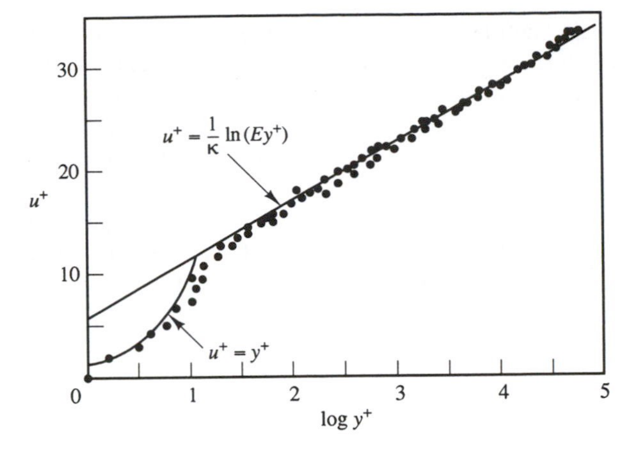

Fig. 3.11 Velocity distribution near a solid wall

Figure 3.12 shows the mean velocity and turbulence property distribution data for a flat plate boundary layer with a constant imposed pressure (Klebanoff, 1955).

The mean velocity is at a maximum far away from the wall and sharply decreases in the region $y/\delta \leq 0.2$ due to the no-slip condition. High values of $\overline{u'^2}$, $\overline{v'^2}$, $\overline{w'^2}$ and $-\overline{u'v'}$ are found adjacent to the wall where the large mean velocity gradients ensure that turbulence production is high. The eddying motions and associated velocity fluctuations are, however, also subject to the no-slip condition at the wall. Therefore all turbulent stresses decrease sharply to zero in this region. The turbulence is anisotropic near the wall since the production process mainly creates component $\overline{u'^2}$. This is borne out by the fact that this is the largest of the mean-squared fluctuations in Figure 3.12.

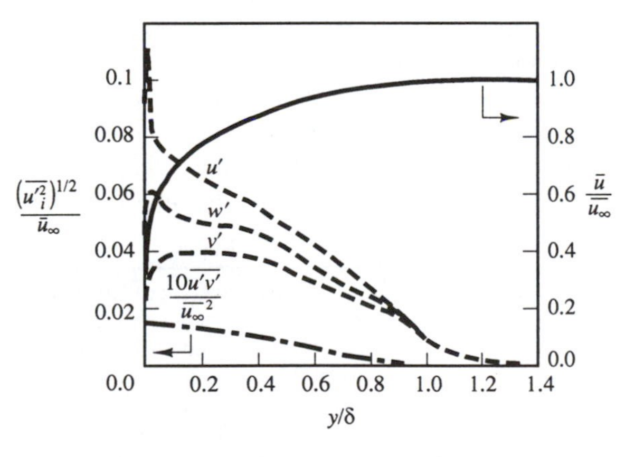

Fig. 3.12 Mean velocity distribution and turbulence properties for a flat plate boundary layer at zero pressure gradient

In the case of the flat plate boundary layer the turbulence properties asymptotically tend towards zero as y/δ increases above a value of 0.8. The rms values of all fluctuating velocities become almost equal here indicating that the turbulence structure becomes more isotropic far away from the wall.

In pipe flows the eddying motions transport turbulence across the centreline from areas of high production. Therefore, the rms fluctuations remain comparatively large in the centre of a pipe. By symmetry the value of $-\overline{u'v'}$ has to go to zero and change sign at the centreline.

The multi-layer structure is a universal feature of turbulent boundary layers near solid surfaces. Monin and Yaglom (1971) plotted data from Klebanoff and Laufer in the near wall region and found not only the universal mean velocity distribution but also that data for the Reynolds stresses for flat plates and pipes collapse onto a single curve if they are non-dimensionalised with the proper velocity scale u_τ.

Between these distinct layers there are intermediate zones which ensure that the various velocity distribution laws merge smoothly. Interested readers may find further details including formulae which cover the whole inner region and the log-law/law-of-the-wake layer in Schlichting (1979) and White (1991).

3.4.3 Summary

In these sections we have reviewed the characteristics of a number of two-dimensional turbulent flows. Although many common features were found it has become clear that, even in these relatively simple thin shear layers, the details of the turbulence structure are very much dependent on the flow itself. In particular the geometry of the boundaries which create and maintain the turbulence is important. Viscous shear stresses depend on the viscosity, a fluid property, but turbulent Reynolds stresses are also affected by the flow itself. Computational procedures must be able to cope with this complication.

3.5 Turbulence models

A turbulence model is a computational procedure to close the system of mean flow equations (3.15), (3.16a–c) and (3.17) so that a more or less wide variety of flow problems can be calculated. For most engineering purposes it is unnecessary to resolve the details of the turbulent fluctuations. Only the effects of the turbulence on the mean flow are usually sought. In particular, we always need expressions for the Reynolds stresses in equations (3.16a–c) and the turbulent scalar transport terms in equation (3.17). For a turbulence model to be useful in a general purpose CFD code it must have wide applicability, be accurate, simple and economical to run. The most common turbulence models are classified in Table 3.2.

Table 3.2 Turbulence models

Classical models	Based on (time-averaged) Reynolds equations
	1. zero equation model – mixing length model
	2. two-equation model – k–ε model
	3. Reynolds stress equation model
	4. algebraic stress model
Large eddy simulation	Based on space-filtered equations

The **classical models** use the Reynolds equations developed in section 3.3 and form the basis of turbulence calculations in currently available commercial CFD codes. **Large eddy simulations** are turbulence models where the time-dependent flow equations are solved for the mean flow and the largest eddies and where the effects of the smaller eddies are modelled. It was argued earlier that the largest eddies interact strongly with the mean flow and contain most of the energy so this approach results in a good model of the main effects of turbulence. Large eddy simulations are at present at the research stage and the calculations are too costly to merit consideration in general purpose computation at present. Although anticipated improvements in computer hardware may change this perspective in the future we will not discuss these models further; the interested reader may find a brief introduction to these and other more advanced turbulence models in Abbott and Basco (1989).

Of the classical models the mixing length and k–ε models are presently by far the most widely used and validated. They are based on the presumption that there exists an analogy between the action of viscous stresses and Reynolds stresses on the mean flow. Both stresses appear on the right hand side of the momentum equation and in Newton's law of viscosity the viscous stresses are taken to be proportional to the rate of deformation of fluid elements. For an incompressible fluid this gives

$$\tau_{ij} = \mu e_{ij} = \mu\left(\frac{\partial u_i}{\partial x_j} + \frac{\partial u_j}{\partial x_i}\right) \tag{2.31}$$

In order to simplify the notation the so-called suffix notation has been used here. The convention of this notation is that i or $j = 1$ corresponds to the x-direction, i or $j = 2$ the y-direction and i or $j = 3$ the z-direction. So for example

$$\tau_{12} = \tau_{xy} = \mu\left(\frac{\partial u_1}{\partial x_2} + \frac{\partial u_2}{\partial x_1}\right) = \mu\left(\frac{\partial u}{\partial y} + \frac{\partial v}{\partial x}\right)$$

It is experimentally observed that turbulence decays unless there is shear in isothermal incompressible flows. Furthermore, turbulent stresses are found to increase as the mean rate of deformation increases. It was proposed by Boussinesq in 1877 that Reynolds stresses could be linked to mean rates of deformation. Using the suffix notation we get

$$\boxed{\tau_{ij} = -\rho\overline{u_i'u_j'} = \mu_t\left(\frac{\partial U_i}{\partial x_j} + \frac{\partial U_j}{\partial x_i}\right)} \tag{3.23}$$

The right hand side is analogous to formula (2.31) above except for the appearance of the turbulent or eddy viscosity μ_t (dimensions Pa s). There is also a kinematic turbulent or eddy viscosity denoted by $\nu_t = \mu_t/\rho$, with dimensions m^2/s.

Turbulent transport of heat, mass and other scalar properties is modelled similarly. Formula (3.23) shows that turbulent momentum transport is assumed to be proportional to mean gradients of velocity (i.e. gradients of momentum per unit mass). By analogy turbulent transport of a scalar is taken to be proportional to the gradient of the mean value of the transported quantity. In suffix notation we get

$$-\rho\overline{u_i'\varphi'} = \Gamma_t \frac{\partial \Phi}{\partial x_i} \tag{3.24}$$

where Γ_t is the turbulent diffusivity.

Since turbulent transport of momentum and heat or mass is due to the same mechanism – eddy mixing – we expect that the value of the turbulent diffusivity Γ_t is close to that of the turbulent viscosity μ_t. We introduce a turbulent Prandtl/Schmidt number defined as follows:

$$\sigma_t = \frac{\mu_t}{\Gamma_t} \tag{3.25}$$

Experiments in many flows have established that this ratio is often nearly constant. Most CFD procedures assume this to be the case and use values of σ_t around 1.

It has become clear from our discussions of simple turbulent flows in section 3.4 that turbulence levels and turbulent stresses vary from point to point in a flow. **Mixing length models** attempt to describe the stresses by means of simple algebraic formulae for μ_t as a function of position. The ***k*–ε model** is a more sophisticated and general, but also more costly, description of turbulence which allows for the effects of transport of turbulence properties by the mean flow and diffusion and for the production and destruction of turbulence. Two transport equations (partial differential equations or PDEs), one for the turbulent kinetic energy k and a further one for the rate of dissipation of turbulent kinetic energy ε, are solved.

The underlying assumption of both these models is that the turbulent viscosity μ_t is isotropic, in other words that the ratio between Reynolds stress and mean rate of deformation is the same in all directions. This assumption fails in many categories of flow where it leads to inaccurate flow predictions. Here it is necessary to derive and solve transport equations for the Reynolds stresses themselves. It may at first seem strange to think that a stress can be subject to transport. However, it is only necessary to remember that the Reynolds stresses initially appeared on the left hand side of the momentum equations and are physically due to convective momentum exchanges as a consequence of turbulent velocity fluctuations. Fluid momentum – mean momentum as well as fluctuating momentum – can be transported by fluid particles and therefore the Reynolds stresses can also be transported.

The six transport equations, one for each Reynolds stress, contain diffusion, pressure–strain and dissipation terms whose individual effects are unknown and cannot be measured. In **Reynolds stress equation models** (also known in the literature as second-order or second-moment closure models) assumptions are made about these unknown terms and the resulting PDEs are solved in conjunction with the transport equation for the rate of dissipation of turbulent kinetic energy ε. The design of Reynolds stress equation models is an area of vigorous research and the models have not been validated as widely as the mixing length and k–ε model. Solving the seven extra PDEs gives rise to a substantial increase in the cost of CFD simulations when compared to the k–ε model, so the application of Reynolds stress equation models outside the academic fraternity is relatively recent.

A much more far-reaching set of modelling assumptions reduces the PDEs describing Reynolds stress transport to algebraic equations to be solved alongside the k and ε equations of the k–ε model. This approach leads to the **algebraic stress models** that are the most economical form of Reynolds stress model able to introduce anisotropic turbulence effects into CFD simulations.

In the following sections the mixing length and k–ε models will be discussed in detail and the main features of the Reynolds stress equation and algebraic stress models will be outlined. Moreover, the results of some of the current research which is likely to impact on industrial turbulence modelling in the immediate future are briefly considered.

3.5.1 Mixing length model

On dimensional grounds we assume that the kinematic turbulent viscosity ν_t, which has dimensions m^2/s, can be expressed as a product of a turbulent velocity scale ϑ (m/s) and a length scale ℓ (m). If *one velocity scale and one length scale* suffice to describe the effects of turbulence dimensional analysis yields

$$\nu_t = C\vartheta\ell \tag{3.26}$$

where C is a dimensionless constant of proportionality. Of course the dynamic turbulent viscosity is given by

$$\mu_t = C\rho\vartheta\ell$$

Most of the kinetic energy of turbulence is contained in the largest eddies and the turbulence length scale ℓ is therefore characteristic of these eddies which interact with the mean flow. If we accept that there is a strong connection between the mean flow and the behaviour of the largest eddies we can attempt to link the characteristic velocity scale of the eddies with the mean flow properties. This has been found to work well in simple two-dimensional turbulent flows where the only significant Reynolds stress is $\tau_{xy} = \tau_{yx} = -\rho\overline{u'v'}$ and the only significant mean velocity gradient is $\partial U/\partial y$. For such flows it is at least dimensionally correct to state that, if the eddy length scale is ℓ,

$$\vartheta = c\ell\left|\frac{\partial U}{\partial y}\right| \tag{3.27}$$

where c is a dimensionless constant. The absolute value is taken to ensure that the velocity scale is always a positive quantity irrespective of the sign of the velocity gradient.

Combining (3.26) and (3.27) and absorbing the two constants C and c which appear in these formulae into a new length scale ℓ_m we obtain

$$\boxed{\nu_t = \ell_m^2\left|\frac{\partial U}{\partial y}\right|} \tag{3.28}$$

This is **Prandtl's mixing length model**. Using formula (3.23) and noting that $\partial U/\partial y$ is the only significant mean velocity gradient the turbulent Reynolds stress is described by

$$\boxed{\tau_{xy} = \tau_{yx} = -\rho\overline{u'v'} = \rho\ell_m^2\left|\frac{\partial U}{\partial y}\right|\frac{\partial U}{\partial y}} \tag{3.29}$$

Turbulence is a function of the flow and if the turbulence changes it is necessary to account for this within the mixing length model by varying ℓ_m. For a substantial category of simple turbulent flows which include the free turbulent flows and wall boundary layers discussed in section 3.4 this can be achieved by means of simple algebraic formulae. Some examples (source: Rodi, 1980) are given in Table 3.3.

The mixing length model can also be used to predict turbulent transport of scalar quantities. The only turbulent transport term which matters in the two-dimensional flows for which the mixing length is useful is modelled as follows:

$$\boxed{-\rho\overline{v'\varphi'} = \Gamma_t\frac{\partial \Phi}{\partial y}} \tag{3.30}$$

where $\Gamma_t = \mu_t/\sigma_t$ and ν_t is found from (3.28). Rodi (1980) recommends values for σ_t of 0.9 in near wall flows, 0.5 for jets and mixing layers and 0.7 in axisymmetric jets.

Table 3.3 Mixing lengths for two-dimensional turbulent flows

Flow	*Mixing length l_m*	*L*
Mixing layer	0.07L	Layer width
Jet	0.09L	Jet half width
Wake	0.16L	Wake half width
Axisymmetric jet	0.075L	Jet half width
Boundary layer ($\partial p/\partial x = 0$)		
viscous sub-layer and		
log-law layer ($y/L \leq 0.22$)	$\kappa y[1 - \exp(-y^+/26)]$	Boundary layer thickness
outer layer ($y/L \geq 0.22$)	$0.09L$	
Pipes and channels (fully developed flow)	$L[0.14 - 0.08(1 - y/L)^2 - 0.06(1 - y/L)^4]$	Pipe radius or channel half width

In the table y represents the distance from the wall and $\kappa = 0.41$ is von Karman's constant. The expressions give very good agreement between computed results and experiments for mean velocity distributions, wall friction coefficients and other flow properties such as heat transfer coefficients etc. in simple two-dimensional flows. The results for two flows from Schlichting (1979) are given in Figure 3.13a and b.

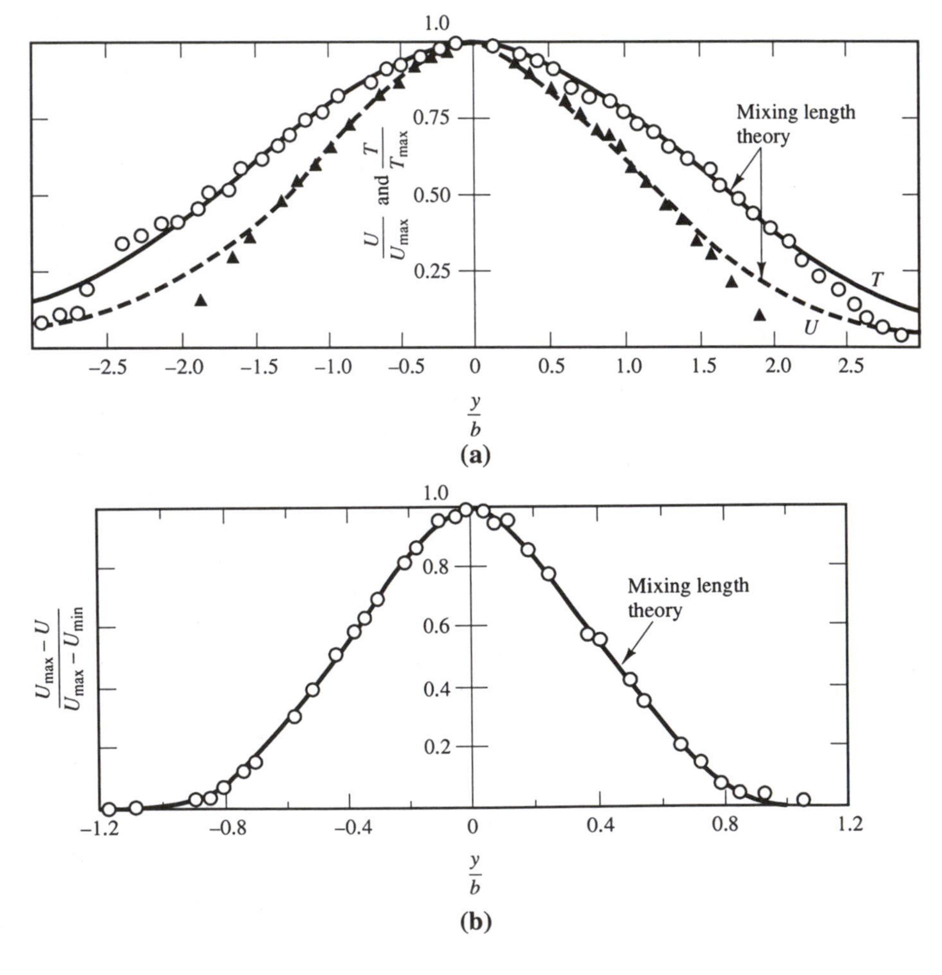

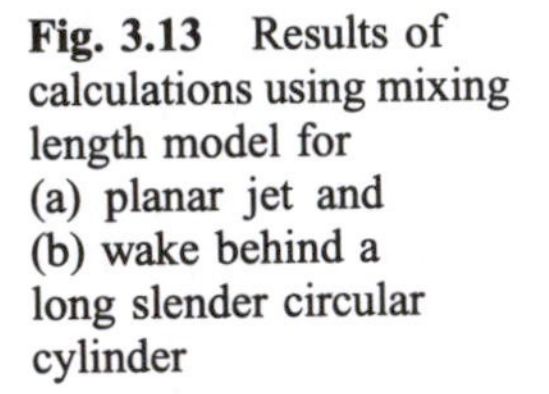

Fig. 3.13 Results of calculations using mixing length model for (a) planar jet and (b) wake behind a long slender circular cylinder

The mixing length is clearly very useful in flows where the turbulence properties develop in proportion to a mean flow length scale, so that ℓ_m can be described as a function of position by means of a simple algebraic formula. This explains its universal popularity in calculations of flows around wing sections. Sophisticated modifications of the formulae for ℓ_m to describe the effects of pressure gradients, small scale separation and boundary layer blowing or suction are available. Mixing length models such as those developed by Baldwin and Lomax (1978) and Cebeci and Smith (1974) are the most widely used turbulence models in external aerodynamics calculations in the aerospace industry.

An overall assessment of the mixing length model is given in Table 3.4.

Table 3.4 Mixing length model assessment

Advantages

- easy to implement and cheap in terms of computing resources
- good predictions for thin shear layers: jets, mixing layers, wakes and boundary layers
- well established

Disadvantages

- completely incapable of describing flows with separation and recirculation
- only calculates mean flow properties and turbulent shear stress

3.5.2 The *k*–ε model

In two-dimensional thin shear layers the changes in the flow direction are always so slow that the turbulence can adjust itself to local conditions. If the convection and diffusion of turbulence properties can be neglected it is possible to express the influence of turbulence on the mean flow in terms of the mixing length. If convection and diffusion are not negligible – as is the case for example in recirculating flows – a compact algebraic prescription for the mixing length is no longer feasible. The mixing length model lacks this kind of generality. The way forward is to consider statements regarding the dynamics of turbulence. The k–ε model focuses on the mechanisms that affect the turbulent kinetic energy.

Some preliminary definitions are required first. The instantaneous kinetic energy $k(t)$ of a turbulent flow is the sum of the mean kinetic energy $K = \frac{1}{2}(U^2 + V^2 + W^2)$ and the turbulent kinetic energy $k = \frac{1}{2}(\overline{u'^2} + \overline{v'^2} + \overline{w'^2})$:

$$k(t) = K + k$$

In the developments below we extensively need to use the rate of deformation and the turbulent stresses. To facilitate the subsequent calculations it is common to write the components of the rate of deformation e_{ij} and the stresses τ_{ij} in tensor (matrix) form:

$$e_{ij} = \begin{pmatrix} e_{xx} & e_{xy} & e_{xz} \\ e_{yx} & e_{yy} & e_{yz} \\ e_{zx} & e_{zy} & e_{zz} \end{pmatrix} \quad \text{and} \quad \tau_{ij} = \begin{pmatrix} \tau_{xx} & \tau_{xy} & \tau_{xz} \\ \tau_{yx} & \tau_{yy} & \tau_{yz} \\ \tau_{zx} & \tau_{zy} & \tau_{zz} \end{pmatrix}$$

Decomposition of the rate of deformation of a fluid element in a turbulent flow into a mean and a fluctuating component, $e_{ij}(t) = E_{ij} + e'_{ij}$, gives the following matrix elements:

$$e_{xx}(t) = E_{xx} + e'_{xx} = \frac{\partial U}{\partial x} + \frac{\partial u'}{\partial x}; \qquad e_{yy}(t) = E_{yy} + e'_{yy} = \frac{\partial V}{\partial y} + \frac{\partial v'}{\partial y};$$

$$e_{zz}(t) = E_{zz} + e'_{zz} = \frac{\partial W}{\partial z} + \frac{\partial w'}{\partial z}$$

$$e_{xy}(t) = E_{xy} + e'_{xy} = e_{yx}(t) = E_{yx} + e'_{yx} = \tfrac{1}{2}\left[\frac{\partial U}{\partial y} + \frac{\partial V}{\partial x}\right] + \tfrac{1}{2}\left[\frac{\partial u'}{\partial y} + \frac{\partial v'}{\partial x}\right]$$

$$e_{xz}(t) = E_{xz} + e'_{xz} = e_{zx}(t) = E_{zx} + e'_{zx} = \tfrac{1}{2}\left[\frac{\partial U}{\partial z} + \frac{\partial W}{\partial x}\right] + \tfrac{1}{2}\left[\frac{\partial u'}{\partial z} + \frac{\partial w'}{\partial x}\right]$$

$$e_{yz}(t) = E_{yz} + e'_{yz} = e_{zy}(t) = E_{zy} + e'_{zy} = \tfrac{1}{2}\left[\frac{\partial V}{\partial z} + \frac{\partial W}{\partial y}\right] + \tfrac{1}{2}\left[\frac{\partial v'}{\partial z} + \frac{\partial w'}{\partial y}\right]$$

The product of a vector $\mathbf{a}$ and a tensor b_{ij} is a vector c whose components can be calculated by application of the ordinary rules of matrix algebra:

$$\mathbf{a}b_{ij} \equiv a_i b_{ij} = \begin{pmatrix} a_1 & a_2 & a_3 \end{pmatrix} \begin{pmatrix} b_{11} & b_{12} & b_{13} \\ b_{21} & b_{22} & b_{23} \\ b_{31} & b_{32} & b_{33} \end{pmatrix}$$

$$= \begin{pmatrix} a_1 b_{11} + a_2 b_{21} + a_3 b_{31} \\ a_1 b_{12} + a_2 b_{22} + a_3 b_{32} \\ a_1 b_{13} + a_2 b_{23} + a_3 b_{33} \end{pmatrix}^T = \begin{pmatrix} c_1 \\ c_2 \\ c_3 \end{pmatrix}^T = c_j = \mathbf{c}$$

The scalar product of two tensors a_{ij} and b_{ij} is evaluated as follows

$$a_{ij} \,.\, b_{ij} = a_{11}b_{11} + a_{12}b_{12} + a_{13}b_{13} + a_{21}b_{21} + a_{22}b_{22} + a_{23}b_{23} + a_{31}b_{31} + a_{32}b_{32} + a_{33}b_{33}$$

We have used the convention of the suffix notation where the x-direction is denoted by subscript 1, the y-direction by 2 and the z-direction by 3. It can be seen that products are formed by taking the sum over all possible values of every repeated suffix.

Governing equation for mean flow kinetic energy K

An equation for the mean kinetic energy K can be obtained by multiplying the x-component Reynolds equation (3.12a) by U, the y-component equation (3.12b) by V and the z-component equation (3.12c) by W. After adding the results together and a fair amount of algebra it can be shown that the time-average equation governing the mean kinetic energy of the flow is as follows (Tennekes and Lumley, 1972):

$$\frac{\partial(\rho K)}{\partial t} + div(\rho K\mathbf{U}) = div\left(-P\mathbf{U} + 2\mu\mathbf{U}E_{ij} - \rho\mathbf{U}\overline{u'_i u'_j}\right) - 2\mu E_{ij} \,.\, E_{ij} + \rho\overline{u'_i u'_j} \,.\, E_{ij} \tag{3.31}$$

(I) (II) (III) (IV) (V) (VI) (VII)

Or in words, for the mean kinetic energy K, we have

Rate of change of K	+	Transport of K by convection	=	Transport of K by pressure	+	Transport of K by viscous stresses	+	Transport of K by Reynolds stress
			−	Rate of dissipation of K	+	Turbulence production		

The transport terms (III), (IV) and (V) are all characterised by the appearance of the *div* and it is common practice to place them together inside one pair of brackets. The effects of the viscous stresses on K have been split into two parts: term (IV), the transport of K due to viscous stresses, and term (VI), the viscous dissipation of mean kinetic energy K. The two terms that contain the Reynolds stresses $-\rho\overline{u_i'u_j'}$ account for the turbulence effects: term (V) is the turbulent transport of K by means of Reynolds stresses and (VII) is the turbulence production term or the net decrease of K due to deformation work by Reynolds stresses production. In high Reynolds number flows the turbulent terms (V) and (VII) – are always much larger than their viscous counterparts (IV) and (VI).

Governing equation for turbulent kinetic energy k

Multiplication of each of the instantaneous Navier–Stokes equations (3.9a–c) by the appropriate fluctuating velocity components (i.e. x-component equation multiplied by u' etc.) and addition of all the results, followed by a repeat of this process on the Reynolds equations (3.12a–c), subtraction of the two resulting equations and very substantial re-arrangement yields the equation for turbulent kinetic energy k (Tennekes and Lumley, 1972):

$$\underset{\text{(I)}}{\frac{\partial(\rho k)}{\partial t}} + \underset{\text{(II)}}{div(\rho k\mathbf{U})} = div\Big(\underset{\text{(III)}}{-\overline{p'\mathbf{u}'}} + \underset{\text{(IV)}}{2\mu\overline{\mathbf{u}'e_{ij}'}} - \underset{\text{(V)}}{\rho\tfrac{1}{2}\overline{u_i' . u_i'u_j'}}\Big) - \underset{\text{(VI)}}{2\mu\overline{e_{ij}' . e_{ij}'}} - \underset{\text{(VII)}}{\rho\overline{u_i'u_j'} . E_{ij}} \qquad (3.32)$$

In words, for the turbulent kinetic energy k, we have

Rate of change of k	+	Transport of k by convection	=	Transport of k by pressure	+	Transport of k by viscous stresses	+	Transport of k by Reynolds stress
			−	Rate of dissipation of k	+	Turbulence production		

Equations (3.31) and (3.32) look very similar in many respects; however, the appearance of primed quantities on the right hand side of the k-equation shows that changes to the turbulent kinetic energy are mainly governed by turbulent interactions. Terms (VII) in both equations are equal in magnitude, but opposite in sign. In two-dimensional thin shear layers we found (see section 3.4) that the only significant Reynolds stress $-\rho\overline{u'v'}$ was usually positive if the main term of E_{ij} in such a flow, the mean velocity gradient $\partial U/\partial y$, is positive. Hence term (VII) gives a positive contribution in the k-equation and represents a production term. In the K-equation, however, the term is negative, so there it destroys mean flow kinetic energy. This expresses mathematically the conversion of mean kinetic energy into turbulent kinetic energy.

The viscous dissipation term (V)

$$-2\mu\overline{e'_{ij}\,.\,e'_{ij}} = -2\mu\left(\overline{e'^2_{11}} + \overline{e'^2_{22}} + \overline{e'^2_{33}} + 2\overline{e'^2_{12}} + 2\overline{e'^2_{13}} + 2\overline{e'^2_{23}}\right)$$

gives a negative contribution to (3.32) due to the appearance of the sum of squared fluctuating deformation rates e'_{ij}. The dissipation of turbulent kinetic energy is caused by work done by the smallest eddies against viscous stresses. The rate of dissipation per unit mass, whose dimensions are m^2/s^3, is of vital importance in the study of turbulence dynamics and is denoted by

$$\boxed{\varepsilon = 2\nu\overline{e'_{ij}\,.\,e'_{ij}}} \tag{3.33}$$

It is always the main destruction term in the turbulent kinetic energy equation, of a similar order of magnitude to the production term and never negligible. In contrast, when the Reynolds number is high, the viscous transport term (IV) in (3.32) is always very small compared to the turbulent transport term (VI).

The k–ε model equations

It is possible to develop similar transport equations for all other turbulence quantities including the rate of viscous dissipation ε (see Bradshaw *et al*, 1981). The exact ε-equation, however, contains many unknown and unmeasurable terms. The **standard *k*–ε model** (Launder and Spalding, 1974) has two model equations, one for k and one for ε, based on our best understanding of the relevant processes causing changes to these variables.

We use k and ε to define velocity scale ϑ and length scale ℓ representative of the large scale turbulence as follows:

$$\vartheta = k^{1/2} \qquad \ell = \frac{k^{3/2}}{\varepsilon}$$

One might question the validity of using the 'small eddy' variable ε to define the 'large eddy' scale ℓ. We are permitted to do this because at high Reynolds numbers the rate at which large eddies extract energy from the mean flow is precisely matched to the rate of transfer of energy across the energy spectrum to small, dissipating, eddies. If this was not the case the energy at some scales of turbulence could grow or diminish without limit. This does not occur in practice and justifies the use of ε in the definition of ℓ.

Applying the same approach as in the mixing length model we specify the eddy viscosity as follows:

$$\mu_t = C\rho\vartheta\ell = \rho C_\mu \frac{k^2}{\varepsilon} \tag{3.34}$$

where C_μ is a dimensionless constant.

The standard model uses the following transport equations used for k and ε:

$$\frac{\partial(\rho k)}{\partial t} + div(\rho k\mathbf{U}) = div\left[\frac{\mu_t}{\sigma_k} grad\ k\right] + 2\mu_t E_{ij} \cdot E_{ij} - \rho\varepsilon \tag{3.35}$$

$$\frac{\partial(\rho\varepsilon)}{\partial t} + div(\rho\varepsilon\mathbf{U}) = div\left[\frac{\mu_t}{\sigma_\varepsilon} grad\ \varepsilon\right] + C_{1\varepsilon}\frac{\varepsilon}{k} 2\mu_t E_{ij} \cdot E_{ij} - C_{2\varepsilon}\rho\frac{\varepsilon^2}{k} \tag{3.36}$$

In words the equations are

Rate of change of k or ε	+	Transport of k or ε by convection	=	Transport of k or ε by diffusion	+	Rate of production of k or ε
			−	Rate of destruction of k or ε		

The equations contain five adjustable constants $C_\mu, \sigma_k, \sigma_\varepsilon, C_{1\varepsilon}$ and $C_{2\varepsilon}$. The standard k–ε model employs values for the constants that are arrived at by comprehensive data fitting for a wide range of turbulent flows:

$$C_\mu = 0.09; \quad \sigma_k = 1.00; \quad \sigma_\varepsilon = 1.30; \quad C_{1\varepsilon} = 1.44; \quad C_{2\varepsilon} = 1.92 \tag{3.37}$$

The production term in the model k-equation is derived from the exact production term in (3.32) by substitution of (3.23). A modelled form of the principal transport processes in the k- and ε-equation appears on the right hand side. The turbulent transport terms are represented using the gradient diffusion idea introduced earlier in the context of scalar transport (see equation 3.24). Prandtl numbers σ_k and σ_ε connect the diffusivities of k and ε to the eddy viscosity μ_t. The pressure term (III) of the exact k-equation cannot be measured directly. Its effect is accounted for in equation (3.35) within the gradient diffusion term.

Production and destruction of turbulent kinetic energy are always closely linked. The dissipation rate ε is large where production of k is large. The model equation (3.36) for ε assumes that its production and destruction terms are proportional to the production and destruction terms of the k-equation (3.35). Adoption of such forms ensures that ε increases rapidly if k increases rapidly and that it decreases sufficiently fast to avoid (non-physical) negative values of turbulent kinetic energy if k decreases.

The factor ε/k in the production and destruction terms makes these terms dimensionally correct in the ε-equation.

To compute the Reynolds stresses with the k–ε model (3.34–3.36) an extended Boussinesq relationship is used:

$$-\rho\overline{u_i'u_j'} = \mu_t\left(\frac{\partial U_i}{\partial x_j} + \frac{\partial U_j}{\partial x_i}\right) - \frac{2}{3}\rho k\delta_{ij} = 2\mu_t E_{ij} - \frac{2}{3}\rho k\delta_{ij} \tag{3.38}$$

Comparison with equation (3.23) shows that this form has an extra term on the right hand side which involves δ_{ij}, the Kronecker delta ($\delta_{ij} = 1$ if $i = j$ and $\delta_{ij} = 0$ if $i \neq j$). The term serves to make the formula applicable to the normal Reynolds stresses for which $i = j$, and hence for $\tau_{xx} = -\rho\overline{u'^2}, \tau_{yy} = -\rho\overline{v'^2}$ and $\tau_{zz} = -\rho\overline{w'^2}$. We consider an incompressible flow and explore the behaviour of the first part of (3.38) by itself. If we sum this over all the normal stresses (i.e. let $i = 1, 2$ and 3 whilst keeping $i = j$) we find, using continuity, that it is zero, since

$$2\mu_t E_{ii} = 2\mu_t\left[\frac{\partial U}{\partial x} + \frac{\partial V}{\partial y} + \frac{\partial W}{\partial z}\right] = 2\mu_t \, div\, \mathbf{U} = 0$$

Clearly in any flow the sum of the normal stresses $-\rho(\overline{u'^2} + \overline{v'^2} + \overline{w'^2})$ is equal to minus twice the turbulence kinetic energy per unit volume ($-2\rho k$). An equal third is allocated to each normal stress component to ensure that their sum always has its physically correct value. It should be noted that this implies an isotropic assumption for the normal Reynolds stresses which the data in section 3.4 have shown is inaccurate even in simple two-dimensional flows.

Boundary conditions

The model equations for k and ε are elliptic by virtue of the gradient diffusion term. Their behaviour is similar to the other elliptic flow equations, which gives rise to the need for the following boundary conditions:

- inlet: distributions of k and ε must be given
- outlet or symmetry axis: $\partial k/\partial n = 0$ and $\partial\varepsilon/\partial n = 0$
- free stream: $k = 0$ and $\varepsilon = 0$
- solid walls: approach depends on Reynolds number (see below)

In exploratory design calculations the detailed boundary condition information required to operate the model may not be available. Industrial CFD users rarely have measurements of k and ε at their disposal. Progress can be made by entering values of k and ε from the literature (e.g. publications referred to in section 3.4) and subsequently exploring the sensitivity of the results to these inlet distributions. If no information is available at all, crude approximations for the inlet distributions for k and ε in internal flows can be obtained from the turbulence intensity T_i and a characteristic length L of the equipment (equivalent pipe radius) by means of the following simple assumed forms:

$$k = \tfrac{3}{2}(U_{ref}T_i)^2; \qquad \varepsilon = C_\mu^{3/4}\frac{k^{3/2}}{\ell}; \qquad \ell = 0.07L$$

The formulae are closely related to the mixing length formulae in section 3.5.1 and the universal distributions near a solid wall given below.

At **high Reynolds number** the standard k–ε model (Launder and Spalding, 1974) avoids the need to integrate the model equations right through to the wall by making use of the universal behaviour of near wall flows discussed in section 3.4. If y is the co-ordinate direction normal to a solid wall, the mean velocity at a point y_p with $30 < y_p^+ < 500$ satisfies the log-law (3.21) and measurements of turbulent kinetic energy budgets indicate that the rate of turbulence production equals the rate of dissipation. Using these assumptions and the eddy viscosity formula (3.34) it is possible to develop the following wall functions:

$$u^+ = \frac{U}{u_\tau} = \frac{1}{\kappa}\ln\left(Ey_P^+\right); \qquad k = \frac{u_\tau^2}{\sqrt{C_\mu}}; \qquad \varepsilon = \frac{u_\tau^3}{\kappa y} \tag{3.39}$$

Von Karman's constant $\kappa = 0.41$ and the wall roughness parameter $E = 9.8$ for smooth walls. Schlichting (1979) gives values of E that are valid for rough walls.

For heat transfer we use the universal near wall temperature distribution valid at high Reynolds numbers (Launder and Spalding, 1974):

$$T^+ \equiv -\frac{(T - T_w)C_P\rho u_\tau}{q_w} = \sigma_{T,t}\left[u^+ + P\left(\frac{\sigma_{T,l}}{\sigma_{T,t}}\right)\right] \tag{3.40}$$

with T_P = temperature at near wall point y_P
T_w = wall temperature
q_w = wall heat flux
C_p = fluid specific heat at constant pressure
$\sigma_{T,t}$ = turbulent Prandtl number
$\sigma_{T,l} = \mu C_p/\Gamma_T$ = Prandtl number
Γ_T = thermal conductivity

Finally P is the 'pee-function', a correction function dependent on the ratio of laminar to turbulent Prandtl numbers (Launder and Spalding, 1974).

At **low Reynolds numbers** the log-law is not valid so the above-mentioned boundary conditions cannot be used. Modifications to the k–ε model to enable it to cope with low Reynolds number flows are reviewed in Patel *et al* (1985). Wall damping needs to be applied to ensure that viscous stresses take over from turbulent Reynolds stresses at low Reynolds numbers and in the viscous sub-layer adjacent to solid walls.

The equations of the low Reynolds number k–ε model, which replace (3.34–3.36), are given below:

$$\mu_t = \rho C_\mu f_\mu \frac{k^2}{\varepsilon} \tag{3.41}$$

$$\frac{\partial(\rho k)}{\partial t} + div(\rho k\mathbf{U}) = div\left[\left(\mu + \frac{\mu_t}{\sigma_k}\right) grad\ k\right] + 2\mu_t E_{ij}\,.\,E_{ij} - \rho\varepsilon \tag{3.42}$$

$$\frac{\partial(\rho\varepsilon)}{\partial t} + div(\rho\varepsilon\mathbf{U}) = div\left[\left(\mu + \frac{\mu_t}{\sigma_\varepsilon}\right) grad\ \varepsilon\right] + C_{1\varepsilon} f_1 \frac{\varepsilon}{k} 2\mu_t E_{ij}\,.\,E_{ij} - C_{2\varepsilon} f_2 \rho \frac{\varepsilon^2}{k} \tag{3.43}$$

The most obvious modification, which is universally made, is to include a viscous contribution in the diffusion terms in (3.42–3.43). The constants C_μ, $C_{1\varepsilon}$ and $C_{2\varepsilon}$ in the standard k–ε model are multiplied by wall-damping functions, f_μ, f_1 and f_2 respectively, that are themselves functions of the turbulence Reynolds number ($= \vartheta\ell/\nu = k^2/(\varepsilon\nu)$) and/or similar parameters. As an example we quote the Lam and Bremhorst (1981) wall-damping functions which are particularly successful:

$$f_\mu = \left[1 - \exp(-0.0165 Re_y)\right]^2 \left(1 + \frac{20.5}{Re_t}\right);$$
$$f_1 = \left(1 + \frac{0.05}{f_\mu}\right)^3; \qquad f_2 = 1 - \exp(-Re_t^2) \tag{3.44}$$

In function f_μ the parameter Re_y is defined by $k^{1/2}y/\nu$. Lam and Bremhorst use $\partial\varepsilon/\partial y = 0$ as a boundary condition.

Assessment of performance

The k–ε model is the most widely used and validated turbulence model. It has achieved notable successes in calculating a wide variety of thin shear layer and recirculating flows without the need for case-by case adjustment of the model constants. The model performs particularly well in confined flows where the Reynolds shear stresses are most important. This includes a wide range of flows with industrial engineering applications, which explains its popularity. Examples of the application of the k–ε model to a range of industrially relevant flows are given in Chapter 10. Versions of the model are available which incorporate effects due to buoyancy (Rodi, 1980). Such models are used to study environmental flows such as pollutant dispersion in the atmosphere and in lakes and the modelling of fires.

In spite of the numerous successes the standard k–ε model shows only moderate agreement in unconfined flows. The model is reported not to perform well in weak shear layers (far wakes and mixing layers) and the spreading rate of axisymmetric jets in stagnant surroundings is severely overpredicted. In large parts of these flows the rate of production of turbulent kinetic energy is much less than the rate of dissipation and the difficulties can only be overcome by making *ad hoc* adjustments to model constants C.

Bradshaw *et al* (1981) state that the practice of incorporating the pressure transport term of the exact k-equation in the gradient diffusion expression of the model equation is deemed to be acceptable on the grounds that the pressure term is sometimes so small that measured turbulent kinetic energy budgets balance without it. They note, however, that many of these measurements contain substantial errors and it is certainly not generally true that pressure diffusion effects are negligible.

The model also has problems in swirling flows and flows with large, rapid, extra strains (e.g. highly curved boundary layers and diverging passages) since it is does not contain a description of the subtle effects of streamline curvature on turbulence. Secondary flows in long non-circular ducts, which are driven by anisotropic normal Reynolds stresses, can also not be predicted owing to the deficiencies of the treatment of normal stresses within the k–ε model. Finally, the model is oblivious to body forces due to rotation of the frame of reference.

A summary of the performance assessment for the standard k–ε model is given in Table 3.5

Table 3.5 k–ε model assessment

Advantages

- simplest turbulence model for which only initial and/or boundary conditions need to be supplied
- excellent performance for many industrially relevant flows
- well established; the most widely validated turbulence model

Disadvantages

- more expensive to implement than mixing length model (two extra PDEs)
- poor performance in a variety of important cases such as
  (i) some unconfined flows
  (ii) flows with large extra strains (e.g. curved boundary layers, swirling flows)
  (iii) rotating flows
  (iv) fully developed flows in non-circular ducts

3.5.3 Reynolds stress equation models

The most complex classical turbulence model is the **Reynolds stress equation model (RSM),** also called the second-order or second-moment closure model. Several major drawbacks of the k–ε model emerge when it is attempted to predict flows with complex strain fields or significant body forces. Under such conditions the individual Reynolds stresses are poorly represented by formula (3.38) even if the turbulent kinetic energy is computed to reasonable accuracy. The exact Reynolds stress transport equation on the other hand can account for the directional effects of the Reynolds stress field.

The modelling strategy originates from work reported in Launder *et al* (1975). We follow established practice in the literature by calling $R_{ij} = -\tau_{ij}/\rho = \overline{u_i'u_j'}$ the Reynolds stress, although the term kinematic Reynolds stress would be more precise. The exact equation for the transport of R_{ij} takes the following form:

$$\frac{DR_{ij}}{Dt} = P_{ij} + D_{ij} - \varepsilon_{ij} + \Pi_{ij} + \Omega_{ij} \tag{3.45}$$

Rate of change of $R_{ij} = \overline{u_i'u_j'}$ + Transport of R_{ij} by convection = Rate of production of R_{ij} + Transport of R_{ij} by diffusion − Rate of dissipation of R_{ij} + Transport of R_{ij} due to turbulent pressure-strain interactions + Transport of R_{ij} due to rotation

Equation (3.45) describes six partial differential equations: one for the transport of each of the six independent Reynolds stresses ($\overline{u_1'^2}, \overline{u_2'^2}, \overline{u_3'^2}, \overline{u_1'u_2'}, \overline{u_1'u_3'}$ and $\overline{u_2'u_3'}$, since $\overline{u_2'u_1'} = \overline{u_1'u_2'}, \overline{u_3'u_1'} = \overline{u_1'u_3'}$ and $\overline{u_3'u_2'} = \overline{u_2'u_3'}$,). If it is compared with the exact transport equation for the turbulent kinetic energy (3.32) two new physical processes appear in the Reynolds stress equations: the pressure–strain correlation term Π_{ij}, whose effect on the kinetic energy can be shown to be zero, and the rotation term Ω_{ij}.

CFD computations with the Reynolds stress transport equations retain the production term in its exact form

$$P_{ij} = -\left(R_{im} \frac{\partial U_j}{\partial x_m} + R_{jm} \frac{\partial U_i}{\partial x_m} \right) \tag{3.46}$$

To obtain a solvable form of (3.45) we need models for the diffusion, the dissipation rate and the pressure–strain correlation terms on the right hand side. Launder *et al* (1975) and Rodi (1980) give comprehensive details of the most general models. For the sake of simplicity we quote those models derived from this approach that are used in some commercial CFD codes. These models lack a little detail, but their structure is most easy to understand and the main message is intact in all cases.

The diffusion term D_{ij} can be modelled by the assumption that the rate of transport of Reynolds stresses by diffusion is proportional to the gradients of Reynolds stresses. This gradient diffusion idea recurs throughout turbulence modelling. Commercial CFD codes often favour the simplest form

$$D_{ij} = \frac{\partial}{\partial x_m} \left(\frac{\nu_t}{\sigma_k} \frac{\partial R_{ij}}{\partial x_m} \right) = div\left(\frac{\nu_t}{\sigma_k}\, grad(R_{ij}) \right) \tag{3.47}$$

with $\nu_t = C_\mu \dfrac{k^2}{\varepsilon}$; $\quad C_\mu = 0.09$ and $\sigma_k = 1.0$

The dissipation rate ε_{ij} is modelled by assuming isotropy of the small dissipative eddies. It is set so that it affects the normal Reynolds stresses ($i = j$) only and in equal measure. This can be achieved by

$$\varepsilon_{ij} = \tfrac{2}{3} \varepsilon \delta_{ij} \tag{3.48}$$

where ε is the dissipation rate of turbulent kinetic energy defined by (3.33). The Kronecker delta, δ_{ij} is given by $\delta_{ij} = 1$ if $i = j$ and $\delta_{ij} = 0$ if $i \neq j$.

The pressure–strain interactions constitute at the same time the most difficult term in (3.45) and the most important one to model accurately. Their effect on the Reynolds stresses is caused by two distinct physical processes: pressure fluctuations due to two eddies interacting with each other and pressure fluctuations due to the interaction of an eddy with a region of flow of different mean velocity. The overall effect of the pressure–strain term is to re-distribute energy amongst the normal Reynolds stresses ($i = j$) so as to make them more isotropic and to reduce the Reynolds shear stresses ($i \neq j$).

Corrections are needed to account for the influence of wall proximity on the pressure–strain terms. These corrections are different in nature from the wall-damping functions encountered in the k–ε model and need to be applied irrespective of the value of the mean flow Reynolds number. Measurements indicate that the wall effect increases the anisotropy of normal Reynolds stresses by damping out fluctuations in the directions normal to the wall and decreases the magnitude of the

Reynolds shear stresses. A comprehensive model that accounts for all these effects is given in Launder *et al* (1975). They also give the following simpler form favoured by some commercially available CFD codes:

$$\Pi_{ij} = -C_1 \frac{\varepsilon}{k}\left(R_{ij} - \tfrac{2}{3}k\delta_{ij}\right) - C_2\left(P_{ij} - \tfrac{2}{3}P\delta_{ij}\right) \tag{3.49}$$

with $C_1 = 1.8$ and $C_2 = 0.6$

The rotational term is given by

$$\Omega_{ij} = -2\omega_k\left(R_{jm}e_{ikm} + R_{im}e_{jkm}\right) \tag{3.50}$$

Here ω_k is the rotation vector and e_{ijk} is the alternating symbol; $e_{ijk} = +1$ if i, j and k are different and in cyclic order, $e_{ijk} = -1$ if i, j and k are different and in anti-cyclic order and $e_{ijk} = 0$ if any two indices are the same.

Turbulent kinetic energy k is needed in the above formulae and can be found by adding the three normal stresses together:

$$k = \tfrac{1}{2}(R_{11} + R_{22} + R_{33}) = \tfrac{1}{2}\left(\overline{u_1'^2} + \overline{u_2'^2} + \overline{u_3'^2}\right).$$

The six equations for Reynolds stress transport are solved along with a model equation for the scalar dissipation rate ε. Again a more exact form is found in Launder *et al* (1975), but the equation from the standard k–ε model is used in commercial CFD for the sake of simplicity.

$$\frac{D\varepsilon}{Dt} = div\left(\frac{\nu_t}{\sigma_\varepsilon} grad\ \varepsilon\right) + C_{1\varepsilon}\frac{\varepsilon}{k}\, 2\nu_t E_{ij} \,.\, E_{ij} - C_{2\varepsilon}\frac{\varepsilon^2}{k} \tag{3.51}$$

where $C_{1\varepsilon} = 1.44$ and $C_{2\varepsilon} = 1.92$

Rate of change of ε	+	Transport of ε by convection	=	Transport of ε by diffusion	+	Rate of production of ε	−	Rate of destruction of ε

The usual boundary conditions for elliptic flows are required for the solution of the Reynolds stress transport equations:

- inlet: specified distributions of R_{ij} and ε
- outlet and symmetry: $\partial R_{ij}/\partial n = 0$ and $\partial\varepsilon/\partial n = 0$
- free stream: $R_{ij} = 0$ and $\varepsilon = 0$
- solid wall: wall functions

In the absence of any information approximate inlet distributions for R_{ij} may be calculated from the turbulence intensity T_i and a characteristic length L of the equipment (e.g. equivalent pipe radius) by means of the following assumed relationships:

$$k = \tfrac{3}{2}(U_{ref}T_i)^2; \qquad \varepsilon = C_\mu^{3/4}\frac{k^{3/2}}{\ell}; \qquad \ell = 0.07L; \qquad \overline{u_1'^2} = k;$$

$$\overline{u_2'^2} = \overline{u_3'^2} = \tfrac{1}{2}k; \qquad \overline{u_i'u_j'} = 0(i \neq j)$$

Expressions such as these should not be used without a subsequent test of the sensitivity of results to the assumed inlet boundary conditions.

For computations at high Reynolds numbers wall-function-type boundary conditions can be used which are very similar to those of the k–ε model. Near wall Reynolds stress values are computed from formulae such as $R_{ij} = \overline{u_i'u_j'} = c_{ij}k$ where c_{ij} are obtained from measurements.

Low Reynolds number modifications to the models can be incorporated to add the effects of molecular viscosity to the diffusion terms and to account for anisotropy in the dissipation rate term in the R_{ij}-equations. Wall-damping functions to adjust the constants of the ε-equation and a modified dissipation rate variable $\tilde{\varepsilon}$ $(\equiv \varepsilon - 2\nu(\partial k^{1/2}/\partial y)^2)$ give more realistic modelling near solid walls. So *et al* (1991) give a recent review of the performance of near wall treatments where details may be found.

Similar models, involving three further model partial differential equations – one for every turbulent scalar flux $\overline{u_i'\varphi'}$ of equation (3.17) – are available for scalar transport. The interested reader is referred to Rodi (1980) for further material. Commercial CFD codes use the simple expedient of adding a turbulent diffusion coefficient $\Gamma_t = \mu_t/\sigma_\phi$ to the laminar diffusion coefficient with the Prandtl/Schmidt numbers σ_ϕ for all scalars equal to 0.7. Very little is known about low Reynolds number modifications to the scalar transport equations in near wall flows.

RSMs are clearly quite complex, but it is generally accepted that they are the 'simplest' type of model with the potential to describe all the mean flow properties and Reynolds stresses without case-by-case adjustment (Table 3.6). The RSM is by no means as well validated as the k–ε model and because of the high cost of the computations it is currently not widely used in industrial flow calculations. The extension and improvement of these models is an area of very active research. Once a consensus about the precise form of the component models has been reached it is likely that the availability of more powerful computing hardware will bring this form of turbulence modelling within the reach of the industrial user in the not too distant future.

Table 3.6 Reynolds stress equation model assessment

Advantages

- potentially the most general of all classical turbulence models
- only initial and/or boundary conditions need to be supplied
- very accurate calculation of mean flow properties and *all* Reynolds stresses for many simple and more complex flows including wall jets, asymmetric channel and non-circular duct flows and curved flows

Disadvantages

- very large computing costs (seven extra PDEs)
- not as widely validated as the mixing length and k–ε models
- performs just as poorly as the k–ε model in some flows owing to identical problems with the ε-equation modelling (e.g. axisymmetric jets and unconfined recirculating flows)

3.5.4 Algebraic stress equation models

The algebraic stress model (ASM) is an economical way of accounting for the anisotropy of Reynolds stresses without going to the full length of solving the Reynolds stress transport equations. The huge computational cost of solving the RSM is caused by the fact that gradients of the Reynolds stresses R_{ij} etc. appear in the convective (D/Dt) and diffusive transport terms D_{ij} of (3.47) and (3.49) respectively. Rodi proposed the idea that if the convective and diffusive transport terms are removed or modelled, the Reynolds stress equations reduce to a set of algebraic equations.

The simplest method is to neglect the convection and diffusion terms altogether. In some cases this appears to be sufficiently accurate (Naot and Rodi, 1982; Demuren and Rodi, 1984). A more generally applicable method is to assume that the sum of the convection and diffusion terms of the Reynolds stresses is proportional to the sum of the convection and diffusion terms of turbulent kinetic energy. Hence

$$\begin{aligned}\frac{D\overline{u_i'u_j'}}{Dt} - D_{ij} &\approx \frac{\overline{u_i'u_j'}}{k}\cdot\left(\frac{Dk}{Dt} - [k\text{-}transport\ (i.e.\ div) terms]\right)\\ &= \frac{\overline{u_i'u_j'}}{k}\cdot\left(-\overline{u_i'u_j'}\,.\,E_{ij} - \varepsilon\right)\end{aligned} \quad (3.52)$$

The terms in the brackets on the right hand side comprise the sum of the rate of production and the rate of dissipation of turbulent kinetic energy from the exact k-equation (3.32). The Reynolds stresses and the turbulent kinetic energy are both turbulence properties and are closely related, so (3.52) is likely not to be too bad an approximation provided that the ratio $\overline{u_i'u_j'}/k$ does not vary too rapidly across the flow. Further refinements may be obtained by relating the transport by convection and diffusion independently to the transport of turbulent kinetic energy.

Introducing approximation (3.52) into the Reynolds stress transport equation (3.45) with production term P_{ij} (3.46), modelled dissipation rate term (3.50) and pressure–strain correlation term (3.49) on the right hand side yields after some rearrangement the following **algebraic stress model**:

$$R_{ij} = \overline{u_i'u_j'} = \tfrac{2}{3}k\delta_{ij} + \left(\frac{C_D}{C_1 - 1 + \dfrac{P}{\varepsilon}}\right)\left(P_{ij} - \tfrac{2}{3}P\delta_{ij}\right)\frac{k}{\varepsilon} \quad (3.53)$$

The Reynolds stresses appear on both sides of the equation – on the right hand side they are contained within P_{ij} – so (3.53) is a set of six simultaneous algebraic equations for the six unknown Reynolds stresses R_{ij} that can be solved by matrix inversion or iterative techniques if k and ε are known. Therefore, the formulae are solved in conjunction with the standard k–ε model equations (3.34–3.37).

The constant C_D is adjustable to make up for the physics 'lost' in the approximation. One commercial CFD code recommends ASM for swirling flows with the following constants:

$$C_D = 0.55 \qquad \text{and} \qquad C_1 = 2.2 \quad (3.54)$$

Turbulent scalar transport can also be described by algebraic models derived from their full transport equations that were alluded to in the previous section. Again Rodi (1980) gives further information for the interested reader.

Demuren and Rodi (1984) report the computation of the secondary flow in non-circular ducts with a somewhat more sophisticated version of this model that includes wall corrections for the pressure–strain term and modified values of adjustable constants to get a good match with measured data in nearly homogeneous shear flows and channel flows. They achieved realistic predictions of the primary flow distortions and secondary flow in square and rectangular ducts. These effects are caused by anisotropy of the normal Reynolds stresses and can therefore not be represented in simulations of the same situation with the standard k–ε model.

The algebraic stress model is an economical method of incorporating the effects of anisotropy into the calculations of Reynolds stresses. The model is not as well validated as the k–ε model but can be used in flows where the latter is known to perform poorly and where the transport assumptions made do not compromise too severely the calculation accuracy (Table 3.7). Recent developments in the design of anisotropic eddy viscosity k–ε models have caused a modest loss of popularity of the ASM.

Table 3.7 Algebraic stress model assessment

Advantages

- cheap method to account for Reynolds stress anisotropy
- potentially combines the generality of approach of the RSM (good modelling of buoyancy and rotation effects possible) with the economy of the k–ε model
- successfully applied to isothermal and buoyant thin shear layers
- if convection and diffusion terms are negligible the ASM performs as well as the RSM

Disadvantages

- only slightly more expensive than the k–ε model (two PDEs and a system of algebraic equations)
- not as widely validated as the mixing length and k–ε models
- same disadvantages as RSM apply
- model is severely restricted in flows where the transport assumptions for convective and diffusive effects do not apply – validation is necessary to define the performance limits

3.5.5 Some recent advances

The field of turbulence modelling provides an area of intense research activity for the CFD and fluid engineering community. The previous sections have outlined the modelling strategy of each of the main classical models which are applied in or are under development for commercially available general purpose codes. Now we report a necessarily small sample of recent developments.

Behind much of the current advanced turbulence modelling research lies the belief that, irrespective of boundary conditions and geometry, there exists a (limited) number of universal features of turbulence which, when identified correctly, can form the basis of a complete description of flow variables of interest to an engineer. The emphasis must be on the word belief, because the very existence of a classical model – based on time-averaged equations – of this kind is contested by a number of renowned experts in the field. Encouraged by, for example, the success of the mixing length model in the external aerodynamics field, they favour the development of dedicated models for limited classes of flow. These two viewpoints naturally lead to two distinct lines of research work:

1. The development and optimisation of turbulence models for limited categories of flows.
2. The search for a comprehensive and completely general purpose turbulence model.

Industry has many pressing flow problems to solve that will not wait for the conception of a universal turbulence model. Fortunately many sectors of industry are specifically interested in a limited class of flows only – e.g. pipe flows for the oil transportation sector, turbines and combustors for power engineering. The large majority of turbulence research consists of case-by-case examination and validation of existing turbulence models for such specific problems.

The literature is far too extensive even to begin to review here. The main sources of useful, applications-oriented information are: *Transactions of the American Society of Mechanical Engineers* – in particular the *Journal of Fluids Engineering, Journal of Heat Transfer* and *Journal of Engineering for Gas Turbines and Power* – as well as the *AIAA Journal*, the *International Journal of Heat and Mass Transfer* and the *International Journal of Heat and Fluid Flow.*

More fundamental turbulence modelling research has recently followed various interesting new directions. Much of the work is published in the above-mentioned journals, especially the *AIAA Journal*, but the *Journal of Fluid Mechanics* and *Physics of Fluids A* provide further and often more in-depth coverage.

All turbulence models considered in this book have been initially conceived by intuition combined with a semi-empirical approach. Launder, who played a key part in the development of practically all the current general purpose models of turbulence, gave a fairly recent review (Launder, 1989) of the position of the RSMs, the most sophisticated models. Developments continue in this area, but Amano and Goel (1987) confirm doubts that even the RSM, with its 'crude' modelling of the right hand side terms, may not be a sufficiently general purpose tool for all problems. They suggest that nothing short of a full third-moment closure model – the solution of the transport equations for the quantities $\overline{u_i'u_j'u_k'}$ in the exact Reynolds stress transport equations – will suffice for the accurate description of recirculating flows such as that over a backward-facing step. Such a requirement is not without consequence since third-moment closures increase the computing cost of solutions. Moreover, it becomes progressively more difficult to make measurements to support the development of models of diffusion, production and destruction terms in transport equations of this kind.

It is therefore encouraging that a group of researchers at NASA Langley Research Center led by Speziale have recently developed a framework for non-linear extensions of the k–ε model. In addition to this work they have derived a variant of the ASM (Gatski and Speziale, 1993) that is claimed to be more computationally stable. Here we briefly discuss the new k–ε models. A major drawback of the standard k–ε model is the fact that the eddy viscosity is identical for all the Reynolds stresses. Measurements indicate that this is not the case even in simple turbulent flows. In two-dimensional thin shear layers the discrepancies often do not give cause for concern because only one Reynolds shear stress is important.

The initial interest in the ASM arose from its ability to provide a cheap way of accounting for anisotropy of the Reynolds stresses. The group of researchers

NASA Langley Research Center led by Speziale have recently advanced non-linear k–ε models. The approach consists of deriving asymptotic expansions for the Reynolds stresses which maintain terms that are quadratic in velocity gradients. Invoking some powerful constraints on the mathematical shape of the resulting models, most of which were first compiled and formulated by Lumley (1978), the following non-linear expression emerges (Speziale, 1987, 1991):

$$\tau_{ij} = -\rho\overline{u_i u_j} = -\tfrac{2}{3}\rho k \delta_{ij} + \rho C_\mu \frac{k^2}{\varepsilon} 2E_{ij} - 4C_D C_\mu^2 \frac{k^3}{\varepsilon^2}\left(E_{im} \,.\, E_{mj} - \tfrac{1}{3}E_{mn} \,.\, E_{mn}\delta_{ij} + E^o_{ij} = \tfrac{1}{3}E^o_{mm}\delta_{ij}\right) \quad (3.55)$$

where $E^o_{ij} = \dfrac{\partial E_{ij}}{\partial t} + \mathbf{u} \,.\, grad\left(E_{ij}\right) - \left(\dfrac{\partial U_i}{\partial x_k} \,.\, E_{kj} + \dfrac{\partial U_j}{\partial x_k} \,.\, E_{ki}\right)$

and $C_D = 1.68$

The value of adjustable constant C_D was found by calibration with experimental data.

Equation (3.55) is the non-linear extension of the k–ε model to flows with moderate and large strains. Formula (3.38) is a special case of (3.55) at low rates of deformation when terms that are quadratic in velocity gradients may be dropped. Like the ASM this model can account for the secondary flow in fully developed non-circular duct flows. Horiuti (1990) argues in favour of a variant of this approach which retains terms up to third order in velocity gradients.

The statistical mechanics approach has led to new mathematical formalisms which, in conjunction with a limited number of assumptions regarding the statistics of small scale turbulence, provide a rigorous basis for the extension of eddy viscosity models. The Renormalization Group (RNG) devised by Yakhot and Orszag of Princeton University has to date attracted most interest.

These workers represent the effects of the small scale turbulence by means of a random forcing function in the Navier–Stokes equation. The RNG procedure systematically removes the small scales of motion from the governing equations by expressing their effects in terms of larger scale motions and a modified viscosity. The mathematics is highly abstruse; we only quote the RNG k–ε model equations which result from the work (Yakhot *et al* 1992):

$$\frac{\partial(\rho k)}{\partial t} + div(\rho k \mathbf{U}) = div\left[\alpha_k \mu_{eff}\, grad\; k\right] + 2\mu_t E_{ij} \,.\, E_{ij} - \rho\varepsilon \quad (3.56)$$

$$\frac{\partial(\rho\varepsilon)}{\partial t} + div(\rho\varepsilon\mathbf{U}) = div\left[\alpha_\varepsilon \mu_{eff}\; grad\; \varepsilon\right] + C^*_{1\varepsilon}\frac{\varepsilon}{k} 2\mu_t E_{ij} \,.\, E_{ij} - C_{2\varepsilon}\rho\frac{\varepsilon^2}{k} \quad (3.57)$$

where $\mu_{eff} = \mu + \mu_t; \qquad \mu_t = \rho C_\mu \dfrac{k^2}{\varepsilon}$

with $C_\mu = 0.0845; \qquad \alpha_k = \alpha_\varepsilon = 1.39; \qquad C_{1\varepsilon} = 1.42; \qquad C_{2\varepsilon} = 1.68$

(3.58)

$$\text{and} \quad C^*_{1\varepsilon} = C_{1\varepsilon} - \frac{\eta(1-\eta/\eta_0)}{1+\beta\eta^3}; \qquad \eta = \left(2E_{ij}\,.\,E_{ij}\right)^{1/2}\frac{k}{\varepsilon}; \qquad \eta_0 = 4.377;$$

$$\beta = 0.012$$

Only the constant β is adjustable; the above value is calculated from near wall turbulence data. All other constants are explicitly computed as part of the RNG process.

The ε-equation has long been suspected as one of the main sources of accuracy limitations for the standard version of the k–ε model and the RSM in flows that experience large rates of deformation. It is, therefore, interesting to note that the model contains a strain-dependent correction term in the constant $C_{1\varepsilon}$ of the RNG model ε-equation. The model can be applied with the isotropic Reynolds stress formula (3.38) or with the non-linear form (3.57). Yakhot *et al* report very good predictions of the flow over a backward-facing step, in particular when using the non-linear Reynolds stress expression.

The model is essentially a variant of the k–ε model and computations are only slightly more expensive than those with the standard version. The performance improvements for complex turbulent flows have aroused so much interest that a number of commercial CFD codes now incorporate the RNG version of this model. The user should, however, note that for all its promise the RNG model is only a relative newcomer to turbulence and still needs to be widely validated.

3.6 Final remarks

This chapter provides a first glimpse of the role of turbulence in defining the broad features of the flow and of the practice of turbulence modelling. Turbulence is a phenomenon of great complexity and has puzzled theoreticians for over a hundred years. Its appearance causes radical changes to the flow which can range from the favourable (efficient mixing) to the detrimental (high energy losses) depending on one's point of view. The fluctuations associated with turbulence give rise to the extra Reynolds stresses on the mean flow.

What makes turbulence so difficult to tackle mathematically is the wide range of length and time scales of motion even in flows with very simple boundary conditions. It should therefore be considered as truly remarkable that the two most widely applied models, the mixing length and k–ε models, succeed in expressing the main features of many turbulent flows by means of one length scale and one time scale defining variable. The standard k–ε model still comes highly recommended for general purpose CFD computations. Although many experts argue that the RSM is the only viable way forward towards a truly general purpose classical turbulence model, the recent advances in the area of non-linear k–ε models are very likely to re-invigorate research on two-equation models.

Large eddy simulation (LES) models require large computing resources and are not (yet?) of use as general purpose tools. Nevertheless, in simple flows LES computations can supply values of turbulence properties that cannot be measured in the laboratory owing to the absence of suitable experimental techniques. Hence LES models will increasingly be used to guide the development of classical models through comparative studies.

Although the resulting mathematical expressions of turbulence models may be quite complicated it should never be forgotten that they **all** contain adjustable constants that need to be determined as best-fit values from experimental data that contain experimental uncertainties. Every engineer is aware of the dangers of extrapolating an empirical model beyond its data range. The same risks occur when abusing turbulence models in this fashion. CFD calculations of 'new' turbulent flows should never be accepted without any validation against high quality experiments.

4

The Finite Volume Method for Diffusion Problems

4.1 Introduction

The nature of the transport equations governing fluid flow and heat transfer and the formal control volume integration were described in Chapter 2. Here we develop the numerical method based on this integration, the **finite volume (or control volume) method**, by considering the simplest transport process of all: pure diffusion in the steady state. The governing equation of steady diffusion can easily be derived from the general transport equation (2.39) for property ϕ by deleting the transient and convective terms. This gives

$$div(\Gamma\ grad\ \phi) + S_{\phi} = 0 \tag{4.1}$$

The control volume integration, which forms the key step of the finite volume method that distinguishes it from all other CFD techniques, yields the following form:

$$\int_{CV} div(\Gamma\ grad\ \phi)dV + \int_{CV} S_{\phi}dV = \int_{A} \mathbf{n}\ .\ (\Gamma\ grad\ \phi)dA + \int_{CV} S_{\phi}dV = 0 \tag{4.2}$$

By working with the one-dimensional steady state diffusion equation the approximation techniques that are needed to obtain the so-called discretised equations are introduced. Later the method is extended to two- and three-dimensional diffusion problems. Application of the method to simple one-dimensional steady state heat transfer problems is illustrated through a series of worked examples and the accuracy of the method is gauged by comparing numerical results with analytical solutions.

4.2 Finite volume method for one-dimensional steady state diffusion

Consider the steady state diffusion of a property ϕ in a one-dimensional domain defined in Figure 4.1. The process is governed by

$$\frac{d}{dx}\left(\Gamma \frac{d\phi}{dx}\right) + S = 0 \tag{4.3}$$

where Γ is the diffusion coefficient and S is the source term. Boundary values of ϕ at points A and B are prescribed. An example of this type of process, one-dimensional heat conduction in a rod, is studied in detail in section 4.3.

Fig. 4.1

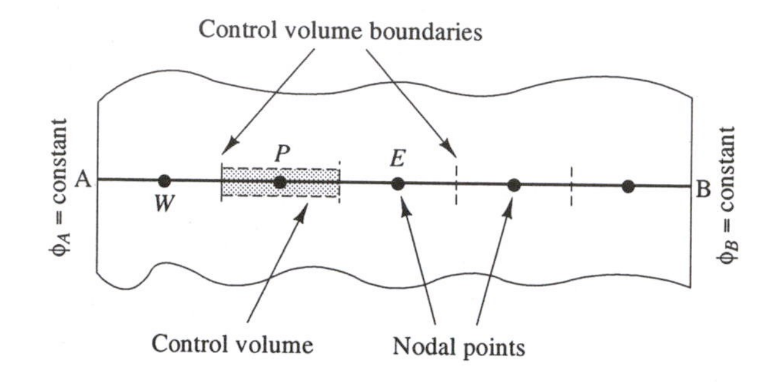

Step 1: Grid generation

The first step in the finite volume method is to divide the domain into discrete control volumes. Let us place a number of nodal points in the space between A and B. The boundaries (or faces) of control volumes are positioned mid-way between adjacent nodes. Thus each node is surrounded by a control volume or cell. It is common practice to set up control volumes near the edge of the domain in such a way that the physical boundaries coincide with the control volume boundaries.

At this point it is appropriate to establish a system of notation that can be used in future developments. The usual convention of CFD methods is shown in Figure 4.2.

A general nodal point is identified by P and its neighbours in a one-dimensional geometry, the nodes to the west and east, are identified by W and E respectively. The west side face of the control volume is referred to by 'w' and the east side control volume face by 'e'. The distances between the nodes W and P, and between nodes P and E, are identified by δx_{WP} and δx_{PE} respectively. Similarly the distances between face w and point P and between P and face e are denoted by δx_{wP} and δx_{Pe} respectively. Figure 4.2 shows that the control volume width is $\Delta x = \delta x_{we}$.

Fig. 4.2

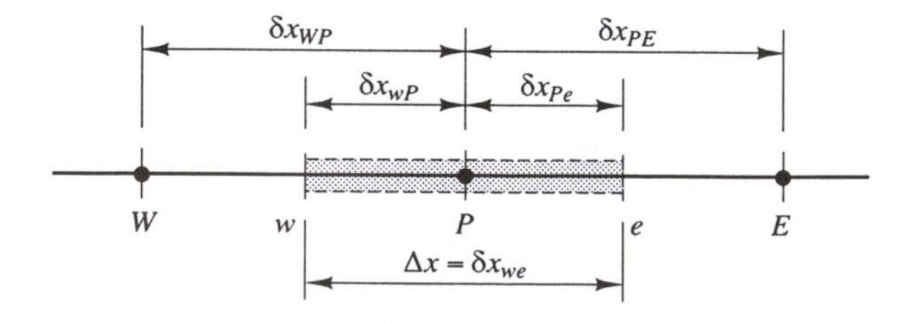

Step 2: discretisation

The key step of the finite volume method is the integration of the governing equation (or equations) over a control volume to yield a discretised equation at its nodal point P. For the control volume defined above this gives

$$\int_{\Delta V} \frac{d}{dx}\left(\Gamma \frac{d\phi}{dx}\right) dV + \int_{\Delta V} S dV = \left(\Gamma A \frac{d\phi}{dx}\right)_e - \left(\Gamma A \frac{d\phi}{dx}\right)_w + \bar{S}\Delta V = 0 \tag{4.4}$$

Here A is the cross-sectional area of the control volume face, ΔV is the volume and $\overline{S}$ is the average value of source S over the control volume. It is a very attractive feature of the finite volume method that the discretised equation has a clear physical interpretation. Equation (4.4) states that the diffusive flux of ϕ leaving the east face minus the diffusive flux of ϕ entering the west face is equal to the generation of ϕ, i.e. it constitutes a balance equation for ϕ over the control volume.

In order to derive useful forms of the discretised equations, the interface diffusion coefficient Γ and the gradient $d\phi/dx$ at east ('e') and west ('w') are required. Following well-established practice, the values of the property ϕ and the diffusion coefficient are defined and evaluated at nodal points. To calculate gradients (and hence fluxes) at the control volume faces an approximate distribution of properties between nodal points is used. Linear approximations seem to be the obvious and simplest way of calculating interface values and the gradients. This practice is called central differencing (see Appendix A). In a uniform grid linearly interpolated values for Γ_e and Γ_w are given by

$$\Gamma_w = \frac{\Gamma_W + \Gamma_P}{2} \tag{4.5a}$$

$$\Gamma_e = \frac{\Gamma_P + \Gamma_E}{2} \tag{4.5b}$$

And the diffusive flux terms are evaluated as

$$\left(\Gamma A \frac{d\phi}{dx}\right)_e = \Gamma_e A_e \left(\frac{\phi_E - \phi_P}{\delta x_{PE}}\right) \tag{4.6}$$

$$\left(\Gamma A \frac{d\phi}{dx}\right)_w = \Gamma_w A_w \left(\frac{\phi_P - \phi_W}{\delta x_{WP}}\right) \tag{4.7}$$

In practical situations, as illustrated later, the source term S may be a function of the dependent variable. In such cases the finite volume method approximates the source term by means of a linear form:

$$\bar{S}\Delta V = S_u + S_p\phi_P \tag{4.8}$$

Substitution of equations (4.6), (4.7) and (4.8) into equation (4.4) gives

$$\Gamma_e A_e \left(\frac{\phi_E - \phi_P}{\delta x_{PE}}\right) - \Gamma_w A_w \left(\frac{\phi_P - \phi_W}{\delta x_{WP}}\right) + (S_u + S_p\phi_P) = 0 \tag{4.9}$$

This can be re-arranged as

$$\boxed{\left(\frac{\Gamma_e}{\delta x_{PE}} A_e + \frac{\Gamma_w}{\delta x_{WP}} A_w - S_p\right)\phi_P = \left(\frac{\Gamma_w}{\delta x_{WP}} A_w\right)\phi_W + \left(\frac{\Gamma_e}{\delta x_{PE}} A_e\right)\phi_E + S_u} \tag{4.10}$$

Identifying the coefficients of ϕ_W and ϕ_E in equation (4.10) as a_W and a_E, and the coefficient of ϕ_P as a_P, the above equation can be written as

$$\boxed{a_P\phi_P = a_W\phi_W + a_E\phi_E + S_u} \quad (4.11)$$

where

a_W	a_E	a_P
$\frac{\Gamma_w}{\delta x_{WP}}A_w$	$\frac{\Gamma_e}{\delta x_{PE}}A_e$	$a_W + a_E - S_P$

The values of S_u and S_p can be obtained from the source model (4.8): $\bar{S}\Delta V = S_u + S_p\phi_P$. Equations (4.11) and (4.8) represent the discretised form of the equation (4.1). This type of discretised equation is central to all further developments.

Step 3: Solution of equations

Discretised equations of the form (4.11) must be set up at each of the nodal points in order to solve a problem. For control volumes that are adjacent to the domain boundaries the general discretised equation (4.11) is modified to incorporate boundary conditions. The resulting system of linear algebraic equations is then solved to obtain the distribution of the property ϕ at nodal points. Any suitable matrix solution technique may be enlisted for this task. In Chapter 7 we describe matrix solution methods that are specially designed for CFD procedures. The techniques of dealing with different types of boundary conditions will be examined in detail in Chapter 9.

4.3 Worked examples: one-dimensional steady state diffusion

The application of the finite volume method to the solution of simple diffusion problems involving conductive heat transfer is presented in this section. The equation governing one-dimensional steady state conductive heat transfer is

$$\frac{d}{dx}\left(k\frac{dT}{dx}\right) + S = 0 \quad (4.12)$$

where thermal conductivity k takes the place of Γ in equation (4.3) and the dependent variable is temperature T. The source term can, for example, be heat generation due to an electrical current passing through the rod. The incorporation of boundary conditions as well as the treatment of source terms will be introduced by means of three worked examples.

Example 4.1 Consider the problem of source-free heat conduction in an insulated rod whose ends are maintained at constant temperatures of 100 °C and 500 °C respectively. The one-

Fig. 4.3

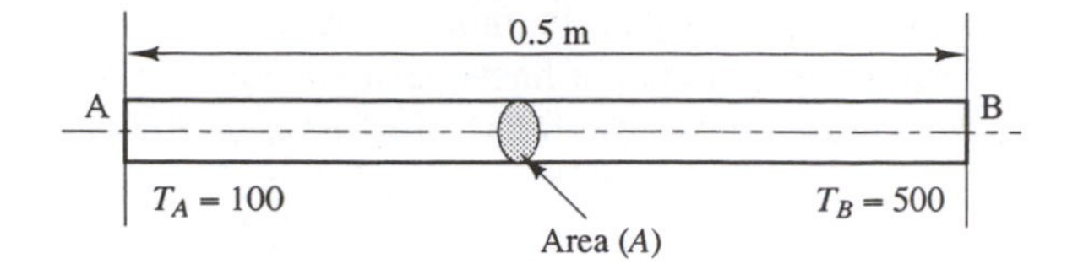

dimensional problem sketched in Figure 4.3 is governed by

$$\frac{d}{dx}\left(k\frac{dT}{dx}\right) = 0 \tag{4.13}$$

Calculate the steady state temperature distribution in the rod. Thermal conductivity k equals 1000 W/m/K, cross-sectional area A is 10×10^{-3} m^2.

Solution Let us divide the length of the rod into five equal control volumes as shown in Figure 4.4. This gives $\delta x = 0.1$ m.

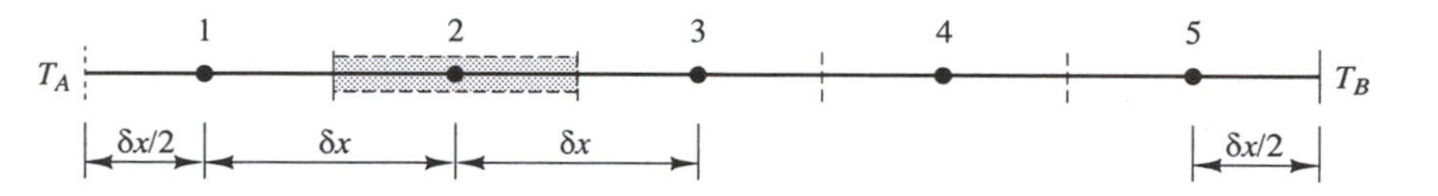

Fig. 4.4 The grid used

The grid consists of five nodes. For each one of nodes 2, 3 and 4 temperature values to the east and west are available as nodal values. Consequently, discretised equations of the form (4.10) can be readily written for control volumes surrounding these nodes:

$$\left(\frac{k_e}{\delta x_{PE}}A_e + \frac{k_w}{\delta x_{WP}}A_w\right)T_P = \left(\frac{k_w}{\delta x_{WP}}A_w\right)T_W + \left(\frac{k_e}{\delta x_{PE}}A_e\right)T_E \tag{4.14}$$

The thermal conductivity ($k_e = k_w = k$), node spacing (δx) and cross-sectional area ($A_e = A_w = A$) are constants. Therefore the **discretised equation for nodal points 2, 3 and 4** is

$$\boxed{a_P T_P = a_W T_W + a_E T_E} \tag{4.15}$$

with

a_W	a_E	a_P
$\frac{k}{\delta x}A$	$\frac{k}{\delta x}A$	$a_W + a_E$

S_u and S_p are zero in this case since there is no source term in the governing equation (4.13).

Nodes 1 and 5 are boundary nodes, and therefore require special attention. Integration of equation (4.13) over the control volume surrounding point 1 gives

$$kA\left(\frac{T_E - T_P}{\delta x}\right) - kA\left(\frac{T_P - T_A}{\delta x/2}\right) = 0 \tag{4.16}$$

This expression shows that the flux through control volume boundary A has been approximated by assuming a linear relationship between temperatures at boundary point A and node P. We can re-arrange (4.16) as follows:

$$\left(\frac{k}{\delta x}A + \frac{2k}{\delta x}A\right)T_P = 0.T_W + \left(\frac{k}{\delta x}A\right)T_E + \left(\frac{2k}{\delta x}A\right)T_A \tag{4.17}$$

Comparing equation (4.17) with equation (4.10) it can be easily identified that the fixed temperature boundary condition enters the calculation as a source term $(S_u + S_p T_P)$ with $S_u = (2kA/\delta x)T_A$ and $S_p = -2kA/\delta x$ and that the link to the (west) boundary side has been suppressed by setting coefficient a_W to zero.

Equation (4.17) may be cast in the same form as (4.11) to yield the **discretised equation for boundary node 1**:

$$\boxed{a_P T_P = a_W T_W + a_E T_E + S_u} \tag{4.18}$$

with

a_W	a_E	a_P	S_P	S_u
0	$\frac{kA}{\delta x}$	$a_W + a_E - S_p$	$-\frac{2kA}{\delta x}$	$\frac{2kA}{\delta x}T_A$

The control volume surrounding node 5 can be treated in a similar manner. Its discretised equation is given by

$$kA\left(\frac{T_B - T_P}{\delta x/2}\right) - kA\left(\frac{T_P - T_W}{\delta x}\right) = 0 \tag{4.19}$$

As before we assume a linear temperature distribution between node P and boundary point B to approximate the heat flux through the control volume boundary. Equation (4.19) can be re-arranged as

$$\left(\frac{k}{\delta x}A + \frac{2k}{\delta x}A\right)T_P = \left(\frac{k}{\delta x}A\right)T_W + 0\ .\ T_E + \left(\frac{2k}{\delta x}A\right)T_B \tag{4.20}$$

The **discretised equation for boundary node 5** is

$$\boxed{a_P T_P = a_W T_W + a_E T_E + S_u} \tag{4.21}$$

where

a_W	a_E	a_P	S_P	S_u
$\frac{kA}{\delta x}$	0	$a_W + a_E - S_p$	$-\frac{2kA}{\delta x}$	$\frac{2kA}{\delta x}T_B$

The discretisation process has yielded one equation for each of the nodal points 1 to 5. Substitution of numerical values gives $kA/\delta x = 100$ and the coefficients of each discretised equation can easily be worked out. Their values are given in Table 4.1.

Table 4.1

Node	a_W	a_E	S_u	S_P	$a_P = a_W + a_E - S_P$
1	0	100	$200T_A$	−200	300
2	100	100	0	0	200
3	100	100	0	0	200
4	100	100	0	0	200
5	100	0	$200T_B$	−200	300

The resulting set of algebraic equations for this example is

$$\begin{aligned} 300T_1 &= 100T_2 + 200T_A \\ 200T_2 &= 100T_1 + 100T_3 \\ 200T_3 &= 100T_2 + 100T_4 \\ 200T_4 &= 100T_3 + 100T_5 \\ 300T_5 &= 100T_4 + 200T_B \end{aligned} \tag{4.22}$$

This set of equations can be re-arranged as

$$\begin{bmatrix} 300 & -100 & 0 & 0 & 0 \\ -100 & 200 & -100 & 0 & 0 \\ 0 & -100 & 200 & -100 & 0 \\ 0 & 0 & -100 & 200 & -100 \\ 0 & 0 & 0 & -100 & 300 \end{bmatrix} \begin{bmatrix} T_1 \\ T_2 \\ T_3 \\ T_4 \\ T_5 \end{bmatrix} = \begin{bmatrix} 200T_A \\ 0 \\ 0 \\ 0 \\ 200T_B \end{bmatrix} \tag{4.23}$$

The above set of equations yields the steady state temperature distribution of the given situation. For simple problems involving a small number of nodes the resulting matrix equation can easily be solved with a software package such as MATLAB (*The Student Edition of MATLAB*, The Math Works Inc., 1992). For $T_A = 100$ and $T_B = 500$ the solution of (4.23) can obtained by using, for example, Gaussian elimination:

$$\begin{bmatrix} T_1 \\ T_2 \\ T_3 \\ T_4 \\ T_5 \end{bmatrix} = \begin{bmatrix} 140 \\ 220 \\ 300 \\ 380 \\ 460 \end{bmatrix} \tag{4.24}$$

The exact solution is a linear distribution between the specified boundary temperatures: $T = 800x + 100$. Figure 4.5 shows that the exact solution and the numerical results coincide.

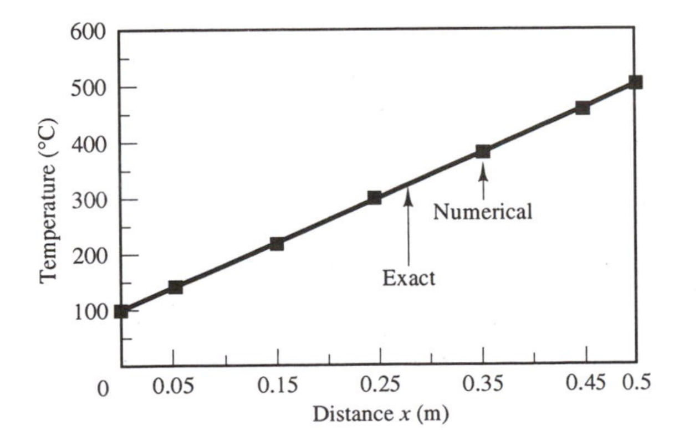

Fig. 4.5 Comparison of the numerical result with the analytical solution

Example 4.2 Now we discuss a problem that includes sources other than those arising from boundary conditions.

Figure 4.6 shows a large plate of thickness $L = 2$ cm with constant thermal conductivity $k = 0.5$ W/m/K and uniform heat generation $q = 1000$ kW/m^3. The faces A and B are at temperatures of 100 °C and 200 °C respectively. Assuming that the dimensions in the y- and z-directions are so large that temperature gradients are significant in the x-direction only, calculate the steady state temperature distribution. Compare the numerical result with the analytical solution. The governing equation is

$$\frac{d}{dx}\left(k\frac{dT}{dx}\right) + q = 0 \tag{4.25}$$

Fig. 4.6

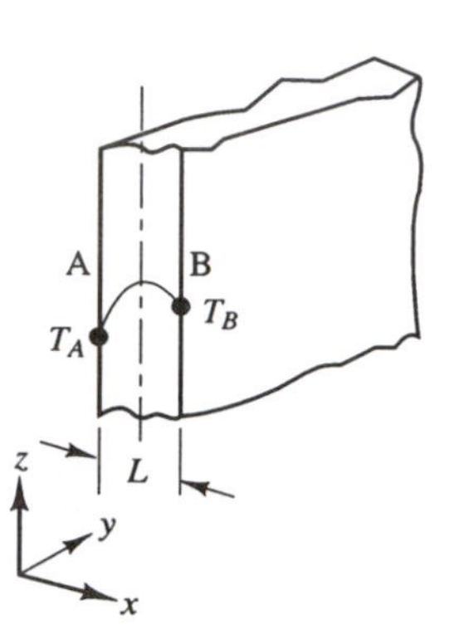

Solution As before the method of solution is demonstrated using a simple grid. The domain is divided into five control volumes (see Figure 4.7) giving $\delta x = 0.004$ m; a unit area is considered in the y–z plane.

Fig. 4.7 The grid used

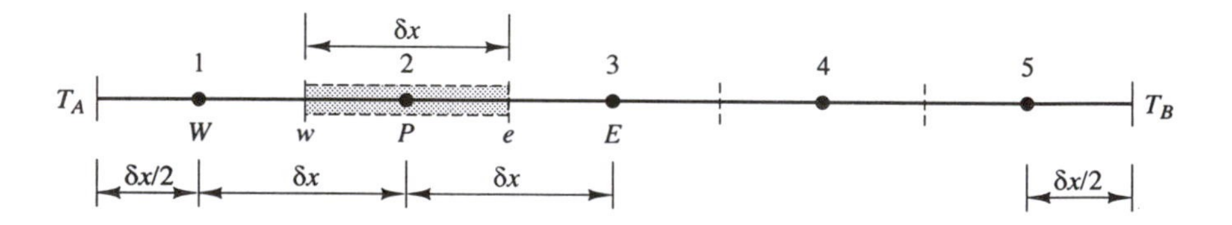

Formal integration of the governing equation over a control volume gives

$$\int_{\Delta V} \frac{d}{dx}\left(k\frac{dT}{dx}\right)dV + \int_{\Delta V} q dV = 0 \tag{4.26}$$

We treat the first term of the above equation as in the previous example. The second integral, the source term of the equation, is evaluated by calculating the average generation (i.e. $\overline{S}\Delta V = q\Delta V$) within each control volume. Equation (4.26) can be written as

$$\left[\left(kA\frac{dT}{dx}\right)_e - \left(kA\frac{dT}{dx}\right)_w\right] + q\Delta V = 0 \tag{4.27}$$

$$\left[k_e A\left(\frac{T_E - T_P}{\delta x}\right) - k_w A\left(\frac{T_P - T_W}{\delta x}\right)\right] + qA\delta x = 0 \tag{4.28}$$

The above equation can be re-arranged as

$$\left(\frac{k_e\,A}{\delta x}+\frac{k_w\,A}{\delta x}\right)T_P=\left(\frac{k_e\,A}{\delta x}\right)T_W+\left(\frac{k_w\,A}{\delta x}\right)T_E+qA\delta x \tag{4.29}$$

This equation is written in the general form of (4.11):

$$\boxed{a_P T_P = a_W T_W + a_E T_E + S_u} \tag{4.30}$$

Since $k_e = k_w = k$ we have the following coefficients:

a_W	a_E	a_p	S_p	S_u
$\frac{kA}{\delta x}$	$\frac{kA}{\delta x}$	$a_W + a_E - S_p$	0	$qA\delta x$

Equation (4.30) is valid for control volumes at **nodal points 2, 3 and 4**.

To incorporate the boundary conditions at nodes 1 and 5 we apply the linear approximation for temperatures between a boundary point and the adjacent nodal point. At node 1 the temperature at the west boundary is known. Integration of equation (4.25) at the control volume surrounding node 1 gives

$$\left[\left(kA\frac{dT}{dx}\right)_e-\left(kA\frac{dT}{dx}\right)_w\right]+q\Delta V=0 \tag{4.31}$$

Introduction of the linear approximation for temperatures between A and P yields

$$\left[k_eA\left(\frac{T_E-T_P}{\delta x}\right)-k_AA\left(\frac{T_P-T_A}{\delta x/2}\right)\right]+qA\delta x=0 \tag{4.32}$$

The above equation can be re-arranged, using $k_e = k_A = k$, to yield the discretised equation for **boundary node 1**:

$$\boxed{a_P T_P = a_W T_W + a_E T_E + S_u} \tag{4.33}$$

where

a_W	a_E	a_P	S_p	S_u
0	$\frac{kA}{\delta x}$	$a_W + a_E - S_p$	$-\frac{2kA}{\delta x}$	$qA\delta x + \frac{2kA}{\delta x}T_A$

At nodal point 5, the temperature on the east face of the control volume is known. The node is treated in a similar way to boundary node 1. At boundary point 5 we have

$$\left[\left(k\,A\frac{dT}{dx}\right)_e-\left(k\,A\frac{dT}{dx}\right)_w\right]+q\Delta V=0 \tag{4.34}$$

$$\left[k_B\,A\left(\frac{T_B-T_P}{\delta x/2}\right)-k_w\,A\left(\frac{T_P-T_W}{\delta x}\right)\right]+qA\delta x=0 \tag{4.35}$$

The above equation can be re-arranged, noting that $k_B = k_w = k$, to give the discretised equation for **boundary node 5**:

$$a_P T_P = a_W T_W + a_E T_E + S_u \tag{4.36}$$

where

a_W	a_E	a_P	S_p	S_u
$\frac{kA}{\delta x}$	0	$a_W + a_E - S_p$	$-\frac{2kA}{\delta x}$	$qA\delta x + \frac{2kA}{\delta x} T_B$

Substitution of numerical values for $A = 1$, $k = 0.5$ W/m/K, $q = 1000$ kW/m^3 and $\delta x = 0.004$ m everywhere gives the coefficients of the discretised equations summarised in Table 4.2.

Table 4.2

Node	a_W	a_E	S_u	S_p	$a_P = a_W + a_E - S_p$
1	0	125	$4000 + 250T_A$	−250	375
2	125	125	4000	0	250
3	125	125	4000	0	250
4	125	125	4000	0	250
5	125	0	$4000 + 250T_B$	−250	375

Given directly in matrix form the equations are

$$\begin{bmatrix} 375 & -125 & 0 & 0 & 0 \\ -125 & 250 & -125 & 0 & 0 \\ 0 & -125 & 250 & -125 & 0 \\ 0 & 0 & -125 & 250 & -125 \\ 0 & 0 & 0 & -125 & 375 \end{bmatrix} \begin{bmatrix} T_1 \\ T_2 \\ T_3 \\ T_4 \\ T_5 \end{bmatrix} = \begin{bmatrix} 29000 \\ 4000 \\ 4000 \\ 4000 \\ 54000 \end{bmatrix} \tag{4.37}$$

The solution to the above set of equations is

$$\begin{bmatrix} T_1 \\ T_2 \\ T_3 \\ T_4 \\ T_5 \end{bmatrix} = \begin{bmatrix} 150 \\ 218 \\ 254 \\ 258 \\ 230 \end{bmatrix} \tag{4.38}$$

Comparison with the analytical solution

The analytical solution to this problem may be obtained by integrating equation (4.25) twice with respect to x and by subsequent application of the boundary conditions. This gives

$$T = \left[\frac{T_B - T_A}{L} + \frac{q}{2k}(L - x)\right]x + T_A \tag{4.39}$$

The comparison between the finite volume solution and the exact solution is shown in Table 4.3 and Figure 4.8 and it can be seen that, even with a coarse grid of five nodes, the agreement is very good.

Table 4.3

Node number	*1*	*2*	*3*	*4*	*5*
x (m)	0.002	0.006	0.01	0.014	0.018
Finite volume solution	150	218	254	258	230
Exact solution	146	214	250	254	226
Percentage error	2.73	1.86	1.60	1.57	1.76

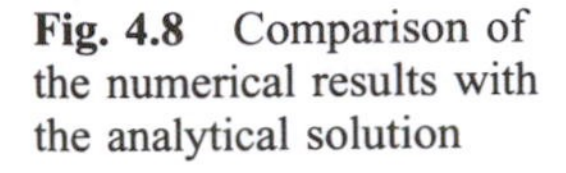

Fig. 4.8 Comparison of the numerical results with the analytical solution

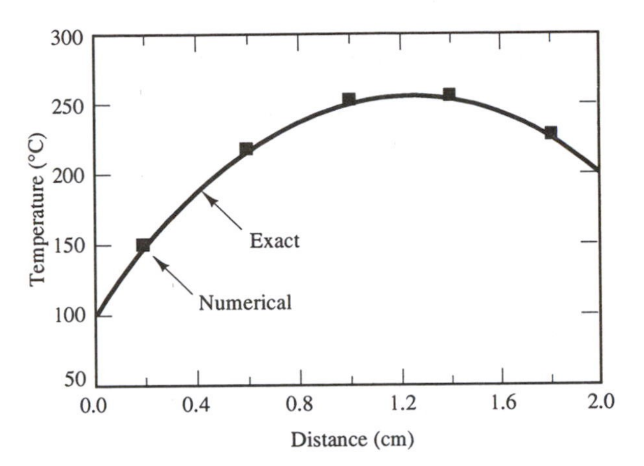

Example 4.3 In the final worked example of this chapter we discuss the cooling of a circular fin by means of convective heat transfer along its length. Convection gives rise to a temperature-dependent heat loss or sink term in the governing equation.

Shown in Figure 4.9 is a cylindrical fin with uniform cross-sectional area A. The base is at a temperature of 100 °C (T_B) and the end is insulated. The fin is exposed to an ambient temperature of 20 °C. One-dimensional heat transfer in this situation is governed by

$$\frac{d}{dx}\left(kA\frac{dT}{dx}\right) - hP(T - T_\infty) = 0 \tag{4.40}$$

where h is the convective heat transfer coefficient, P the perimeter, k the thermal conductivity of the material and T_∞ the ambient temperature. Calculate the temperature distribution along the fin and compare the results with the analytical solution given by

$$\frac{T - T_\infty}{T_B - T_\infty} = \frac{\cosh[n(L - x)]}{\cosh(nL)} \tag{4.41}$$

where $n^2 = hP/(kA)$, L is the length of the fin and x the distance along the fin. Data: $L = 1$ m, $hP/(kA) = 25\ m^{-2}$ (note kA is constant).

Fig. 4.9 The geometry for Example 4.3

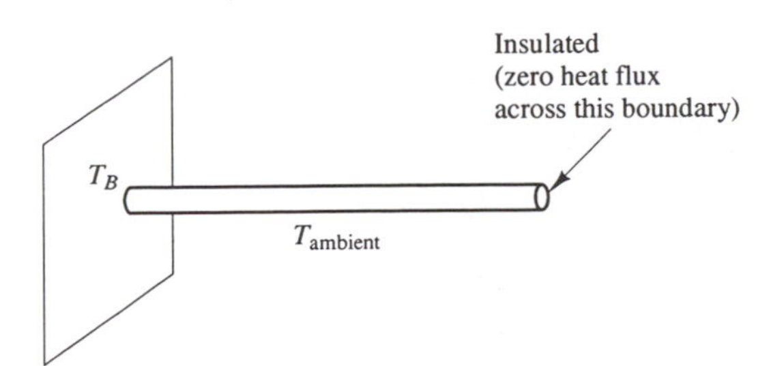

Solution The governing equation in the example contains a sink term, $-hP(T - T_\infty)$, the convective heat loss, which is a function of the local temperature T. As before the first step in solving the problem by the finite volume method is to set up a grid. We use a uniform grid and divide the length into five control volumes so that $\delta x = 0.2$ m. The grid is shown in Figure 4.10.

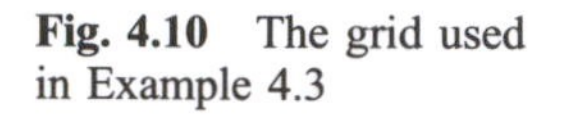
Fig. 4.10 The grid used in Example 4.3

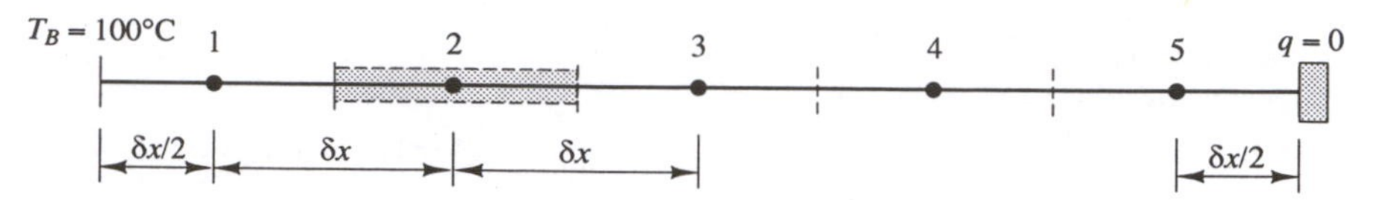

When kA = constant, the governing equation (4.40) can be written as

$$\frac{d}{dx}\left(\frac{dT}{dx}\right) - n^2(T - T_\infty) = 0 \text{ where } n^2 = hp/(kA) \tag{4.42}$$

Integration of the above equation over a control volume gives

$$\int_{\Delta V} \frac{d}{dx}\left(\frac{dT}{dx}\right)dV - \int_{\Delta V} n^2(T - T_\infty)dV = 0 \tag{4.43}$$

The first integral of the above equation is treated as in Examples 4.1 and 4.2; the second integral due to the source term in the equation is evaluated by assuming that the integral is locally constant within each control volume.

$$\left[\left(A\frac{dT}{dx}\right)_e - \left(A\frac{dT}{dx}\right)_w\right] - \left[n^2(T_P - T_\infty)A\delta x\right] = 0$$

First we develop a formula valid for nodal points 2, 3 and 4 by introducing the usual linear approximations for the temperature gradient. Subsequent division by cross-sectional area A gives

$$\left[\left(\frac{T_E - T_P}{\delta x}\right) - \left(\frac{T_P - T_W}{\delta x}\right)\right] - \left[n^2(T_p - T_\infty)\delta x\right] = 0 \tag{4.44}$$

This can be re-arranged as

$$\left(\frac{1}{\delta x} + \frac{1}{\delta x}\right)T_p = \left(\frac{1}{\delta x}\right)T_w + \left(\frac{1}{\delta x}\right)T_E + n^2\delta x T_\infty - n^2\delta x T_P \tag{4.45}$$

For **interior nodal points 2, 3 and 4** we write, using general form (4.11),

$$\boxed{a_P T_P = a_W T_W + a_E T_E + S_u} \tag{4.46}$$

with

a_W	a_E	a_P	S_p	S_u
$\frac{1}{\delta x}$	$\frac{1}{\delta x}$	$a_W + a_E - S_p$	$-n^2\delta x$	$n^2\delta x T_\infty$

Next we apply the boundary conditions at node points 1 and 5. At node 1 the west control volume boundary is kept at a specified temperature. It is treated in the same way as Example 4.1, i.e.

$$\left[\left(\frac{T_E - T_P}{\delta x}\right) - \left(\frac{T_P - T_B}{\delta x/2}\right)\right] - \left[n^2(T_P - T_\infty)\delta x\right] = 0 \tag{4.47}$$

The coefficients of the discretised equation at **boundary node 1** are

a_W	a_E	a_P	S_p	S_u
0	$\frac{1}{\delta x}$	$a_W + a_E - S_p$	$-n^2\delta x - \frac{2}{\delta x}$	$n^2\delta x T_\infty + \frac{2}{\delta x}T_B$

At node 5 the flux across the east boundary is zero since the east side of the control volume is an insulated boundary:

$$\left[0 - \left(\frac{T_P - T_W}{\delta x}\right)\right] - \left[n^2(T_P - T_\infty)\delta x\right] = 0 \tag{4.48}$$

Hence the east coefficient is set to zero. There are no additional source terms associated with the zero flux boundary condition. The coefficients at **boundary node 5** are given by

a_W	a_E	a_P	S_p	S_u
$\frac{1}{\delta x}$	0	$a_W + a_E - S_p$	$-n^2\delta x$	$n^2\delta x T_\infty$

Substituting numerical values gives the coefficients in Table 4.4

Table 4.4

Node	a_W	a_E	S_u	S_p	$a_P = a_W + a_E - S_p$
1	0	5	$100 + 10T_B$	−15	20
2	5	5	100	−5	15
3	5	5	100	−5	15
4	5	5	100	−5	15
5	5	0	100	−5	10

The matrix form of the equations set is

$$\begin{bmatrix} 20 & -5 & 0 & 0 & 0 \\ -5 & 15 & -5 & 0 & 0 \\ 0 & -5 & 15 & -5 & 0 \\ 0 & 0 & -5 & 15 & -5 \\ 0 & 0 & 0 & -5 & 10 \end{bmatrix} \begin{bmatrix} T_1 \\ T_2 \\ T_3 \\ T_4 \\ T_5 \end{bmatrix} = \begin{bmatrix} 1100 \\ 100 \\ 100 \\ 100 \\ 100 \end{bmatrix} \tag{4.49}$$

The solution to the above system is

$$\begin{bmatrix} T_1 \\ T_2 \\ T_3 \\ T_4 \\ T_5 \end{bmatrix} = \begin{bmatrix} 64.22 \\ 36.91 \\ 26.50 \\ 22.60 \\ 21.30 \end{bmatrix} \tag{4.50}$$

Comparison with the analytical solution

Table 4.5 compares the finite volume solution with analytical expression (4.41). The maximum percentage error ((analytical solution – finite volume solution)/analytical solution) is around 6%. Given the coarseness of the grid used in the calculation the numerical solution is reasonably close to the exact solution.

Table 4.5

Node	*Distance*	*Finite volume solution*	*Analytical solution*	*Difference*	*Percentage Error*
1	0.1	64.22	68.52	4.30	6.27
2	0.3	36.91	37.86	0.95	2.51
3	0.5	26.50	26.61	0.11	0.41
4	0.7	22.60	22.53	−0.07	−0.31
5	0.9	21.30	21.21	−0.09	−0.42

The numerical solution can be improved by employing a finer grid. Let us consider the same problem with the rod length subdivided into 10 control volumes. The derivation of the discretised equations is the same as before, but the numerical values of the coefficients and source terms are different owing to the smaller grid spacing of $\delta x = 0.1$ m. The comparison of results of the second calculation with the analytical solution is shown in Figure 4.11 and Table 4.6. The second numerical results show better agreement with the analytical solution; now the maximum deviation is only 2%.

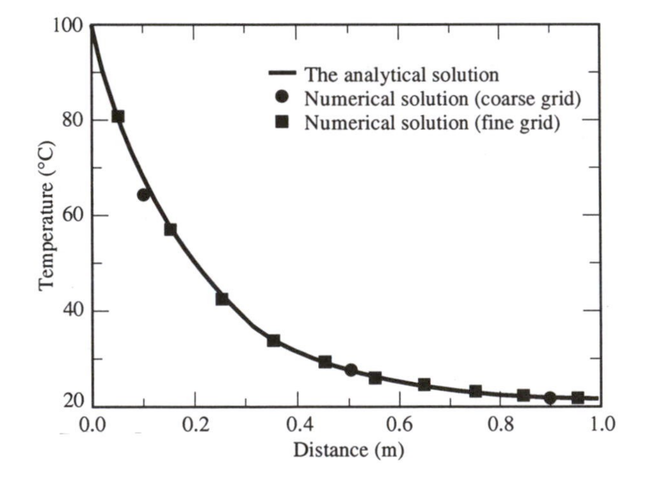

Fig. 4.11 Comparison of numerical and analytical results

Table 4.6

Node	*Distance*	*Finite volume solution*	*Analytical solution*	*Difference*	*Percentage error*
1	0.05	80.59	82.31	1.72	2.08
2	0.15	56.94	57.79	0.85	1.47
3	0.25	42.53	42.93	0.40	0.93
4	0.35	33.74	33.92	0.18	0.53
5	0.45	28.40	28.46	0.06	0.21
6	0.55	25.16	25.17	0.01	0.03
7	0.65	23.21	23.19	−0.02	−0.08
8	0.75	22.06	22.03	−0.03	−0.13
9	0.85	21.47	21.39	−0.08	−0.37
10	0.95	21.13	21.11	−0.02	−0.09

4.4 Finite volume method for two-dimensional diffusion problems

The methodology used in deriving discretised equations in the one-dimensional case can be easily extended to two-dimensional problems. To illustrate the technique let us consider the two-dimensional steady state diffusion equation given by

$$\frac{\partial}{\partial x}\left(\Gamma\frac{\partial\phi}{\partial x}\right)+\frac{\partial}{\partial y}\left(\Gamma\frac{\partial\phi}{\partial y}\right)+S=0 \tag{4.51}$$

A portion of the two-dimensional grid used for the discretisation is shown in Figure 4.12.

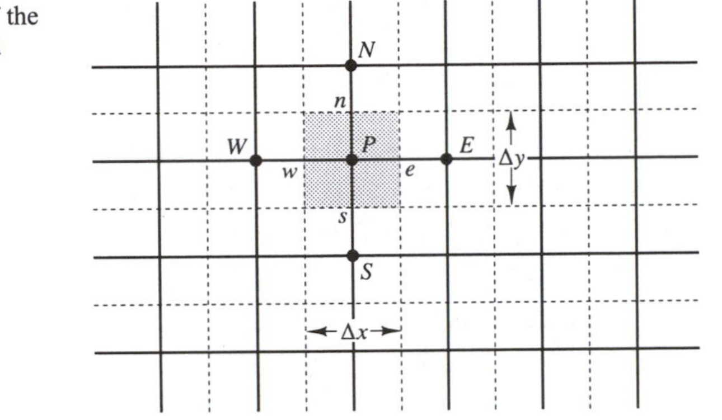

Fig. 4.12 A part of the two-dimensional grid

In addition to the east (E) and west (W) neighbours a general grid node P now also has north (N) and south (S) neighbours. The same notation as in the one-dimensional analysis is used for faces and cell dimensions. When the above equation is formally integrated over the control volume we obtain

$$\int_{\Delta V}\frac{\partial}{\partial x}\left(\Gamma\frac{\partial\phi}{\partial x}\right)dx\,.\,dy+\int_{\Delta V}\frac{\partial}{\partial y}\left(\Gamma\frac{\partial\phi}{\partial y}\right)dx\,.\,dy+\int_{\Delta V}S_\phi dV=0 \tag{4.52}$$

So, noting that $A_e = A_w = \Delta y$ and $A_n = A_s = \Delta x$, we obtain:

$$\left[\Gamma_e A_e\left(\frac{\partial\phi}{\partial x}\right)_e-\Gamma_w A_w\left(\frac{\partial\phi}{\partial x}\right)_w\right] + \left[\Gamma_n A_n\left(\frac{\partial\phi}{\partial y}\right)_n-\Gamma_s A_s\left(\frac{\partial\phi}{\partial y}\right)_s\right]+\bar{S}\Delta V=0 \tag{4.53}$$

As before this equation represents the balance of the generation of ϕ in a control volume and the fluxes through its cell faces. Using the approximations introduced in the previous section we can write expressions for the flux through control volume faces:

$$\text{Flux across the west face} = \Gamma_w A_w \left.\frac{\partial\phi}{\partial x}\right|_w = \Gamma_w A_w \frac{(\phi_P-\phi_W)}{\delta x_{WP}} \tag{4.54a}$$

$$\text{Flux across the east face} = \Gamma_e A_e \left.\frac{\partial\phi}{\partial x}\right|_e = \Gamma_e A_e \frac{(\phi_E-\phi_P)}{\delta x_{PE}} \tag{4.54b}$$

$$\text{Flux across the south face} = \Gamma_s A_s \left.\frac{\partial\phi}{\partial y}\right|_s = \Gamma_s A_s \frac{(\phi_P-\phi_S)}{\delta y_{SP}} \tag{4.54c}$$

$$\text{Flux across the north face} = \Gamma_n A_n \left.\frac{\partial\phi}{\partial y}\right|_n = \Gamma_n A_n \frac{(\phi_N-\phi_P)}{\delta y_{PN}} \tag{4.54d}$$

By substituting the above expressions into equation (4.53) we obtain

$$\Gamma_e A_e \frac{(\phi_E - \phi_P)}{\delta x_{PE}} - \Gamma_w A_w \frac{(\phi_P - \phi_W)}{\delta x_{WP}} + \Gamma_n A_n \frac{(\phi_N - \phi_P)}{\delta y_{PN}} - \Gamma_s A_s \frac{(\phi_P - \phi_S)}{\delta y_{SP}} + \bar{S}\Delta V = 0 \quad (4.55)$$

When the source term is represented in linearised form $\bar{S}\Delta V = S_u + S_p\phi_P$, this equation can be re-arranged as

$$\left(\frac{\Gamma_w A_w}{\delta x_{WP}} + \frac{\Gamma_e A_e}{\delta x_{PE}} + \frac{\Gamma_s A_s}{\delta y_{SP}} + \frac{\Gamma_n A_n}{\delta y_{PN}} - S_p\right)\phi_P = \left(\frac{\Gamma_w A_w}{\delta x_{WP}}\right)\phi_W + \left(\frac{\Gamma_e A_e}{\delta x_{PE}}\right)\phi_E + \left(\frac{\Gamma_s A_s}{\delta y_{SP}}\right)\phi_S + \left(\frac{\Gamma_n A_n}{\delta y_{PN}}\right)\phi_N + S_u \quad (4.56)$$

Equation (4.56) is now cast in the general discretised equation form for interior nodes:

$$\boxed{a_P\phi_P = a_W\phi_W + a_E\phi_E + a_S\phi_S + a_N\phi_N + S_u} \quad (4.57)$$

where

a_W	a_E	a_S	a_N	a_P
$\frac{\Gamma_w A_w}{\delta x_{WP}}$	$\frac{\Gamma_e A_e}{\delta x_{PE}}$	$\frac{\Gamma_s A_s}{\delta y_{SP}}$	$\frac{\Gamma_n A_n}{\delta y_{PN}}$	$a_W + a_E + a_S + a_N - S_p$

The face areas in a two-dimensional case are $A_w = A_e = \Delta y; A_n = A_s = \Delta x$.

We obtain the distribution of the property ϕ in a given two-dimensional situation by writing discretised equations of the form (4.57) at each grid node of the subdivided domain. At the boundaries where the temperatures or fluxes are known the discretised equations are modified to incorporate boundary conditions in the manner demonstrated in Examples 4.1 and 4.2. The boundary side coefficient is set to zero (cutting the link with the boundary) and the flux crossing the boundary is introduced as a source which is appended to any existing S_u and S_p terms. Subsequently, the resulting set of equations is solved to obtain the two-dimensional distribution of the property ϕ. Example 7.2 in Chapter 7 shows how the method can be applied to calculate conductive heat transfer in two-dimensional situations.

4.5 Finite volume method for three-dimensional diffusion problems

Steady state diffusion in a three-dimensional situation is governed by

$$\frac{\partial}{\partial x}\left(\Gamma\frac{\partial\phi}{\partial x}\right) + \frac{\partial}{\partial y}\left(\Gamma\frac{\partial\phi}{\partial y}\right) + \frac{\partial}{\partial z}\left(\Gamma\frac{\partial\phi}{\partial z}\right) + S = 0 \quad (4.58)$$

Now a three-dimensional grid is used to subdivide the domain. A typical control volume is shown in Figure 4.13.

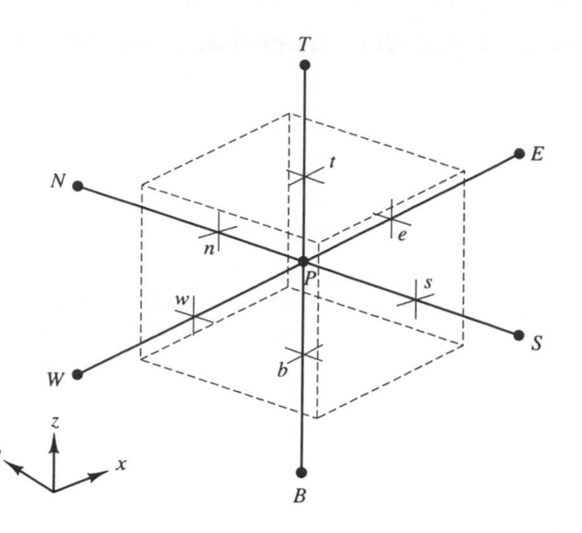

Fig. 4.13 A cell in three dimensions and neighbouring nodes

A cell containing node P now has six neighbouring nodes identified as west, east, south, north, bottom and top nodes (W, E, S, N, B, T). As before, the notation, w, e, s, n, b and t are used to refer to the west, east, south, north, bottom and top cell faces respectively.

Integration of Equation (4.58) over the control volume shown gives

$$\left[\Gamma_e A_e\left(\frac{\partial\phi}{\partial x}\right)_e - \Gamma_w A_w\left(\frac{\partial\phi}{\partial x}\right)_w\right] + \left[\Gamma_n A_n\left(\frac{\partial\phi}{\partial y}\right)_n - \Gamma_s A_s\left(\frac{\partial\phi}{\partial y}\right)_s\right] + \left[\Gamma_t A_t\left(\frac{\partial\phi}{\partial z}\right)_t - \Gamma_b A_b\left(\frac{\partial\phi}{\partial z}\right)_b\right] + \bar{S}\Delta V = 0 \quad (4.59)$$

Following the procedure developed for one- and two-dimensional cases the discretised form of the equation (4.59) is obtained:

$$\left[\Gamma_e \frac{(\phi_E - \phi_P)A_e}{\delta x_{PE}} - \Gamma_w \frac{(\phi_P - \phi_W)A_w}{\delta x_{WP}}\right] + \left[\Gamma_n \frac{(\phi_N - \phi_P)A_n}{\delta y_{PN}} - \Gamma_s \frac{(\phi_P - \phi_S)A_s}{\delta y_{SP}}\right] + \left[\Gamma_t \frac{(\phi_T - \phi_P)A_t}{\delta z_{PT}} - \Gamma_b \frac{(\phi_P - \phi_B)A_b}{\delta z_{BP}}\right] + (S_u + S_p\phi_P) = 0 \quad (4.60)$$

As before this can be re-arranged as to give the discretised equation for interior nodes:

$$a_P\phi_P = a_W\phi_W + a_E\phi_E + a_S\phi_S + a_N\phi_N + a_B\phi_B + a_T\phi_T + S_u \quad (4.61)$$

where

a_W	a_E	a_S	a_N	a_B	a_T	a_P
$\dfrac{\Gamma_w A_w}{\delta x_{WP}}$	$\dfrac{\Gamma_e A_e}{\delta x_{PE}}$	$\dfrac{\Gamma_s A_s}{\delta y_{SP}}$	$\dfrac{\Gamma_n A_n}{\delta y_{PN}}$	$\dfrac{\Gamma_b A_b}{\delta z_{BP}}$	$\dfrac{\Gamma_t A_t}{\delta z_{PT}}$	$a_W + a_E + a_S + a_N + a_B + a_T - S_p$

Boundary conditions can be introduced by cutting links with the appropriate face(s) and modifying the source term as described in section 4.3.

4.6 Summary of discretised equations for diffusion problems

- The discretised equations for one-, two- and three-dimensional diffusion problems have been found to take the following general form:

$$a_P\phi_P = \sum a_{nb}\phi_{nb} + S_u \tag{4.62}$$

where Σ indicates summation over all neighbouring nodes (nb), and a_{nb} are the neighbouring coefficients, a_W, a_E in 1D, a_W, a_E, a_S, a_N in 2D and $a_W, a_E, a_S, a_N, a_B, a_T$ in 3D; ϕ_{nb} are the values of the property ϕ at the neighbouring nodes and $(S_u + S_p\phi_p)$ is the linearised source term.
- In all cases the coefficients around point P satisfy the following relation:

$$a_P = \sum a_{nb} - S_p \tag{4.63}$$

- A summary of the neighbour coefficients for one-, two- and three-dimensional diffusion problems is given in Table 4.7.

Table 4.7

	a_W	a_E	a_S	a_N	a_B	A_T
1D	$\frac{\Gamma_w A_w}{\delta x_{WP}}$	$\frac{\Gamma_e A_e}{\delta x_{PE}}$	–	–	–	–
2D	$\frac{\Gamma_w A_w}{\delta x_{WP}}$	$\frac{\Gamma_e A_e}{\delta x_{PE}}$	$\frac{\Gamma_s A_s}{\delta y_{SP}}$	$\frac{\Gamma_n A_n}{\delta y_{PN}}$	–	–
3D	$\frac{\Gamma_w A_w}{\delta x_{WP}}$	$\frac{\Gamma_e A_e}{\delta x_{PE}}$	$\frac{\Gamma_s A_s}{\delta y_{SP}}$	$\frac{\Gamma_n A_n}{\delta y_{PN}}$	$\frac{\Gamma_b A_b}{\delta y_{BP}}$	$\frac{\Gamma_t A_t}{\delta z_{PT}}$

- Source terms can be included by identifying their linearised form $\overline{S}\Delta V = S_u + S_p\phi_P$ and specifying values for S_u and S_p.
- Boundary conditions are incorporated by cutting the link to the boundary side and introducing the boundary side flux – exact or linearly approximated–through additional source terms S_u and S_p. For a one-dimensional control volume of width $\Delta\zeta$ with a boundary B:

$$\text{link cutting : set coefficient } a_B = 0 \tag{4.64}$$

$$\text{source contributions : fixed value } \phi_B : S_u = \frac{2k_B A_B}{\Delta\zeta}\phi_B; \quad S_p = -\frac{2k_B A_B}{\Delta\zeta} \tag{4.65}$$

$$\text{fixed flux } q_B : \quad S_u + S_p\phi_P = q_B \tag{4.66}$$

5

The Finite Volume Method for Convection–Diffusion Problems

5.1 Introduction

In problems where fluid flow plays a significant role we must account for the effects of convection. Diffusion always occurs alongside convection in nature so here we examine methods to predict combined convection and diffusion. The steady convection–diffusion equation can be derived from the transport equation (2.39) for a general property ϕ by deleting the transient term

$$div(\rho \mathbf{u} \phi) = div(\Gamma \; grad \; \phi) + S_\phi \tag{5.1}$$

Formal integration over a control volume gives

$$\int_A \mathbf{n} \,.\, (\rho \phi \mathbf{u}) dA = \int_A \mathbf{n} \,.\, (\Gamma \; grad \; \phi) dA + \int_{CV} S_\phi dV \tag{5.2}$$

This equation represents the flux balance in a control volume. The left hand side gives the net convective flux and the right hand side contains the net diffusive flux and the generation or destruction of the property ϕ within the control volume.

The principal problem in the discretisation of the convective terms is the calculation of the value of transported property ϕ at control volume faces and its convective flux across these boundaries. In Chapter 4 we introduced the central differencing method of obtaining discretised equations for the diffusion and source terms on the right hand side of equation (5.2). It would seem obvious to try out this practice, which worked so well for diffusion problems, on the convective terms. However, the diffusion process affects the distribution of a transported quantity along its gradients in all directions, whereas convection spreads influence only in the flow direction. This crucial difference manifests itself in a stringent upper limit to the grid size, that is dependent on the relative strength of convection and diffusion, for stable convection–diffusion calculations with central differencing.

Naturally, we also present the case for a number of alternative discretisation practices for the convective effects which enable stable computations under less restrictive conditions. In the current analysis no reference will be made to the evaluation of face velocities. It is assumed that they are 'somehow' known. The method of computing velocities will be discussed in Chapter 6.

5.2 Steady one-dimensional convection and diffusion

In the absence of sources, the steady convection and diffusion of a property ϕ in a given one-dimensional flow field u is governed by

$$\frac{d}{dx}(\rho u\phi) = \frac{d}{dx}\left(\Gamma\frac{d\phi}{dx}\right) \tag{5.3}$$

The flow must also satisfy continuity so

$$\frac{d(\rho u)}{dx} = 0 \tag{5.4}$$

We consider the one-dimensional control volume shown in Figure 5.1 and use the notation introduced in Chapter 4. Our attention is focused on a general node P; the neighbouring nodes are identified by W and E and the control volume faces by w and e.

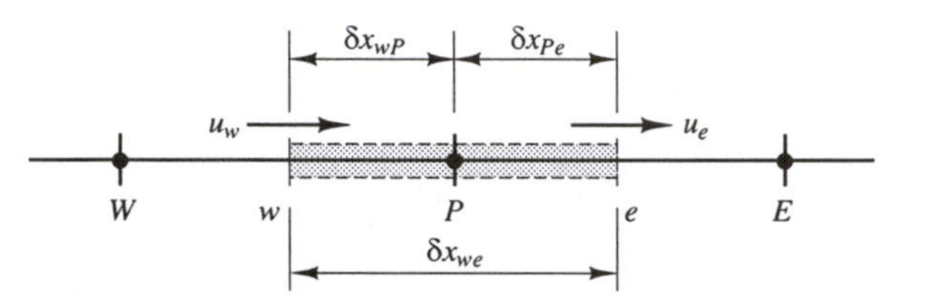

Fig. 5.1 A control volume around node P

Integration of transport equation (5.3) over the control volume of Figure 5.1 gives

$$(\rho uA\phi)_e - (\rho uA\phi)_w = \left(\Gamma A\frac{\partial\phi}{\partial x}\right)_e - \left(\Gamma A\frac{\partial\phi}{\partial x}\right)_w \tag{5.5}$$

And integration of continuity equation (5.4) yields

$$(\rho uA)_e - (\rho uA)_w = 0 \tag{5.6}$$

To obtain discretised equations for the convection–diffusion problem we must approximate the terms in equation (5.5). It is convenient to define two variables F and D to represent the convective mass flux per unit area and diffusion conductance at cell faces:

$$F = \rho u \qquad \text{and} \qquad D = \frac{\Gamma}{\delta x} \tag{5.7}$$

The cell face values of the variables F and D can be written as

$$F_w = (\rho u)_w, \qquad F_e = (\rho u)_e \tag{5.8a}$$

$$D_w = \frac{\Gamma_w}{\delta x_{WP}}, \qquad D_e = \frac{\Gamma_e}{\delta x_{PE}} \tag{5.8b}$$

We develop our techniques assuming that $A_w = A_e = A$ and employ the central differencing approach to represent the contribution of the diffusion terms on the right hand side. The integrated convection–diffusion equation (5.5) can now be written as

$$\boxed{F_e\phi_e - F_w\phi_w = D_e(\phi_E - \phi_P) - D_w(\phi_P - \phi_W)} \tag{5.9}$$

and the integrated continuity equation (5.6) as

$$\boxed{F_e - F_w = 0} \tag{5.10}$$

We also assume that the velocity field is 'somehow known', which takes care of the values of F_e and F_w. In order to solve equation (5.9) we need to calculate the transported property ϕ at the e and w faces. Schemes for this purpose are assessed in the following sections.

5.3 The central differencing scheme

The central differencing approximation has been used to represent the diffusion terms which appear on the right hand side of equation (5.9) and it seems logical to try linear interpolation to compute the cell face values for the convective terms on the left hand side of this equation. For a uniform grid we can write the cell face values of property ϕ as

$$\phi_e = (\phi_P + \phi_E)/2 \tag{5.11a}$$

$$\phi_w = (\phi_W + \phi_P)/2 \tag{5.11b}$$

Substitution of the above expressions into the convection terms of (5.9) yields

$$\frac{F_e}{2}(\phi_P + \phi_E) - \frac{F_w}{2}(\phi_W + \phi_P) = D_e(\phi_E - \phi_P) - D_w(\phi_P - \phi_W) \tag{5.12}$$

This can be re-arranged to give

$$\left[\left(D_w - \frac{F_w}{2}\right) + \left(D_e + \frac{F_e}{2}\right)\right]\phi_P = \left(D_w + \frac{F_w}{2}\right)\phi_W + \left(D_e - \frac{F_e}{2}\right)\phi_E$$

$$\left[\left(D_w + \frac{F_w}{2}\right) + \left(D_e - \frac{F_e}{2}\right) + (F_e - F_w)\right]\phi_P = \left(D_w + \frac{F_w}{2}\right)\phi_W + \left(D_e - \frac{F_e}{2}\right)\phi_E \tag{5.13}$$

Identifying the coefficients of ϕ_W and ϕ_E as a_W and a_E the **central differencing** expressions for the discretised convection–diffusion equation are

$$\boxed{a_P\phi_P = a_W\phi_W + a_E\phi_E} \tag{5.14}$$

where

a_W	a_E	a_P
$D_w + \frac{F_w}{2}$	$D_e - \frac{F_e}{2}$	$a_W + a_E + (F_e - F_w)$

It can be easily recognised that equation (5.14) for steady convection–diffusion problems takes the same general form as equation (4.11) for pure diffusion problems. The difference is that the coefficients of the former contain additional terms to account for convection. To solve a one-dimensional convection–diffusion problem we write discretised equations of the form (5.14) for all grid nodes. This yields a set of algebraic equations that is solved to obtain the distribution of the transported property ϕ. The process is now illustrated by means of a worked example.

Example 5.1 A property ϕ is transported by means of convection and diffusion through the one-dimensional domain sketched in Figure 5.2. The governing equation is (5.3); boundary conditions are $\phi_0 = 1$ at $x = 0$ and $\phi_L = 0$ at $x = L$. Using five equally spaced cells and the central differencing scheme for convection and diffusion calculate the distribution of ϕ as a function of x for (i) Case 1: $u = 0.1$ m/s, (ii) Case 2: $u = 2.5$ m/s, and compare the results with the analytical solution

$$\frac{\phi - \phi_0}{\phi_L - \phi_0} = \frac{\exp(\rho u x/\Gamma) - 1}{\exp(\rho u L/\Gamma) - 1} \tag{5.15}$$

(iii) Case 3: recalculate the solution for $u = 2.5$ m/s with 20 grid nodes and compare the results with the analytical solution. The following data apply: length $L = 1.0$ m, $\rho = 1.0$ kg/m^3, $\Gamma = 0.1$ kg/m/s.

Fig. 5.2

u

$\phi = 1$ at $x = 0$; $\phi = 0$ at $x = L$

Solution The method of solution is demonstrated using the simple grid shown in Figure 5.3. The domain has been divided into five control volumes giving $\delta x = 0.2$ m. Note that $F = \rho u, D = \Gamma/\delta x, F_e = F_w = F$ and $D_e = D_w = D$ everywhere. The boundaries are denoted by A and B.

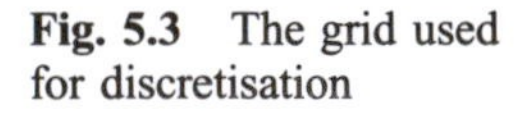

Fig. 5.3 The grid used for discretisation

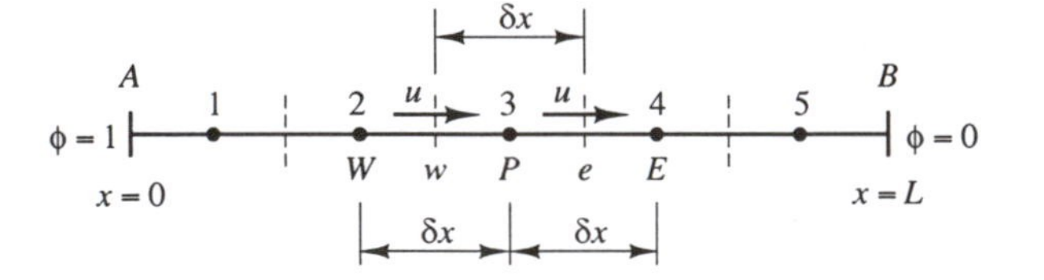

The discretisation equation (5.14) and its coefficients apply at internal nodal points 2, 3 and 4, but control volumes 1 and 5 need special treatment since they are adjacent to the domain boundaries. We integrate governing equation (5.3) and use central differencing both for the diffusion terms and the convective flux through the east face of cell 1. The value of ϕ is given at the west face of this cell ($\phi_w = \phi_A = 1$) so we do not need to make any approximations in the convective flux term at this boundary. This yields the following equation for node 1:

$$\frac{F_e}{2}(\phi_P + \phi_E) - F_A\phi_A = D_e(\phi_E - \phi_P) - D_A(\phi_P - \phi_A) \tag{5.16}$$

For control volume 5, the ϕ-value at the east face is known ($\phi_e = \phi_B = 0$). We obtain

$$F_B\phi_B - \frac{F_w}{2}(\phi_P + \phi_W) = D_B(\phi_B - \phi_P) - D_w(\phi_P - \phi_W) \tag{5.17}$$

Re-arrangement of equations (5.16) and (5.17), noting that $D_A = D_B = 2\Gamma/\delta x = 2D$ and $F_A = F_B = F$, gives discretised equations at boundary nodes of the following form:

$$a_P\phi_P = a_W\phi_W + a_E\phi_E + S_u \tag{5.18}$$

with central coefficient

$$a_P = a_W + a_E + (F_e - F_w) - S_p$$

and

Node	a_W	a_E	S_p	S_u
1	0	$D - F/2$	$-(2D + F)$	$(2D + F)\phi_A$
2, 3, 4	$D + F/2$	$D - F/2$	0	0
5	$D + F/2$	0	$-(2D - F)$	$(2D - F)\phi_B$

To introduce the boundary conditions we have suppressed the link to the boundary side and entered the boundary flux through the source terms.

(i) Case 1

$u = 0.1$ m/s: $F = \rho u = 0.1, D = \Gamma/\delta x = 0.1/0.2 = 0.5$ gives the coefficients as summarised in Table 5.1.

Table 5.1

Node	a_W	a_E	S_u	S_p	$a_pP = a_W + a_E - S_p$
1	0	0.45	$1.1\phi_A$	−1.1	1.55
2	0.55	0.45	0	0	1.0
3	0.55	0.45	0	0	1.0
4	0.55	0.45	0	0	1.0
5	0.55	0	$0.9\phi_B$	−0.9	1.45

The matrix form of the equation set, using $\phi_A = 1$ and $\phi_B = 0$ is

$$\begin{bmatrix} 1.55 & -0.45 & 0 & 0 & 0 \\ -0.55 & 1.0 & -0.45 & 0 & 0 \\ 0 & -0.55 & 1.0 & -0.45 & 0 \\ 0 & 0 & -0.55 & 1.0 & -0.45 \\ 0 & 0 & 0 & -0.55 & 1.45 \end{bmatrix} \begin{bmatrix} \phi_1 \\ \phi_2 \\ \phi_3 \\ \phi_4 \\ \phi_5 \end{bmatrix} = \begin{bmatrix} 1.1 \\ 0 \\ 0 \\ 0 \\ 0 \end{bmatrix} \tag{5.19}$$

The solution to the above system is

$$\begin{bmatrix} \phi_1 \\ \phi_2 \\ \phi_3 \\ \phi_4 \\ \phi_5 \end{bmatrix} = \begin{bmatrix} 0.9421 \\ 0.8006 \\ 0.6276 \\ 0.4163 \\ 0.1579 \end{bmatrix} \tag{5.20}$$

Comparison with the analytical solution. Substitution of the data into equation (5.15) gives the exact solution of the problem:

$$\phi(x) = \frac{2.7183 - \exp(x)}{1.7183}$$

The numerical and analytical solutions are compared in Table 5.2 and in Figure 5.4.

Table 5.2

Node	*Distance*	*Finite volume solution*	*Analytical solution*	*Difference*	*Percentage error*
1	0.1	0.9421	0.9387	−0.003	−0.36
2	0.3	0.8006	0.7963	−0.004	−0.53
3	0.5	0.6276	0.6224	−0.005	−0.83
4	0.7	0.4163	0.4100	−0.006	−1.53
5	0.9	0.1579	0.1505	−0.007	−4.91

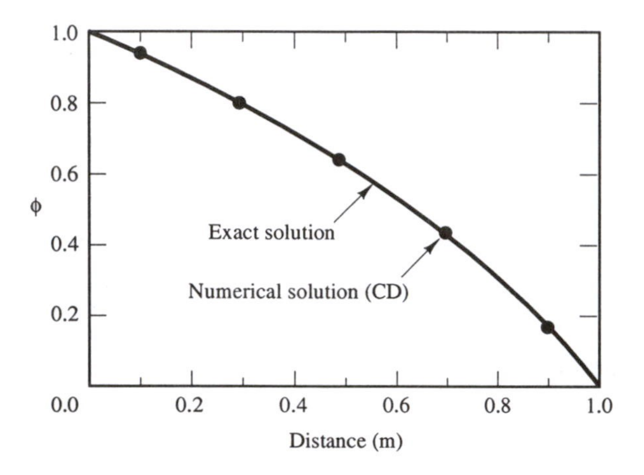

Fig. 5.4 Comparison of numerical and analytical solutions for Case 1

Given the coarseness of the grid the central differencing scheme gives reasonable agreement with the analytical solution.

(ii) Case 2

$u = 2.5$ m/s: $F = \rho u = 2.5, D = \Gamma/\delta x = 0.1/0.2 = 0.5$ gives the coefficients as summarised in Table 5.3.

Comparison of numerical and analytical solutions The matrix equations are formed from the coefficients in Table 5.3 by the same method used in Case 1 and subsequently solved. The analytical solution for the data that apply here is

$$\phi(x) = 1 + \frac{1 - \exp(25x)}{7.20 \times 10^{10}}$$

Table 5.3

Node	a_W	a_E	S_u	S_p	$a_P = a_W + a_E - S_p$
1	0	−0.75	$3.5\phi_A$	−3.5	2.75
2	1.75	−0.75	0	0	1.0
3	1.75	−0.75	0	0	1.0
4	1.75	−0.75	0	0	1.0
5	1.75	0	$-1.5\phi_B$	1.5	0.25

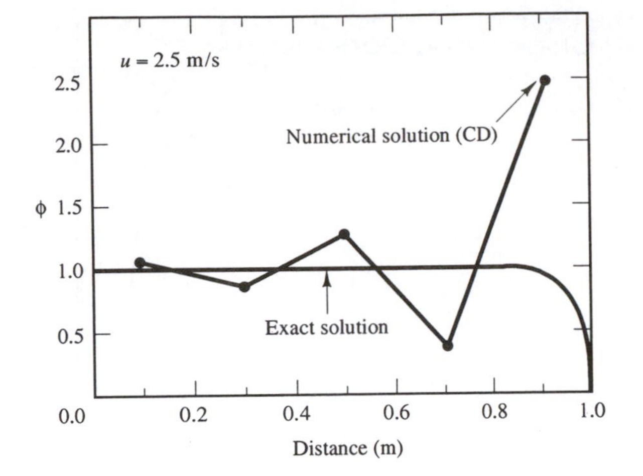

Fig. 5.5 Comparison of numerical and analytical solutions for Case 2

Table 5.4

Node	*Distance*	*Finite volume solution*	*Analytical solution*	*Difference*	*Percentage error*
1	0.1	1.0356	1.0000	−0.035	−3.56
2	0.3	0.8694	0.9999	0.131	13.05
3	0.5	1.2573	0.9999	−0.257	−25.74
4	0.7	0.3521	0.9994	0.647	64.70
5	0.9	2.4644	0.9179	−1.546	−168.48

The numerical and analytical solutions are compared in Table 5.4 and shown in Figure 5.5. The central differencing scheme produces a solution that appears to oscillate about the exact solution. These oscillations are often called 'wiggles' in the literature; the agreement with the analytical solution is clearly not very good.

(iii) Case 3

$u = 2.5$ m/s: a grid of 20 nodes gives $\delta x = 0.05$, $F = \rho u = 2.5$, $D = \Gamma/\delta x = 0.1/0.05 = 2.0$. The coefficients are summarised in Table 5.5 and the resulting solution is compared with the analytical solution in Figure 5.6.

Table 5.5

Node	a_W	a_E	S_u	S_p	$a_P = a_W + a_E - S_p$
1	0	0.75	$6.5\phi_A$	−6.5	7.25
2–19	3.25	0.75	0	0	4.00
20	3.25	0	$5\phi_B$	−1.5	4.75

The agreement between the numerical results and the analytical solution is now good. Comparison of the data for this case with the one computed on the five-point grid of Case 2 shows that grid refinement has reduced the *F/D* ratio from 5 to 1.25. The central differencing scheme seems to yield accurate results when the *F/D* ratio is low. The influence of the *F/D* ratio and the reasons for the appearance of 'wiggles' in central difference solutions when this ratio is high will be discussed below.

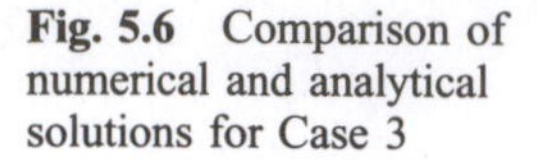

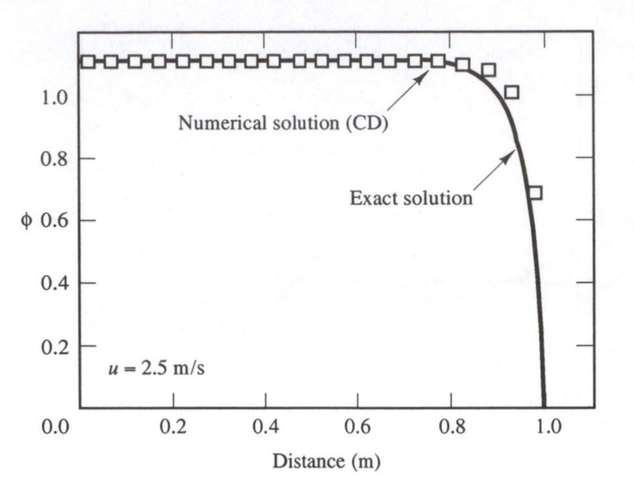

Fig. 5.6 Comparison of numerical and analytical solutions for Case 3

5.4 Properties of discretisation schemes

The failure of central differencing in certain cases involving combined convection and diffusion forces us to take a more in-depth look at the properties of discretisation schemes. In theory numerical results may be obtained that are indistinguishable from the 'exact' solution of the transport equation when the number of computational cells is infinitely large irrespective of the differencing method used. However, in practical calculations we can only use a finite – sometimes quite small – number of cells and our numerical results will only be physically realistic when the discretisation scheme has certain fundamental properties. The most important ones are:

- Conservativeness
- Boundedness
- Transportiveness

5.4.1 Conservativeness

Integration of the convection–diffusion equation over a finite number of control volumes yields a set of discretised conservation equations involving fluxes of the transported property ϕ through control volume faces. To ensure conservation of ϕ for the whole solution domain the flux of ϕ leaving a control volume across a certain face must be equal to the flux of ϕ entering the adjacent control volume through the same face. To achieve this the **flux** through a common face must be represented in a **consistent** manner – by one and the same expression – in adjacent control volumes.

For example, consider the one-dimensional steady state diffusion problem without source terms shown in Figure 5.7.

The fluxes across the domain boundaries are denoted by q_A and q_B. Let us consider four control volumes and apply central differencing to calculate the diffusive flux across the cell faces. The expression for the flux leaving the element around node 2 across its west face is $\Gamma_{w_2}(\phi_2 - \phi_1)/\delta x$ and the flux entering across its east face is $\Gamma_{e_2}(\phi_3 - \phi_2)/\delta x$. An overall flux balance may be obtained by summing the net flux through each control volume taking into account the boundary

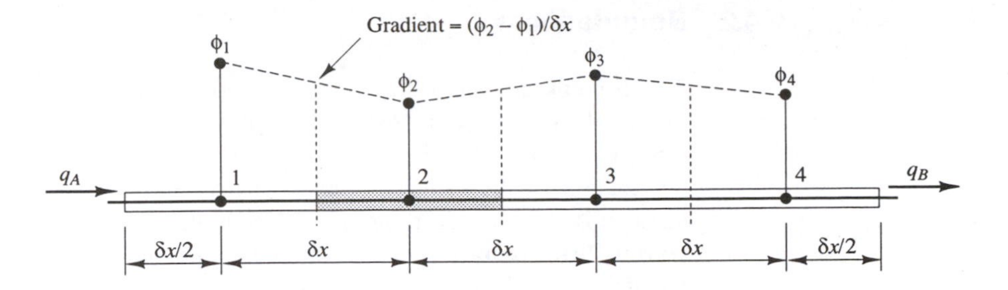

Fig. 5.7 Example of consistent specification of diffusive fluxes

fluxes for the control volumes around nodes 1 and 4.

$$\left[\Gamma_{e_1}\frac{(\phi_2-\phi_1)}{\delta x}-q_A\right]+\left[\Gamma_{e_2}\frac{(\phi_3-\phi_2)}{\delta x}-\Gamma_{w_2}\frac{(\phi_2-\phi_1)}{\delta x}\right]$$
$$+\left[\Gamma_{e_3}\frac{(\phi_4-\phi_3)}{\delta x}-\Gamma_{w_3}\frac{(\phi_3-\phi_2)}{\delta x}\right]$$
$$+\left[q_B-\Gamma_{w_4}\frac{(\phi_4-\phi_3)}{\delta x}\right]=q_B-q_A \quad (5.21)$$

Since $\Gamma_{e_1}=\Gamma_{w_2},\Gamma_{e_2}=\Gamma_{w_3}$ and $\Gamma_{e_3}=\Gamma_{w_4}$ the fluxes across control volume faces are expressed in a consistent manner and cancel out in pairs when summed over the entire domain. Only the two boundary fluxes q_A and q_B remain in the overall balance so equation (5.21) expresses overall conservation of property ϕ. Flux consistency ensures conservation of ϕ over the entire domain for the central difference formulation of the diffusion flux.

Inconsistent flux interpolation formulae give rise to unsuitable schemes that do not satisfy overall conservation. For example, let us consider the situation where a quadratic interpolation formula, based on values at 1, 2 and 3, is used for the control volume 2, and a quadratic profile, based on values at points 2, 3 and 4, is used for control volume 3.

As shown in Figure 5.8 the resulting quadratic profiles can be quite different. Consequently, the flux values calculated at the east face of control volume 2 and the west face of control volume 3 may be unequal if the gradients of the two curves are different at the cell face. If this is the case the two fluxes do not cancel out when summed and overall conservation is not satisfied. The example should not suggest to the reader that quadratic interpolation is entirely bad. Further on we shall meet a popular quadratic discretisation practice – the so-called QUICK scheme – that *is* consistent.

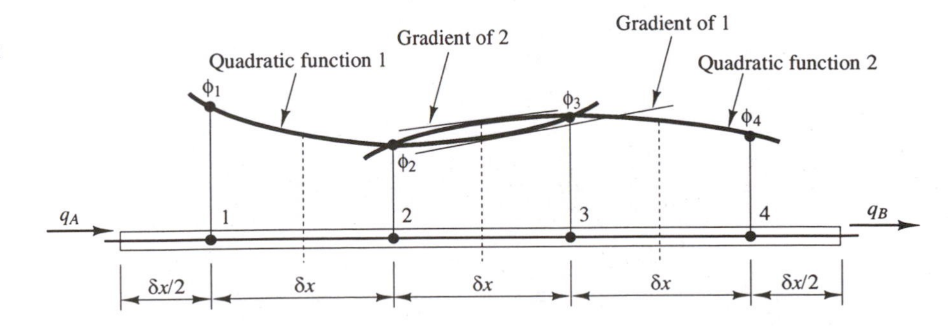

Fig. 5.8 Example of inconsistent specification of diffusive fluxes

5.4.2 Boundedness

The discretised equations at each nodal point represent a set of algebraic equations that needs to be solved. Normally iterative numerical techniques are used to solve large equation sets. These methods start the solution process from a guessed distribution of the variable ϕ and perform successive updates until a converged solution is obtained. Scarborough (1958) has shown that a **sufficient condition for a convergent iterative method** can be expressed in terms of the values of the coefficients of the discretised equations:

$$\frac{\sum |a_{nb}|}{|a'_P|} \begin{cases} \leq 1 \text{ at all nodes} \\ < 1 \text{ at one node at least} \end{cases} \tag{5.22}$$

Here a'_P is the net coefficient of the central node P (i.e. $a_P - S_p$) and the summation in the numerator is taken over all the neighbouring nodes (nb). If the differencing scheme produces coefficients that satisfy the above criterion the resulting matrix of coefficients is **diagonally dominant**. To achieve diagonal dominance we need large values of net coefficient ($a_P - S_P$) so the **linearisation** practice of **source terms** should ensure that S_p is always **negative**. If this is the case $-S_p$ is always positive and adds to a_P.

Diagonal dominance is a desirable feature for satisfying the ‘boundedness’ criterion. This states that in the **absence of sources** the internal nodal values of the **property** ϕ should be **bounded by its boundary values**. Hence in a steady state conduction problem without sources and with boundary temperatures of 500 °C and 200 °C all interior values of T should be less than 500 °C and greater than 200 °C. Another essential requirement for boundedness is that **all coefficients of the discretised equations should have the same sign** (usually all positive). Physically this implies that an increase in the variable ϕ at one node should result in an increase in ϕ at neighbouring nodes. If the discretisation scheme does not satisfy the boundedness requirements it is possible that the solution does not converge at all or, if it does, that it contains ‘wiggles’. This is powerfully illustrated by the results of Case 2 of Example 5.1. In all other worked examples we have developed discretised equations with positive coefficients a_P and a_{nb}, but in Case 2 most of the east coefficients were negative (see Table 5.3) and the solution contained large under- and overshoots!

5.4.3 Transportiveness

The transportiveness property of a fluid flow (Roache, 1976) can be illustrated by considering a constant source of ϕ at a point P as shown in Figure 5.9. We define the non-dimensional cell Peclet number as a measure of the relative strengths of convection and diffusion:

$$Pe = \frac{F}{D} = \frac{\rho u}{\Gamma/\delta x} \tag{5.23}$$

where δx = characteristic length (cell width). The lines in Figure 5.9 indicate the general shape of contours of a constant ϕ (say $\phi = 1$) for different values of Pe.

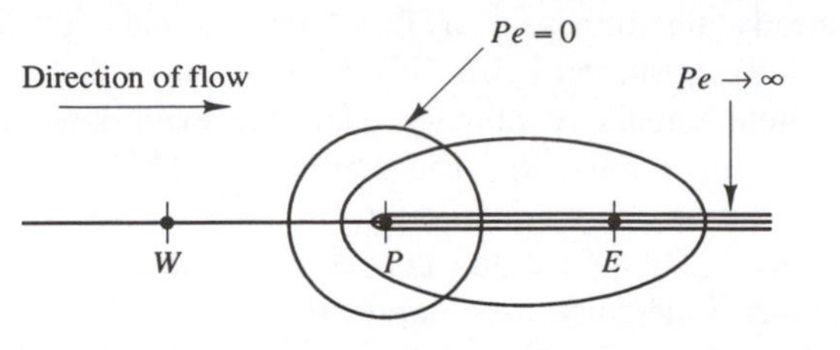

Fig. 5.9 Distribution of ϕ in the vicinity of a source at different Peclet numbers

Let us consider two extreme cases to identify the extent of the influence of the upstream node P at the downstream node E:

- no convection and pure diffusion ($Pe = 0$)
- no diffusion and pure convection ($Pe \to \infty$)

In the case of pure diffusion the fluid is stagnant ($Pe = 0$) and the contours of constant ϕ will be concentric circles with P at their centre since the diffusion process tends to spread ϕ equally in all directions. Conditions at the east node E will be influenced by those upstream at P and also by conditions further downstream. As Pe increases the contours change shape from circular to elliptical and are shifted in the direction of the flow as indicated in Figure 5.9. Influencing becomes increasingly biased towards the upstream direction at large values of Pe so that the node E is strongly influenced by conditions at P, but conditions at P will experience weak influence or no influence at all from E. In the case of pure convection ($Pe \to \infty$) the elliptical contours are completely stretched out in the flow direction. All of property ϕ emanating from the source at P is immediately transported downstream towards E. Thus the value of ϕ at E is affected only by upstream conditions and since there is no diffusion ϕ_E is equal to ϕ_P. It is very important that the relationship between the magnitude of the Peclet number and the directionality of influencing, known as the **transportiveness**, is borne out in the discretisation scheme.

5.5 Assessment of the central differencing scheme for convection–diffusion problems

Conservativeness

The central differencing scheme uses consistent expressions to evaluate convective and diffusive fluxes at the control volume faces. The discussions in section 5.4.1 show that the scheme is conservative.

Boundedness

(i) The internal coefficients of discretised scalar transport equation (5.14) are

a_W	a_E	a_P
$D_w + \dfrac{F_w}{2}$	$D_e - \dfrac{F_e}{2}$	$a_W + a_E + (F_e - F_w)$

A steady one-dimensional flow field is also governed by the discretised continuity equation (5.10). This equation states that $(F_e - F_w)$ is zero when the flow field satisfies continuity. Thus the expression for a_P in (5.14) becomes equal to $a_P = a_W + a_E$. The coefficients of the central differencing scheme satisfy the Scarborough criterion (5.22).

(ii) With $a_E = D_e - F_e/2$ the convective contribution to the east coefficient is negative; if the convection dominates it is possible for a_E to be negative. Given that $F_w > 0$ and $F_e > 0$ (i.e. the flow is unidirectional), for a_E to be positive D_e and F_e must satisfy the following condition:

$$F_e/D_e = Pe_e < 2 \tag{5.24}$$

If Pe_e is greater than 2 the east coefficient will be negative. This violates one of the requirements for boundedness and may lead to physically impossible solutions.

In the example of section 5.3 we took $Pe = 5$ in Case 2 so condition (5.24) is violated. The consequences were evident in the results which showed large 'undershoots' and 'overshoots'. Taking Pe less than 2 in Cases 1 and 3 gave bounded answers close to the analytical solution.

Transportiveness

The central differencing scheme introduces influencing at node P from the directions of all its neighbours to calculate the convective and diffusive flux. Thus the scheme does not recognise the direction of the flow or the strength of convection relative to diffusion. It does not possess the transportiveness property at high Pe.

Accuracy

The Taylor series truncation error of the central differencing scheme is second order (see Appendix A for further details). The requirement for positive coefficients in the central differencing scheme as given by formula (5.24) implies that the scheme will be stable and accurate only if $Pe = F/D < 2$. It is important to note that the cell Peclet number, as defined by (5.23), is a combination of fluid properties (ρ and Γ), a flow property (u) and a property of the computational grid (δx). So for given values of ρ and Γ it is only possible to satisfy condition (5.24) if the velocity is small, hence in diffusion-dominated low Reynolds number flows, or if the grid spacing is small. Owing to this limitation central differencing is not a suitable discretisation practice for general purpose flow calculations. This creates the need for discretisation schemes which possess more favourable properties. Below we discuss the upwind, hybrid, power-law and QUICK schemes.

5.6 The upwind differencing scheme

One of the major inadequacies of the central differencing scheme is its inability to identify flow direction. The value of property ϕ at a west cell face is always influenced by both ϕ_P and ϕ_W in central differencing. In a strongly convective flow from west to east, the above treatment is unsuitable because the west cell face should receive much stronger influencing from node W than from node P. The upwind differencing or 'donor cell' differencing scheme takes into account the flow direction when determining the value at a cell face: the convected value of ϕ at a cell face is taken to be equal to the value at the upstream node. In Figure 5.10 we show the nodal values used to calculate cell face values when the flow is in the positive direction (west to east) and in Figure 5.11 those for the negative direction.

Fig. 5.10

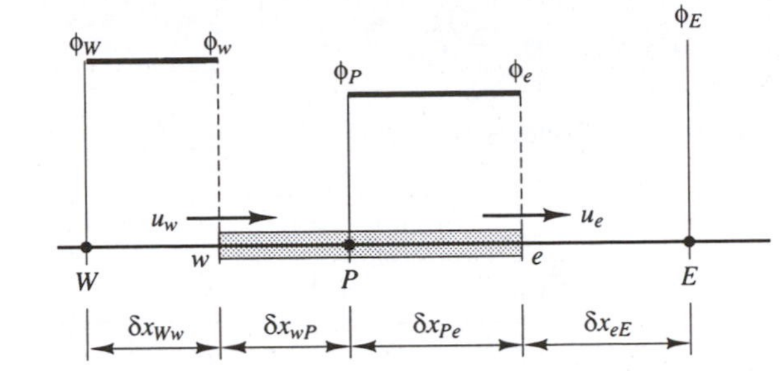

Fig. 5.11

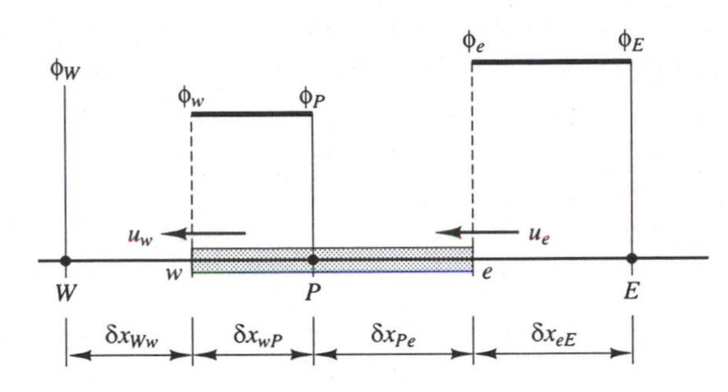

When the flow is in the positive direction, $u_w > 0$, $u_e > 0 (F_w > 0, F_e > 0)$, the upwind scheme sets

$$\phi_w = \phi_W \text{ and } \phi_e = \phi_P \tag{5.25}$$

and the discretised equation (5.9) becomes

$$F_e\phi_P - F_w\phi_W = D_e(\phi_E - \phi_P) - D_w(\phi_P - \phi_W) \tag{5.26}$$

which can be re-arranged as

$$(D_w + D_e + F_e)\phi_P = (D_w + F_w)\phi_W + D_e\phi_E$$

to give

$$[(D_w + F_w) + D_e + (F_e - F_w)]\phi_P = (D_w + F_w)\phi_W + D_e\phi_E \tag{5.27}$$

When the flow is in the negative direction, $u_w < 0$, $u_e < 0 (F_w < 0, F_e < 0)$, the scheme takes

$$\phi_w = \phi_P \text{ and } \phi_e = \phi_E \tag{5.28}$$

Now the discretised equation is

$$F_e\phi_E - F_w\phi_P = D_e(\phi_E - \phi_P) - D_w(\phi_P - \phi_W) \tag{5.29}$$

or

$$[D_w + (D_e - F_e) + (F_e - F_w)]\phi_P = D_w\phi_W + (D_e - F_e)\phi_E \tag{5.30}$$

Identifying the coefficients of ϕ_W and ϕ_E as a_W and a_E the equations (5.27) and (5.30) can be written in the usual general form

$$\boxed{a_P\phi_P = a_W\phi_W + a_E\phi_E} \tag{5.31}$$

with central coefficient

$$\boxed{a_P = a_W + a_E + (F_e - F_w)}$$

and neighbour coefficients

	a_W	a_E
$F_w > 0,\ F_e > 0$	$D_w + F_w$	D_e
$F_w < 0,\ F_e < 0$	D_w	$D_e - F_e$

A form of notation for the **neighbour coefficients of the upwind differencing method** that covers both flow directions is given below:

a_W	a_E
$D_w + \max(F_w,\ 0)$	$D_e + \max(0,\ -F_e)$

Example 5.2 Solve the problem considered in Example 5.1 using the upwind differencing scheme for (i) $u = 0.1$ m/s, (ii) $u = 2.5$ m/s with the coarse five-point grid.

Solution The grid shown in Figure 5.3 is again used here for the discretisation. The discretisation equation at internal nodes 2, 3 and 4 and the relevant neighbour coefficients are given by (5.31) and its accompanying tables. Note that in this example $F = F_e = F_w = \rho u$ and $D = D_e = D_w = \Gamma/\delta x$ everywhere.

At the boundary node 1, the use of upwind differencing for the convective terms gives

$$F_e\phi_P - F_A\phi_A = D_e(\phi_E - \phi_P) - D_A(\phi_P - \phi_A) \tag{5.32}$$

And at node 5

$$F_B\phi_P - F_w\phi_W = D_B(\phi_B - \phi_P) - D_w(\phi_P - \phi_W) \tag{5.33}$$

At the boundary nodes we have $D_A = D_B = 2\Gamma/\delta x = 2D$ and $F_A = F_B = F$ and as usual the boundary conditions enter the discretised equations as source contributions:

$$\boxed{a_P\phi_P = a_W\phi_W + a_E\phi_E + S_u} \tag{5.34}$$

with

$$\boxed{a_P = a_W + a_E + (F_e - F_w) - S_P}$$

and

Node	a_W	a_E	S_P	S_u
1	0	D	$-(2D+F)$	$(2D+F)\phi_A$
2, 3, 4	$D+F$	D	0	0
5	$D+F$	0	$-2D$	$2D\phi_B$

The reader will by now be familiar with the process of calculating coefficients and constructing and solving the matrix equation. For the sake of brevity we leave this as an exercise and concentrate on the evaluation of the results. The analytical solution is again given by equation (5.15) and is compared with the numerical, upwind differencing, solution.

Case 1

$u = 0.1$ m/s: $F = \rho u = 0.1, D = \Gamma/\delta x = 0.1/0.2 = 0.5$ so $Pe = F/D = 0.2$. The results are summarised in Table 5.6 and Figure 5.12 shows that the upwind differencing scheme produces good results at this cell Peclet number.

Table 5.6

Node	*Distance*	*Finite volume solution*	*Analytical solution*	*Difference*	*Percentage error*
1	0.1	0.9337	0.9387	0.005	0.53
2	0.3	0.7879	0.7963	0.008	1.05
3	0.5	0.6130	0.6224	0.009	1.51
4	0.7	0.4031	0.4100	0.007	1.68
5	0.9	0.1512	0.1505	−0.001	−0.02

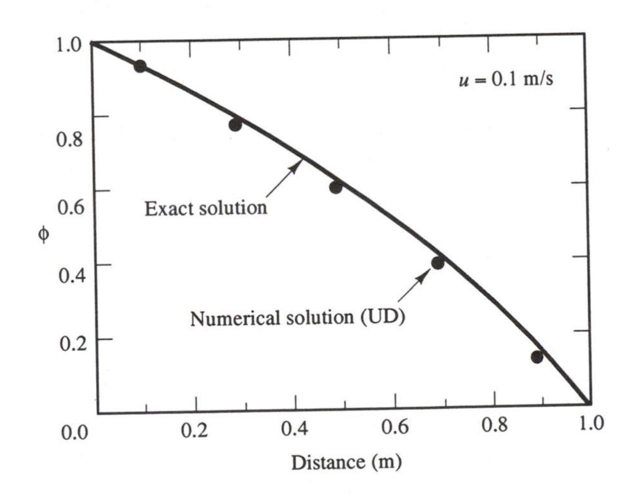

Fig. 5.12 Comparison of the upwind difference numerical results and the analytical solution for Case 1

Case 2

$u = 2.5$ m/s: $F = \rho u = 2.5, D = \Gamma/\delta x = 0.1/0.2 = 0.5$ now $Pe = 5$. The numerical results are compared with the analytical solution in Table 5.7 and Figure 5.13.

Table 5.7

Node	Distance	Finite volume solution	Analytical solution	Difference	Percentage error
1	0.1	0.9998	0.9999	0.0001	0.00
2	0.3	0.9987	0.9999	0.001	0.01
3	0.5	0.9921	0.9999	0.007	0.70
4	0.7	0.9524	0.9994	0.047	4.70
5	0.9	0.7143	0.8946	0.180	20.15

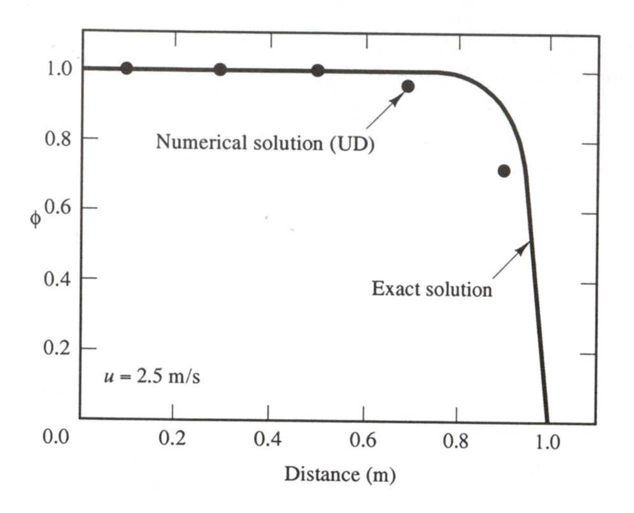

Fig. 5.13 Comparison of the upwind difference numerical results and the analytical solution for Case 2

The central differencing scheme failed to produce a reasonable result with the same grid resolution. The upwind scheme produces a much more realistic solution that is, however, not very close to the exact solution near boundary B.

5.6.1 Assessment of the upwind differencing scheme

Conservativeness

The upwind differencing scheme utilises consistent expressions to calculate fluxes through cell faces: therefore it can be easily shown that the formulation is conservative.

Boundedness

The coefficients of the discretised equation are always positive and satisfy the requirements for boundedness. When the flow satisfies continuity the term $(F_e - F_w)$ in a_P (see (5.31)) is zero and gives $a_P = a_W + a_E$, which is desirable for stable iterative solutions. All the coefficients are positive and the coefficient matrix is diagonally dominant; hence no 'wiggles' occur in the solution.

Transportiveness

The scheme accounts for the direction of the flow so transportiveness is built into the formulation.

Accuracy

The scheme is based on the backward differencing formula so the accuracy is only first order on the basis of the Taylor series truncation error (see Appendix A).

Because of its simplicity the upwind differencing scheme has been widely applied in early CFD calculations. It can be easily extended to multi-dimensional problems by repeated application of the upwind strategy embodied in the coefficients of (5.31) in each co-ordinate direction. A major drawback of the scheme is that it produces erroneous results when the flow is not aligned with the grid lines. The upwind differencing scheme causes the distributions of the transported properties to become smeared in such problems. The resulting error has a diffusion-like appearance and is referred to as **false diffusion**. The effect can be illustrated by calculating the transport of scalar property ϕ using upwind differencing in a domain where the flow is at an angle to a Cartesian grid.

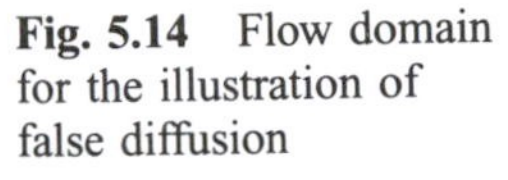
Fig. 5.14 Flow domain for the illustration of false diffusion

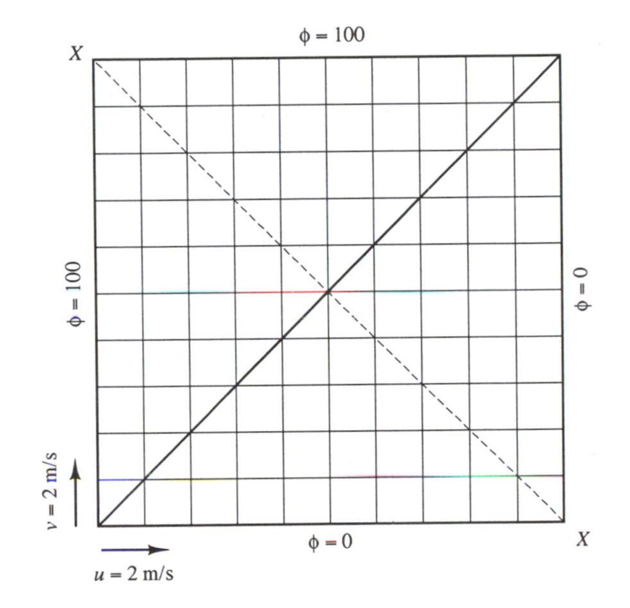

In Figure 5.14 we have a domain where $u = v = 2$ m/s everywhere so the velocity field is uniform and parallel to the diagonal (solid line) across the grid. The boundary conditions for the scalar are $\phi = 0$ along the south and east boundaries, and $\phi = 100$ on the west and north boundaries. At the first and the last nodes where the diagonal intersects the boundary grid nodes a value of 50 is assigned to the property ϕ.

To identify the false diffusion due to the upwind scheme, a pure convection process is considered without physical diffusion. There are no source terms for ϕ and a steady state solution is sought. The correct solution is known in this case. As the flow is parallel to the solid diagonal the value of ϕ at all nodes above the diagonal should be 100 and below the diagonal it should be zero. The degree of false diffusion can be illustrated by calculating the distribution of ϕ and plotting the results along the diagonal (X–X). Since there is no physical diffusion the exact solution exhibits a step change of ϕ from 100 to zero when the diagonal X–X crosses the solid diagonal. The calculated results for different grids are shown in Figure 5.15 together with the exact solution. The numerical results show badly smeared profiles.

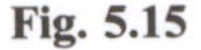

Fig. 5.15

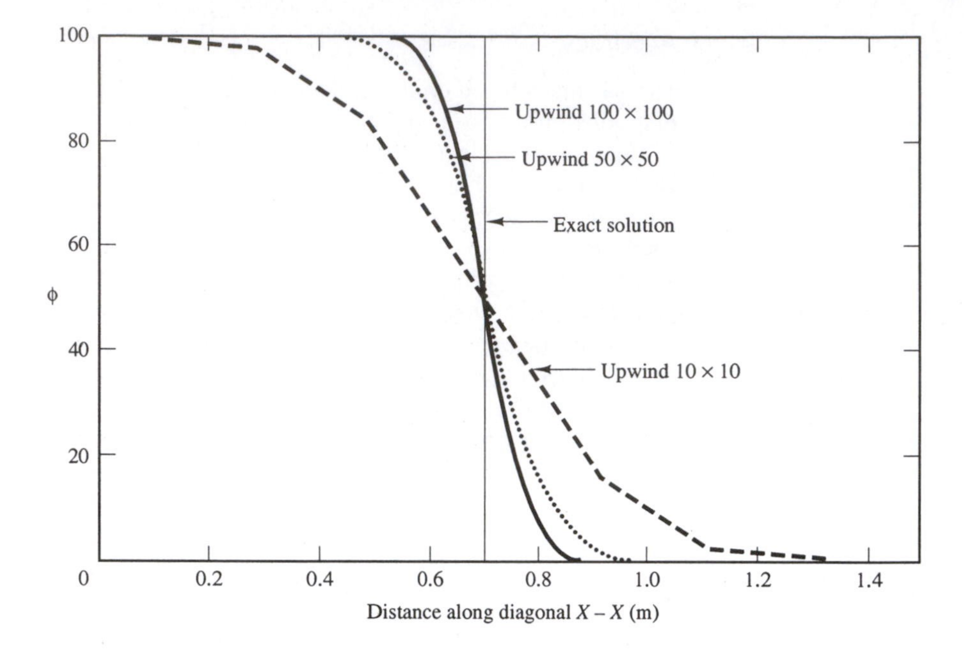

The error is largest for the coarsest grid and the figure shows that refinement of the grid can, in principle, overcome the problem of false diffusion. The results for 50×50 and 100×100 grids show profiles that are closer to the exact solution. In practical flow calculations, however, the degree of grid refinement required to eliminate false diffusion can be prohibitively expensive. Trials have shown that in high Reynolds number flows, false diffusion can be large enough to give physically incorrect results (Leschziner, 1980; Huang *et al*, 1985). Therefore, the upwind differencing scheme is not entirely suitable for accurate flow calculations and considerable research has been directed towards finding improved discretisation schemes.

5.7 The hybrid differencing scheme

The hybrid differencing scheme of Spalding (1972) is based on a combination of central and upwind differencing schemes. The central differencing scheme, which is accurate to second-order, is employed for small Peclet numbers ($Pe < 2$) and the upwind scheme, which is accurate to first order but accounts for transportiveness, is employed for large Peclet numbers ($Pe \geq 2$). As before we develop the discretisation of the one-dimensional convection–diffusion equation without source terms. This equation can be interpreted as a flux balance equation. The hybrid differencing scheme uses piecewise formulae based on the local Peclet number to evaluate the net flux through each control volume face. The Peclet number is evaluated at the face of the control volume. For example, for a west face

$$Pe_w = \frac{F_w}{D_w} = \frac{(\rho u)_w}{\Gamma_w/\delta x_{WP}} \tag{5.35}$$

The hybrid differencing formula for the net flux per unit area through the west face is as follows:

$$q_w = F_w\left[\frac{1}{2}\left(1+\frac{2}{Pe_w}\right)\phi_W + \frac{1}{2}\left(1-\frac{2}{Pe_w}\right)\phi_P\right] \quad \text{for} \quad -2 < Pe_w < 2$$
$$q_w = F_w A_w \phi_W \quad \text{for} \quad Pe_w \geq 2$$
$$q_w = F_w A_w \phi_P \quad \text{for} \quad Pe_w \leq -2 \tag{5.36}$$

It can be easily seen that for low Peclet numbers this is equivalent to using central differencing for the convection and diffusion terms, but when $|Pe| > 2$ it is equivalent to upwinding for convection and setting the diffusion to zero. The general form of the discretised equation is

$$a_P\phi_P = a_W\phi_W + a_E\phi_E \tag{5.37}$$

The central coefficient is given by

$$a_P = a_W + a_E + (F_e - F_w)$$

After some re-arrangement it is easy to establish that the neighbour coefficients for the **hybrid differencing** scheme for steady **one-dimensional convection–diffusion** can be written as follows:

a_W	a_E
$\max\left[F_w, \left(D_w + \frac{F_w}{2}\right), 0\right]$	$\max\left[-F_e, \left(D_e - \frac{F_e}{2}\right), 0\right]$

Example 5.3 Solve the problem considered in Case 2 of Example 5.1 using the hybrid scheme for $u = 2.5$ m/s. Compare a five-node solution with a 25 node solution.

Solution If we use the five node grid and the data of Case 2 of Example 5.1 and $u = 2.5$ m/s we have: $F = F_e = F_w = \rho u = 2.5$ and $D = D_e = D_w = \Gamma/\delta x = 0.5$ and hence a Peclet number $Pe_w = Pe_e = \rho u \delta x/\Gamma = 5$. Since the cell Peclet number Pe is greater than 2 the hybrid scheme uses the upwind expression for the convective terms and sets the diffusion to zero.

The discretisation equation at internal nodes 2, 3 and 4 is defined by (5.37) and its coefficients. We also need to introduce boundary conditions at nodes 1 and 5 which need special treatment. At the boundary node 1 we write

$$F_e\phi_P - F_A\phi_A = 0 - D_A(\phi_P - \phi_A) \tag{5.38}$$

and at node 5

$$F_B\phi_P - F_w\phi_W = D_B(\phi_B - \phi_P) - 0 \tag{5.39}$$

It can be seen that the diffusive flux at the boundary is entered on the right hand side and the convective fluxes are given by means of the upwind method. We note that

$F_A = F_B = F$ and $D_B = 2\Gamma/\delta x = 2D$ so the discretised equation can be written as

$$a_P\phi_P = a_W\phi_W + a_E\phi_E + S_u \tag{5.40}$$

with

$$a_P = a_W + a_E + (F_e - F_w) - S_p$$

and

Node	a_W	a_E	S_p	S_u
1	0	0	$-(2D+F)$	$(2D+F)\phi_A$
2, 3, 4	F	0	0	0
5	F	0	$-2D$	$2D\phi_B$

Substitution of numerical values gives the coefficients summarised in Table 5.8.

Table 5.8

Node	a_W	a_E	S_u	S_p	$a_P = a_W + a_E - S_p$
1	0	0	$3.5\phi_A$	−3.5	3.5
2	2.5	0	0	0	2.5
3	2.5	0	0	0	2.5
4	2.5	0	0	0	2.5
5	2.5	0	$1.0\phi_B$	−1.0	3.5

The matrix form of the equation set is

$$\begin{bmatrix} 3.5 & 0 & 0 & 0 & 0 \\ -2.5 & 2.5 & 0 & 0 & 0 \\ 0 & -2.5 & 2.5 & 0 & 0 \\ 0 & 0 & -2.5 & 2.5 & 0 \\ 0 & 0 & 0 & -2.5 & 3.5 \end{bmatrix} \begin{bmatrix} \phi_1 \\ \phi_2 \\ \phi_3 \\ \phi_4 \\ \phi_5 \end{bmatrix} = \begin{bmatrix} 3.5 \\ 0 \\ 0 \\ 0 \\ 0 \end{bmatrix} \tag{5.41}$$

The solution to the above system is

$$\begin{bmatrix} \phi_1 \\ \phi_2 \\ \phi_3 \\ \phi_4 \\ \phi_5 \end{bmatrix} = \begin{bmatrix} 1.0 \\ 1.0 \\ 1.0 \\ 1.0 \\ 0.7143 \end{bmatrix} \tag{5.42}$$

Comparison with the analytical solution

The numerical results are compared with the analytical solution in Table 5.9 and, since the cell Peclet number is high, they are the same as those for pure upwind differencing. When the grid is refined to an extent that the cell $Pe < 2$ the scheme reverts to central differencing and produces an accurate solution. This is illustrated

Table 5.9

Node	*Distance*	*Finite volume solution*	*Analytical solution*	*Difference*	*Percentage error*
1	0.1	1.0	0.9999	−0.0001	−0.01
2	0.3	1.0	0.9999	−0.0001	−0.01
3	0.3	1.0	0.9999	−0.0001	−0.01
4	0.7	1.0	0.9994	−0.0006	−0.06
5	0.9	0.7143	0.8946	0.1843	20.15

by using a 25 node grid with $\delta x = 0.04$ m so $F = D = 2.5$. Both the results computed on the coarse and the fine grids are shown in Figure 5.16 together with the analytical solution. Now $Pe = 1$, the hybrid scheme reverts to central differencing and it can be seen that the solution obtained with the fine grid is remarkably good.

Fig. 5.16

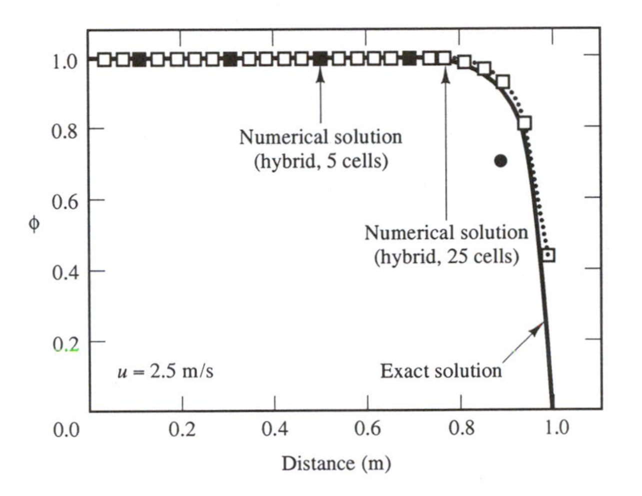

5.7.1 Assessment of the hybrid differencing scheme

The hybrid difference scheme exploits the favourable properties of the upwind and central differencing schemes. It switches to the upwind differencing when the central differencing produces inaccurate results at high *Pe* numbers. The scheme is fully conservative and since the coefficients are always positive it is unconditionally bounded. It satisfies the transportiveness requirement by using an upwind formulation for large values of Peclet number. The scheme produces physically realistic solutions and is highly stable when compared with the higher order schemes to be discussed later in the chapter. Hybrid differencing has been widely used in various computational fluid dynamics (CFD) procedures and has proved to be very useful for predicting practical flows. The disadvantage is that the accuracy in terms of Taylor series truncation error is only first-order.

5.7.2 Hybrid differencing scheme for multi-dimensional convection–diffusion

The hybrid differencing scheme can easily be extended to two- and three-dimensional problems by repeated application of the derivation in each new coordinate direction. The discretised equation that covers all cases is given by

$$a_P\phi_P = a_W\phi_W + a_E\phi_E + a_S\phi_S + a_N\phi_N + a_B\phi_B + a_T\phi_T \tag{5.43}$$

with central coefficient

$$a_P = a_W + a_E + a_S + a_N + a_B + a_T + \Delta F$$

and the coefficients of this equation for the **hybrid differencing scheme** are as follows:

	One-dimensional flow	Two-dimensional flow	Three-dimensional flow
a_W	$\max\left[F_w, \left(D_w + \frac{F_w}{2}\right), 0\right]$	$\max\left[F_w, \left(D_w + \frac{F_w}{2}\right), 0\right]$	$\max\left[F_w, \left(D_w + \frac{F_w}{2}\right), 0\right]$
a_E	$\max\left[-F_e, \left(D_e - \frac{F_e}{2}\right), 0\right]$	$\max\left[-F_e, \left(D_e - \frac{F_e}{2}\right), 0\right]$	$\max\left[-F_e, \left(D_e - \frac{F_e}{2}\right), 0\right]$
a_S	–	$\max\left[F_s, \left(D_s + \frac{F_s}{2}\right), 0\right]$	$\max\left[F_s, \left(D_s + \frac{F_s}{2}\right), 0\right]$
a_N	–	$\max\left[-F_n, \left(D_n - \frac{F_n}{2}\right), 0\right]$	$\max\left[-F_n, \left(D_n - \frac{F_n}{2}\right), 0\right]$
a_B	–	–	$\max\left[F_b, \left(D_b + \frac{F_b}{2}\right), 0\right]$
a_T	–	–	$\max\left[-F_t, \left(D_t - \frac{F_t}{2}\right), 0\right]$
ΔF	$F_e - F_w$	$F_e - F_w + F_n - F_s$	$F_e - F_w + F_n - F_s + F_t - F_b$

In the above expressions the values of F and D are calculated with the following formulae:

Face	w	e	s	n	b	t
F	$(\rho u)_w A_w$	$(\rho u)_e A_e$	$(\rho v)_s A_s$	$(\rho v)_n A_n$	$(\rho w)_b A_b$	$(\rho w)_t A_t$
D	$\frac{\Gamma_w}{\delta x_{WP}} A_w$	$\frac{\Gamma_e}{\delta x_{PE}} A_e$	$\frac{\Gamma_s}{\delta y_{SP}} A_s$	$\frac{\Gamma_n}{\delta y_{PN}} A_n$	$\frac{\Gamma_b}{\delta z_{PN}} A_b$	$\frac{\Gamma_t}{\delta z_{PT}} A_t$

Modifications to these coefficients to cater for boundary conditions in two and three dimensions are available in the form of expressions such as (5.40).

5.8 The power-law scheme

The power-law differencing scheme of Patankar (1980) is a more accurate approximation to the one-dimensional exact solution and produces better results than the hybrid scheme. In this scheme diffusion is set to zero when cell Pe exceeds 10. If $0 < Pe < 10$ the flux is evaluated by using a polynomial expression; for example, the net flux per unit area at the west control volume face is evaluated using

$$q_w = F_w[\phi_W - \beta_w(\phi_P - \phi_W)] \qquad \text{for } 0 < Pe < 10 \tag{5.44a}$$

$$\text{where } \beta_w = (1 - 0.1\text{Pe}_w)^5/\text{Pe}_w$$

and

$$q_w = F_w\phi_W \qquad \text{for } Pe > 10 \tag{5.44b}$$

The coefficients of the one-dimensional discretised equation utilising the **power-law** scheme for steady **one-dimensional convection–diffusion** are given by

$$\text{central coefficient: } a_P = a_W + a_E + (F_e - F_w)$$

and

a_W	a_E
$D_w \max\left[0, (1 - 0.1\lvert Pe_w\rvert)^5\right] + \max[F_w, 0]$	$D_e \max\left[0, (1 - 0.1\lvert Pe_e\rvert)^5\right] + \max[-F_e, 0]$

Properties of the power-law differencing scheme are similar to those of the hybrid scheme. The power-law differencing scheme is more accurate for one-dimensional problems since it attempts to represent the exact solution more closely. The scheme has proved to be useful in practical flow calculations and can be used as an alternative to the hybrid scheme. Some commercial computer codes, for example FLUENT version 4.22, use this scheme as the default scheme for flow calculations (FLUENT Users' Manual, 1992).

5.9 Higher order differencing schemes for convection–diffusion problems

The accuracy of hybrid and upwind schemes is only first-order in terms of Taylor series truncation error (TSTE). The use of upwind quantities ensures that the schemes are very stable and obey the transportiveness requirement but the first-order accuracy makes them prone to numerical diffusion errors. Such errors can be minimised by employing higher order discretisation. Higher order schemes involve more neighbour points and reduce the discretisation errors by bringing in a wider influence. The central differencing scheme which has second-order accuracy proved to be unstable and does not possess the transportiveness property. Formulations that do not take into account the flow direction are unstable and, therefore, more accurate higher order schemes, which preserve upwinding for stability and sensitivity to the flow direction, are needed. Some of the widely used approaches are discussed below.

5.9.1 Quadratic upwind differencing scheme: the QUICK scheme

The quadratic upstream interpolation for convective kinetics (QUICK) scheme of Leonard (1979) uses a three-point upstream-weighted quadratic interpolation for cell face values. The face value of ϕ is obtained from a quadratic function passing

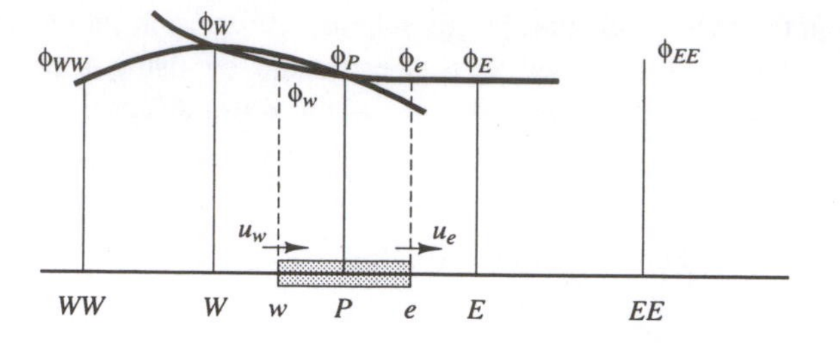

Fig. 5.17 Quadratic profiles used in the QUICK scheme

through two bracketing nodes (on each side of the face) and a node on the upstream side (Figure 5.17).

For example, when $u_w > 0$ and $u_e > 0$ a quadratic fit through *WW*, *W* and *P* is used to evaluate ϕ_w, and a further quadratic fit through *W*, *P* and *E* to calculate ϕ_e. For $u_w < 0$ and $u_e < 0$ values of ϕ at *W*, *P* and *E* are used for ϕ_w, and values at *P*, *E* and *EE* for ϕ_e. It can be shown that for a uniform grid the value of ϕ at the cell face between two bracketing nodes i and $i-1$, and upstream node $i-2$ is given by the following formula:

$$\phi_{face} = \frac{6}{8}\phi_{i-1} + \frac{3}{8}\phi_i - \frac{1}{8}\phi_{i-2} \tag{5.45}$$

When $u_w > 0$, the bracketing nodes for the west face 'w' are *W* and *P*, the upstream node is *WW* (Figure 5.17), and

$$\phi_w = \frac{6}{8}\phi_W + \frac{3}{8}\phi_P - \frac{1}{8}\phi_{WW} \tag{5.46}$$

When $u_e > 0$, the bracketing nodes for the east face 'e' are *P* and *E*, the upstream node is *W*, so

$$\phi_e = \frac{6}{8}\phi_P + \frac{3}{8}\phi_E - \frac{1}{8}\phi_W \tag{5.47}$$

The diffusion terms may be evaluated using the gradient of the appropriate parabola. It is interesting to note that on a uniform grid this practice gives the same expressions as central differencing for diffusion. If $F_w > 0$ and $F_e > 0$ and if we use equations (5.46–5.47) for the convective terms and central differencing for the diffusion terms, the discretised form of the one-dimensional convection–diffusion transport equation (5.9) may be written as

$$\left[F_e\left(\frac{6}{8}\phi_P + \frac{3}{8}\phi_E - \frac{1}{8}\phi_W\right) - F_w\left(\frac{6}{8}\phi_W + \frac{3}{8}\phi_P - \frac{1}{8}\phi_{WW}\right)\right]$$
$$= D_e(\phi_E - \phi_P) - D_w(\phi_P - \phi_W)$$

which can be re-arranged to give

$$\left[D_w - \frac{3}{8}F_w + D_e + \frac{6}{8}F_e\right]\phi_P = \left[D_w + \frac{6}{8}F_w + \frac{1}{8}F_e\right]\phi_W + \left[D_e - \frac{3}{8}F_e\right]\phi_E - \frac{1}{8}F_w\phi_{WW} \tag{5.48}$$

This is now written in the standard form for discretised equations

$$\boxed{a_P\phi_P = a_W\phi_W + a_E\phi_E + a_{WW}\phi_{WW}} \tag{5.49}$$

where

a_W	a_E	a_{WW}	a_P
$D_w + \frac{6}{8}F_w + \frac{1}{8}F_e$	$D_e - \frac{3}{8}F_e$	$-\frac{1}{8}F_w$	$a_W + a_E + a_{WW} + (F_e - F_w)$

For $F_w < 0$ and $F_e < 0$ the flux across the west and east boundaries is given by the expressions

$$\begin{aligned}\phi_w &= \frac{6}{8}\phi_P + \frac{3}{8}\phi_W - \frac{1}{8}\phi_E \\ \phi_e &= \frac{6}{8}\phi_E + \frac{3}{8}\phi_P - \frac{1}{8}\phi_{EE}\end{aligned} \tag{5.50}$$

Substitution of these two formulae for the convective terms in the discretised convection–diffusion equation (5.9) together with central differencing for the diffusion terms leads, after re-arrangement as above, to the following coefficients:

a_W	a_E	a_{EE}	a_P
$D_w + \frac{3}{8}F_w$	$D_e - \frac{6}{8}F_e - \frac{1}{8}F_w$	$\frac{1}{8}F_e$	$a_W + a_E + a_{EE} + (F_e - F_w)$

General expressions, valid for positive and negative flow directions, can be obtained by combining the two sets of coefficients above.

The **QUICK scheme for one-dimensional convection–diffusion problems** can be summarised as follows:

$$a_P\phi_P = a_W\phi_W + a_E\phi_E + a_{WW}\phi_{WW} + a_{EE}\phi_{EE} \tag{5.51}$$

with central coefficient

$$a_P = a_W + a_E + a_{WW} + a_{EE} + (F_e - F_w)$$

and neighbour coefficients

a_W	a_{WW}	a_E	a_{EE}
$D_w + \frac{6}{8}\alpha_w F_w + \frac{1}{8}\alpha_e F_e + \frac{3}{8}(1 - \alpha_w)F_w$	$-\frac{1}{8}\alpha_w F_w$	$D_e - \frac{3}{8}\alpha_e F_e - \frac{6}{8}(1-\alpha_e)F_e - \frac{1}{8}(1 - \alpha_w)F_w$	$\frac{1}{8}(1 - \alpha_e)F_e$

where

$$\alpha_w = 1 \text{ for } F_w > 0 \text{ and } \alpha_e = 1 \text{ for } F_e > 0$$

$$\alpha_w = 0 \text{ for } F_w < 0 \text{ and } \alpha_e = 0 \text{ for } F_e < 0$$

Example 5.4 Using the QUICK scheme solve the problem considered in Example 5.1 for $u = 0.2$ m/s on a five-point grid. Compare the QUICK solution with the exact and the central differencing solution.

Solution As before the five-node grid introduced in Example 5.1 is used for the discretisation. With the data of this example and $u = 0.2$ m/s we have $F = F_e = F_w = 0.2$ and $D = D_e = D_w = 0.5$ everywhere so that the cell Peclet number becomes $Pe_w = Pe_e = \rho u\delta x/\Gamma = 0.4$. The discretisation equation with the QUICK scheme at internal nodes 3 and 4 is given by (5.51) together with its coefficients.

In the QUICK scheme the ϕ-value at cell boundaries is calculated with formulae (5.46–5.47) that use three nodal values. Nodes 1, 2 and 5 are all affected by the proximity of domain boundaries and need to be treated separately. At the boundary node 1 ϕ is given at the west (w) face ($\phi_w = \phi_A$), but there is no west (W) node to evaluate ϕ_e at the east face by (5.47). To overcome this problem Leonard (1979) suggests a linear extrapolation to create a 'mirror' node at a distance $\delta x/2$ to the west of the physical boundary. This is illustrated in Figure 5.18.

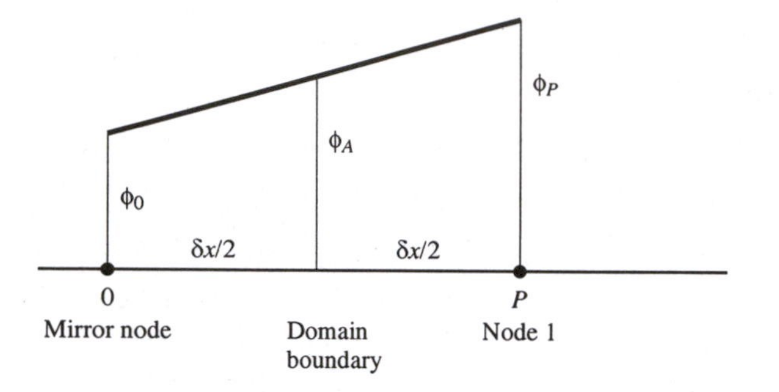

Fig. 5.18 Mirror node treatment at the boundary

It can be easily shown that the linearly extrapolated value at the mirror node is given by

$$\phi_0 = 2\phi_A - \phi_P \tag{5.52}$$

The extrapolation to the 'mirror' node has given us the required W node for the formula (5.47) that calculates ϕ_e at the east face of control volume 1:

$$\begin{aligned}\phi_e &= \frac{6}{8}\phi_P + \frac{3}{8}\phi_E - \frac{1}{8}(2\phi_A - \phi_P) \\ &= \frac{7}{8}\phi_P + \frac{3}{8}\phi_E - \frac{2}{8}\phi_A\end{aligned} \tag{5.53}$$

At the boundary nodes the gradients must be evaluated using an expression consistent with formula (5.53). It can be shown that the diffusive flux through the west boundary is given by

$$\Gamma\frac{\partial\phi}{\partial x}\bigg|_A = \frac{D_A}{3}(9\phi_P - 8\phi_A - \phi_E) \tag{5.54}$$

The discretised equation at node 1 is

$$F_e\left[\frac{7}{8}\phi_P + \frac{3}{8}\phi_E - \frac{2}{8}\phi_A\right] - F_A\phi_A = D_e(\phi_E - \phi_P) - \frac{D_A}{3}(9\phi_P - 8\phi_A - \phi_E) \tag{5.55}$$

At control volume 5, the ϕ-value at the east face is known ($\phi_e = \phi_B$) and the diffusive flux of ϕ through the east boundary is given by

$$\Gamma\frac{\partial\phi}{\partial x}\bigg|_B = \frac{D_B}{3}(8\phi_B - 9\phi_P + \phi_W) \tag{5.56}$$

At node 5 the discretised equation becomes

$$F_B\phi_B - F_w\left[\frac{6}{8}\phi_W + \frac{3}{8}\phi_P - \frac{1}{8}\phi_{WW}\right] = \frac{D_B}{3}(8\phi_B - 9\phi_P + \phi_W) - D_w(\phi_P - \phi_W) \quad (5.57)$$

Since a special expression is used to evaluate ϕ at the east face of the control volume 1 we must use the same expression for ϕ to calculate the convective flux through the west face of control volume 2 to ensure flux consistency. So at node 2 we have

$$F_e\left[\frac{6}{8}\phi_P + \frac{3}{8}\phi_E - \frac{1}{8}\phi_W\right] - F_w\left[\frac{7}{8}\phi_W + \frac{3}{8}\phi_P - \frac{2}{8}\phi_A\right] = D_e(\phi_E - \phi_P) - D_w(\phi_P - \phi_W) \quad (5.58)$$

The discretised equations for nodes 1, 2 and 5 are now written to fit into the standard form to give:

$$a_P\phi_P = a_{WW}\phi_{WW} + a_W\phi_W + a_E\phi_E + S_u \quad (5.59)$$

with

$$a_P = a_{WW} + a_W + a_E + (F_e - F_w) - S_p$$

and

Node	a_{WW}	a_W	a_E	S_p	S_u
1	0	0	$D_e + \frac{1}{3}D_A - \frac{3}{8}F_e$	$-\left(\frac{8}{3}D_A + \frac{2}{8}F_e + F_A\right)$	$\left(\frac{8}{3}D_A + \frac{2}{8}F_e + F_A\right)\phi_A$
2	0	$D_w + \frac{7}{8}F_w + \frac{1}{8}F_e$	$D_e - \frac{3}{8}F_e$	$\frac{1}{4}F_w$	$-\frac{1}{4}F_w\phi_A$
5	$-\frac{1}{8}F_w$	$D_w + \frac{1}{3}D_B + \frac{6}{8}F_w$	0	$-\left(\frac{8}{3}D_B - F_B\right)$	$\left(\frac{8}{3}D_B - F_B\right)\phi_B$

Substitution of numerical values gives the coefficients summarised in Table 5.10.

Table 5.10

Node	a_W	a_E	a_{WW}	S_u	S_p	a_P
1	0	0.592	0	$1.583\phi_A$	−1.583	2.175
2	0.7	0.425	0	$-0.05\phi_A$	0.05	1.075
3	0.675	0.425	−0.025	0	0	1.075
4	0.675	0.425	−0.025	0	0	1.075
5	0.817	0	−0.025	$1.133\phi_B$	−1.133	1.925

The matrix form of the equation set is

$$\begin{bmatrix} 2.175 & -0.592 & 0 & 0 & 0 \\ -0.7 & 1.075 & -0.425 & 0 & 0 \\ 0.025 & -0.675 & 1.075 & -0.425 & 0 \\ 0 & 0.025 & -0.675 & 1.075 & -0.425 \\ 0 & 0 & 0.025 & -0.817 & 1.925 \end{bmatrix} \begin{bmatrix} \phi_1 \\ \phi_2 \\ \phi_3 \\ \phi_4 \\ \phi_5 \end{bmatrix} = \begin{bmatrix} 1.583 \\ -0.05 \\ 0 \\ 0 \\ 0 \end{bmatrix} \quad (5.60)$$

The solution to the above system is

$$\begin{bmatrix} \phi_1 \\ \phi_2 \\ \phi_3 \\ \phi_4 \\ \phi_5 \end{bmatrix} = \begin{bmatrix} 0.9648 \\ 0.8707 \\ 0.7309 \\ 0.5226 \\ 0.2123 \end{bmatrix} \tag{5.61}$$

Comparison with the analytical solution

Figure 5.19 shows that the QUICK solution is almost indistinguishable from the exact solution. Table 5.11 confirms that the errors are very small even with this coarse mesh. Following the steps outlined in Example 5.1 the central differencing solution is computed with the data given above. The sum of absolute errors in Table 5.11 indicates that the QUICK scheme gives a more accurate solution than the central differencing scheme.

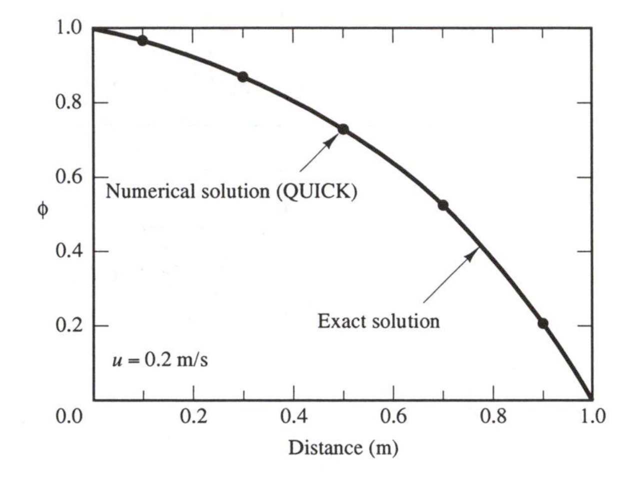

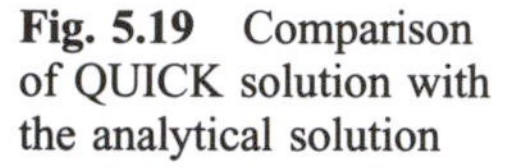
Fig. 5.19 Comparison of QUICK solution with the analytical solution

Table 5.11

Node	*Distance*	*Analytical solution*	*QUICK solution*	*Difference*	*CD solution*	*Difference*
1	0.1	0.9653	0.9648	0.0005	0.9696	0.0043
2	0.3	0.8713	0.8707	0.0006	0.8786	0.0073
3	0.5	0.7310	0.7309	0.0001	0.7421	0.0111
4	0.7	0.5218	0.5226	−0.0008	0.5374	0.0156
5	0.9	0.2096	0.2123	−0.0027	0.2303	0.0207
$\sum$ Absolute error				0.0047		0.059

5.9.2 Assessment of the QUICK scheme

The scheme uses consistent quadratic profiles – the cell face values of fluxes are always calculated by quadratic interpolation between two bracketing nodes and an upstream node – and is therefore conservative. Since the scheme is based on a quadratic function its accuracy in terms of the Taylor series truncation error is third order on a uniform mesh. The transportiveness property is built into the scheme as the quadratic function is based on two upstream and one downstream nodal values. If the flow field satisfies continuity the coefficient a_p equals the sum of all neighbour coefficients which is desirable for boundedness.

On the downside, the main coefficients (E and W) are not guaranteed to be positive and the coefficients a_{WW} and a_{EE} are negative. For example, if $u_w > 0$ and $u_e > 0$ the east coefficient becomes negative at relatively modest cell Peclet numbers ($Pe_e = F_e/D_e > 8/3$). This gives rise to stability problems and unbounded solutions under certain flow conditions. Similarly the west coefficient can become negative when the flow is in the negative direction. The QUICK scheme is therefore conditionally stable.

Another notable feature is the fact that the discretised equations involve not only immediate-neighbour nodes but also nodes further away. Tri-diagonal matrix solution methods (see Chapter 7) are not directly applicable.

5.9.3 Stability problems of the QUICK scheme and remedies

Since the QUICK scheme in the form presented above can be unstable due to the appearance of negative main coefficients it has been re-formulated in different ways that alleviate stability problems. These formulations all involve placing troublesome negative coefficients in the source term so as to retain positive main coefficients. The contributing part is appropriately weighted to give better stability and positive coefficients as far as possible. Some of the better known practical approaches are described in Han *et al* (1981), Pollard and Siu (1982) and Hayase *et al* (1992). The last authors generalised the approach for re-arranging QUICK schemes and derived a stable and fast converging variant.

The **Hayase *et al* (1990) QUICK scheme** can be summarised as follows:

$$\begin{aligned}
\phi_w &= \phi_W + \frac{1}{8}[3\phi_P - 2\phi_W - \phi_{WW}] && \text{for} \quad F_w > 0 \\
\phi_e &= \phi_P + \frac{1}{8}[3\phi_E - 2\phi_P - \phi_W] && \text{for} \quad F_e > 0 \\
\phi_w &= \phi_P + \frac{1}{8}[3\phi_W - 2\phi_P - \phi_E] && \text{for} \quad F_w < 0 \\
\phi_e &= \phi_E + \frac{1}{8}[3\phi_P - 2\phi_E - \phi_{EE}] && \text{for} \quad F_e < 0
\end{aligned} \tag{5.62}$$

The discretisation equation takes the form

$$\boxed{a_P\phi_P = a_W\phi_W + a_E\phi_E + \bar{S}} \tag{5.63}$$

The central coefficient is

$$\boxed{a_P = a_W + a_E + (F_e - F_w)}$$

and

a_W	a_E	$\bar{S}$
$D_w + \alpha_w F_w$	$D_e - (1-\alpha_e)F_e$	$\frac{1}{8}(3\phi_P - 2\phi_W - \phi_{WW})\alpha_w F_w + \frac{1}{8}(\phi_W + 2\phi_P - 3\phi_E)\alpha_e F_e$ $+\frac{1}{8}(3\phi_W - 2\phi_P - \phi_E)(1-\alpha_w)F_w + \frac{1}{8}(2\phi_E + \phi_{EE} - 3\phi_P)(1-\alpha_e)F_e$

where

$$\alpha_w = 1 \text{ for } F_w > 0 \text{ and } \alpha_e = 1 \text{ for } F_e > 0$$

$$\alpha_w = 0 \text{ for } F_w < 0 \text{ and } \alpha_e = 0 \text{ for } F_e < 0$$

The advantage of this approach is that the coefficients are always positive and now satisfy the requirements for conservativeness, boundedness and transportiveness. It should be noted that all variations of QUICK, including the one developed by Hayase *et al.*, give the same solution upon convergence.

5.9.4 General comments on the QUICK differencing scheme

The QUICK differencing scheme has greater formal accuracy than the central differencing or hybrid schemes and it retains the upwind weighted characteristics. The resultant false diffusion is small and solutions achieved with coarse grids are often considerably more accurate than those of the upwind or hybrid schemes. Figure 5.20 shows a comparison between upwind and QUICK for the two-dimensional test case considered in section 5.6.1. It can be seen that the QUICK scheme matches the exact solution much more accurately than the upwind scheme on a 50 × 50 grid.

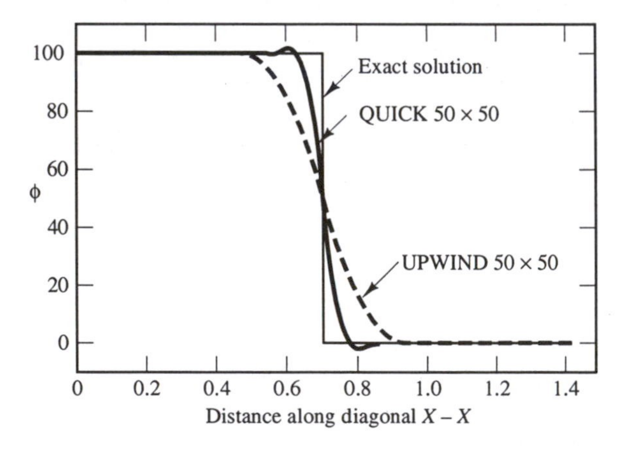

Fig. 5.20 Comparison of QUICK and upwind solutions for the 2D test case considered in section 5.6.1

The QUICK scheme can, however, give (minor) undershoots and overshoots as is evident in Figure 5.20. In complex flow calculations, the use of QUICK can lead to subtle problems caused by such unbounded results: for example, they could give rise to negative turbulence kinetic energy (k) in k–ε model (see Chapter 3) computations. The possibility of undershoots and overshoots needs to be considered when interpreting solutions.

5.10 Other higher order schemes

Schemes of order 3 and above have been developed for the discretisation of convective terms with varying degrees of success. Implementation of boundary conditions can be problematic with such higher order schemes. Computational cost is another factor which needs to be considered in using higher order schemes. The fact that the QUICK scheme can give 'undershoots' and 'overshoots' has led to the development of second-order schemes that avoid these problems. The class of TVD (Total Variation Diminishing) schemes are specially formulated to achieve oscillation-free solutions and have proved to be useful in CFD calculations. A discussion of such schemes is beyond the scope of this book and the reader is referred to Hirsch (1990), Van Leer (1973, 1974, 1979), Boris and Book (1973, 1976), Osher (1984), Osher and Chakravarthy (1984), Zhu (1991) and Alvarez *et al* (1993) for further details.

5.11 Summary

The problems of discretising the convection–diffusion equation, under the assumption that the flow field is known, have been discussed. The crucial issue is the formulation of suitable expressions for the values of the transported property ϕ at cell faces when accounting for the convective contribution in the equation.

- All the finite volume schemes presented in this chapter describe the effects of simultaneous convection and diffusion by means of discretised equations whose coefficients are weighted combinations of the convective mass flux per unit area F and the diffusion conductance D.
- The discretised equations for a general internal node for the central, upwind and hybrid differencing and the power-law schemes of a one-dimensional convection–diffusion problem take the following form:

$$\boxed{a_P\phi_P = a_W\phi_W + a_E\phi_E} \tag{5.64}$$

with

$$\boxed{a_P = a_W + a_E + (F_e - F_w)}$$

- The neighbour coefficients for these schemes are

Scheme	a_W	a_E
Central differencing	$D_w + F_w/2$	$D_e - F_e/2$
Upwind differencing	$D_w + \max(F_w,\ 0)$	$D_e + \max(0,\ -F_e)$
Hybrid differencing	$\max[F_w, (D_w + F_w/2),\ 0]$	$\max[-F_e, (D_e - F_e/2),\ 0]$
Power law	$D_w\ \max\left[0, (1 - 0.1\lvert Pe_w\rvert)^5\right]$ $+\max(F_w,\ 0)$	$D_e\ \max\left[0, (1 - 0.1\lvert Pe_e\rvert)^5\right]$ $+\max(-F_e,\ 0)$

- The boundary conditions enter the discretised equations via source terms. Their treatment is specific to each discretisation scheme.
- Discretisation schemes that possess conservativeness, boundedness and transportiveness give physically realistic results and stable iterative solutions:
  - The central differencing method is not suitable for general purpose convection–diffusion problems because it lacks transportiveness and gives unrealistic solutions at large values of the cell Peclet number.
  - Upwind, hybrid and power-law differencing all possess conservativeness, boundedness and transportiveness and are highly stable, but suffer from false diffusion in multi-dimensional flows if the velocity vector is not parallel to one of the co-ordinate directions.
- The discretised equations of the standard QUICK method of Leonard (1979) have the following form for a general internal node point:

$$\boxed{a_P\phi_P = a_W\phi_W + a_E\phi_E + a_{WW}\phi_{WW} + a_{EE}\phi_{EE}} \tag{5.65}$$

where

$$\boxed{a_P = a_W + a_E + a_{WW} + a_{EE} + (F_e - F_w)}$$

The neighbour coefficients of the standard QUICK scheme are

	Standard QUICK
a_W	$D_w + \frac{6}{8}\alpha_w F_w + \frac{1}{8}\alpha_e F_e + \frac{3}{8}(1-\alpha_w)F_w$
a_{WW}	$-\frac{1}{8}\alpha_w F_w$
a_E	$D_e - \frac{3}{8}\alpha_e F_e - \frac{6}{8}(1-\alpha_e)F_e - \frac{1}{8}(1-\alpha_w)F_w$
a_{EE}	$\frac{1}{8}(1-\alpha_e)F_e$

with $\alpha_w = 1$ for $F_w > 0$ and $\alpha_e = 1$ for $F_e > 0$
$\alpha_w = 0$ for $F_w < 0$ and $\alpha_e = 0$ for $F_e < 0$

- Higher order schemes, such as QUICK, can minimise false diffusion errors but are less computationally stable. This manifests itself as small over- and undershoots in the solution of some problems including those with large gradients of ϕ leading to non-physical behaviour, e.g. negative turbulence properties k and ε, in extreme cases. Nevertheless, if used with care and judgement the QUICK scheme can give very accurate solutions of convection–diffusion problems.

6

Solution Algorithms for Pressure–Velocity Coupling in Steady Flows

6.1 Introduction

The convection of a scalar variable ϕ depends on the magnitude and direction of the local velocity field. To develop our methods in the previous chapter we assumed that the velocity field was somehow known. In general the velocity field is, however, not known and emerges as part of the overall solution process along with all other flow variables. In this chapter we look at the most popular strategies for computing the entire flow field.

Transport equations for each velocity component – momentum equations – can be derived from the general transport equation (2.39) by replacing the variable ϕ by u, v and w respectively. The velocity field must, of course, also satisfy the continuity equation. Let us consider the equations governing a two-dimensional laminar steady flow:

x-momentum equation

$$\frac{\partial}{\partial x}(\rho uu) + \frac{\partial}{\partial y}(\rho vu) = \frac{\partial}{\partial x}\left(\mu \frac{\partial u}{\partial x}\right) + \frac{\partial}{\partial y}\left(\mu \frac{\partial u}{\partial y}\right) - \frac{\partial p}{\partial x} + S_u \tag{6.1}$$

y-momentum equation

$$\frac{\partial}{\partial x}(\rho uv) + \frac{\partial}{\partial y}(\rho vv) = \frac{\partial}{\partial x}\left(\mu \frac{\partial v}{\partial x}\right) + \frac{\partial}{\partial y}\left(\mu \frac{\partial v}{\partial y}\right) - \frac{\partial p}{\partial y} + S_v \tag{6.2}$$

continuity equation

$$\frac{\partial}{\partial x}(\rho u) + \frac{\partial}{\partial y}(\rho v) = 0 \tag{6.3}$$

The pressure gradient term, which forms the main momentum source term in most flows of engineering importance, has been written separately to facilitate the discussion that follows.

The solution of equation set (6.1–6.3) presents us with two new problems:

- The convective terms of the momentum equation contain non-linear quantities, for example the first term of equation (6.1) is the x-derivative of ρu^2.
- All three equations are intricately coupled because every velocity component appears in each momentum equation and the continuity equation. The most complex issue to resolve is the role played by the pressure. It appears in both momentum equations, but there is evidently no (transport or other) equation for pressure.

If the pressure gradient is known, the process of obtaining discretised equations for velocities from the momentum equations is similar to that for any other scalar, and schemes based on those explained in Chapter 5 are applicable. In general purpose flow computations we also wish to calculate the pressure field as part of the solution so its gradient is not normally known beforehand. If the flow is compressible the continuity equation may be used as a transport equation for density and, in addition to (6.1–6.3), the energy equation is a transport equation for temperature. The pressure may then be obtained from the density and temperature by using the equation of state $p = p(\rho, T)$. However, if the flow is incompressible the density is constant and hence by definition not linked to the pressure. In this case coupling between pressure and velocity introduces a constraint on the solution of the flow field: if the correct pressure field is applied in the momentum equations the resulting velocity field should satisfy continuity.

Both the problems associated with the non-linearities in the equation set and the pressure–velocity linkage can be resolved by adopting an iterative solution strategy such as the SIMPLE algorithm of Patankar and Spalding (1972). In this algorithm the convective fluxes per unit mass F through cell faces are evaluated from so-called guessed velocity components. Furthermore, a guessed pressure field is used to solve the momentum equations and a pressure correction equation, deduced from the continuity equation, is solved to obtain a pressure correction field which is in turn used to update the velocity and pressure fields. To start the iteration process we use initial guesses for the velocity and pressure fields. As the algorithm proceeds our aim must be progressively to improve these guessed fields. The process is iterated until convergence of the velocity and pressure fields. The main features of the SIMPLE algorithm and its more recent enhancements will be discussed in this chapter.

6.2 The staggered grid

The solution procedure for the transport of a general property ϕ developed in Chapter 5 will, of course, be enlisted to solve the momentum equations. Matters are, however, not completely straightforward since there are problems associated with the pressure source terms of the momentum equations that need special treatment.

The finite volume method starts, as always, with the discretisation of the flow domain and of the relevant transport equations (6.1–6.3). First we need to decide where to store the velocities. It seems logical to define these at the same locations as the scalar variables such as pressure, temperature etc. However, if the velocities and pressures are both defined at the nodes of an ordinary control volume a highly non-uniform pressure field can act like a uniform field in the discretised momentum equations. This can be demonstrated with the simple two-dimensional situation shown in Figure 6.1, where a uniform grid is used for simplicity. Let us assume that

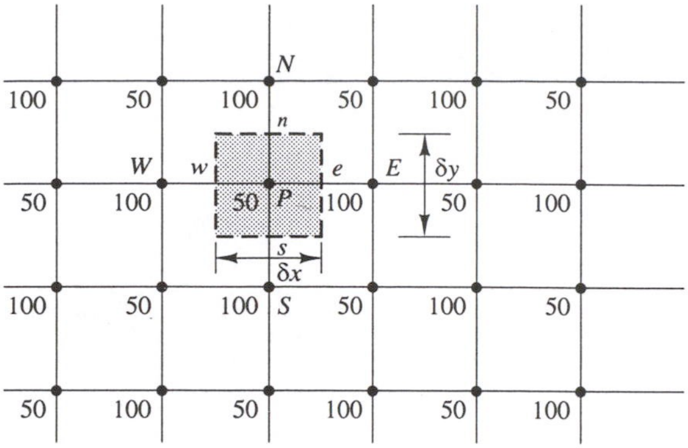

Fig. 6.1 A 'checker-board' pressure field

we have somehow obtained a highly irregular 'checker-board' pressure field with values as shown in Figure 6.1.

If the pressures at 'e' and 'w' are obtained by linear interpolation the pressure gradient term $\partial p/\partial x$ in the u-momentum equation is given by

$$\frac{\partial p}{\partial x} = \frac{p_e - p_w}{\delta x} = \frac{\left(\frac{p_E + p_P}{2}\right) - \left(\frac{p_P + p_W}{2}\right)}{\delta x}$$
$$= \frac{p_E - p_W}{2\delta x} \tag{6.4}$$

Similarly, the pressure gradient $\partial p/\partial y$ for the v-momentum equation is evaluated as

$$\frac{\partial p}{\partial y} = \frac{p_N - p_S}{2\delta y} \tag{6.5}$$

The pressure at the central node (P) does not appear in (6.4) and (6.5). Substituting the appropriate values from the 'checker-board' pressure field in Figure 6.1 into formulae (6.4–6.5) we find that all the discretised gradients are zero at all the nodal points even though the pressure field exhibits spatial oscillations in both directions. As a result, this pressure field would give the same (zero) momentum source in the discretised equations as a uniform pressure field. This behaviour is obviously non-physical.

It is clear that, if the velocities are defined at the scalar grid nodes, the influence of pressure is not properly represented in the discretised momentum equations. A remedy for this problem is to use a **staggered grid** for the velocity components (Harlow and Welch, 1965). The idea is to evaluate scalar variables, such as pressure, density, temperature etc., at ordinary nodal points but to calculate velocity components on staggered grids centred around the cell faces. The arrangement for a two-dimensional flow calculation is shown in Figure 6.2.

The scalar variables, including pressure, are stored at the nodes marked (•). The velocities are defined at the (scalar) cell faces in between the nodes and are indicated by arrows. Horizontal (→) arrows indicate the locations for u-velocities and vertical (↑) ones denote those for v-velocities. In addition to the E, W, N, S notation Figure 6.2 also introduces a new system of notation based on a numbering of grid lines and cell faces. It will be explained and used later on in this chapter.

For the moment we continue to use the original E, W, N, S notation; the u-velocities are stored at scalar cell faces 'e' and 'w' and the v-velocities at faces 'n' and 's'. In a three-dimensional flow the w-component is evaluated at cell faces 't'

Fig. 6.2

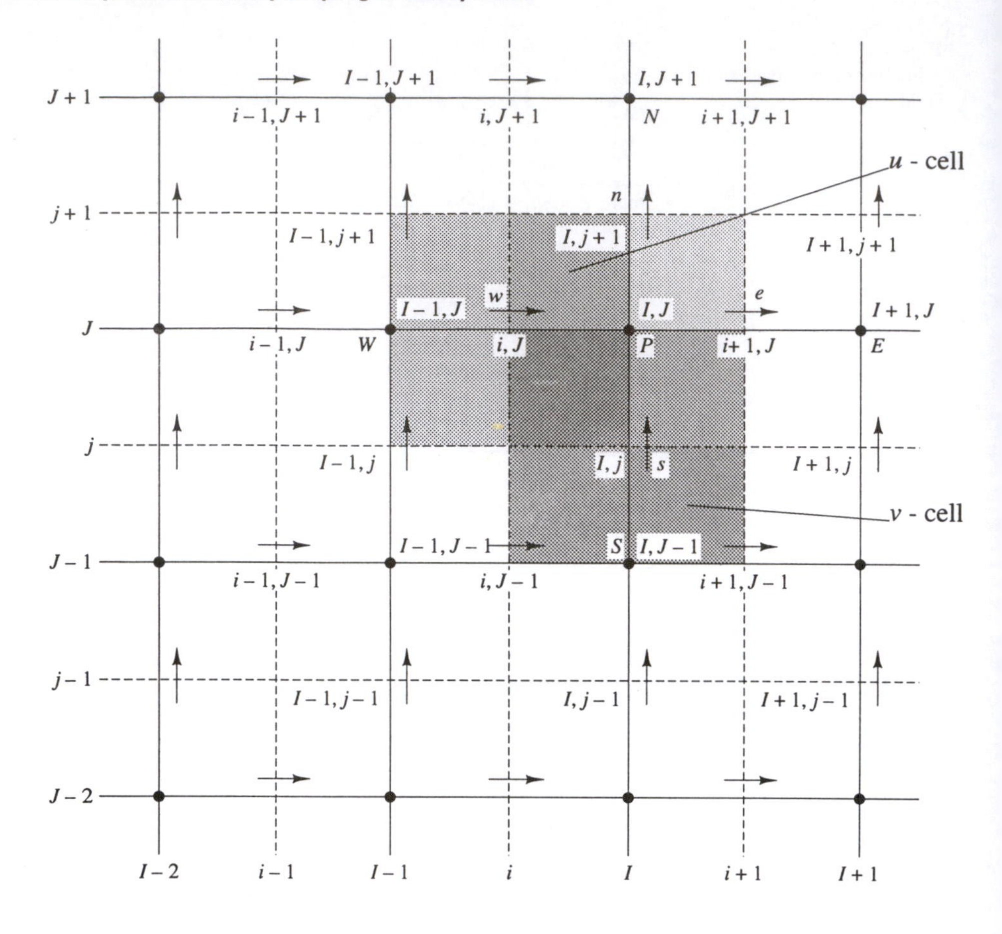

and 'b'. We observe that the control volumes for u and v are different from the scalar control volumes and different from each other. The scalar control volumes are sometimes referred to as the pressure control volumes because, as we shall see later, the discretised continuity equation is turned into a pressure correction equation, which is evaluated on scalar control volumes.

In the staggered grid arrangement, the pressure nodes coincide with the cell faces of the u-control volume. The pressure gradient term $\partial p/\partial x$ is given by

$$\frac{\partial p}{\partial x} = \frac{p_P - p_W}{\delta x_u} \tag{6.6}$$

where δx_u is the width of the u-control volume. Similarly $\partial p/\partial y$ for the v-control volume shown is given by

$$\frac{\partial p}{\partial y} = \frac{p_P - p_S}{\delta y_v} \tag{6.7}$$

where δy_v is width of the v-control volume.

If we consider the 'checker-board' pressure field again, substitution of the appropriate nodal pressure values into equations (6.6) and (6.7) now yields very significant non-zero pressure gradient terms. The staggering of the velocity avoids the unrealistic behaviour of the discretised momentum equation for spatially oscillating pressures like the 'checker-board' field. A further advantage of the staggered grid arrangement is that it generates velocities at exactly the locations where they are required for the scalar transport – convection–diffusion – computations. Hence, no interpolation is needed to calculate velocities at the scalar cell faces.

6.3 The momentum equations

As mentioned earlier, if the pressure field is known, the discretisation of velocity equations and the subsequent solution procedure is similar to that of a scalar equation. Since the velocity grid is staggered the new notation based on grid line and cell face numbering will be used. In Figure 6.2 the unbroken grid lines are numbered by means of capital letters. In the x-direction the numbering is $\ldots, I-1, I, I+1$, ... etc. and in the y-direction $\ldots, J-1, J, J+1, \ldots$ etc. The dashed lines that construct the scalar cell faces are denoted by lower case letters $\ldots, i-1, i, i+1, \ldots$ and $\ldots, j-1, j, j+1, \ldots$ in the x- and y-direction respectively.

A subscript system based on this numbering allows us to define the locations of grid nodes and cell faces with precision. Scalar nodes, located at the intersection of two grid lines, are identified by two capital letters: e.g. point P in Figure 6.2 is denoted by (I, J). The u-velocities are stored at the e- and w-cell faces of a scalar control volume. These are located at the intersection of a line defining a cell boundary and a grid line and are, therefore, defined by a combination of a lower case letter and a capital: e.g. the w-face of the cell around point P is identified by (i, J). For the same reasons the storage locations for the v-velocities are combinations of a capital and a lower case letter: e.g. the s-face is given by (I, j).

We may use forward or backward staggered velocity grids. The uniform grids in Figure 6.2 are backward staggered since the i-location for the u-velocity $u_{i,J}$ is at a distance of $-1/2\delta x_u$ from the scalar node (I, J). Likewise, the j-location for the v-velocity $v_{I,j}$ is $-1/2\delta y_v$ from node (I, J). Expressed in the new co-ordinate system the discretised u-momentum equation for the velocity at location (i, J) is given by

$$a_{i,J}u_{i,J} = \sum a_{nb}u_{nb} - \frac{p_{I,J} - p_{I-1,J}}{\delta x_u}\Delta V_u + \bar{S}\,\Delta V_u$$

or

$$a_{i,J}u_{i,J} = \sum a_{nb}u_{nb} + (p_{I-1,J} - p_{I,J})A_{i,J} + b_{i,J} \tag{6.8}$$

where ΔV_u is the volume of the u-cell, $b_{i,j} = \bar{S}\Delta V_u$ is the momentum source term, $A_{i,J}$ is the (east or west) cell face area of the u-control volume. The pressure gradient source term in (6.8) has been discretised by means of a linear interpolation between the pressure nodes located at the u-control volume boundaries.

In the new numbering system the E, W, N and S neighbours involved in the summation $\sum a_{nb}u_{nb}$ are $(i-1, J)$, $(i+1, J)$, $(i, J+1)$ and $(i, J-1)$. Their locations and the prevailing velocities are shown in more detail in Figure 6.3. The values of coefficients $a_{i,J}$ and a_{nb} may be calculated with any of the differencing methods (upwind, hybrid, QUICK) suitable for convection–diffusion problems. The coefficients contain combinations of the convective flux per unit mass F and the diffusive conductance D at u-control volume cell faces. Applying the new notation

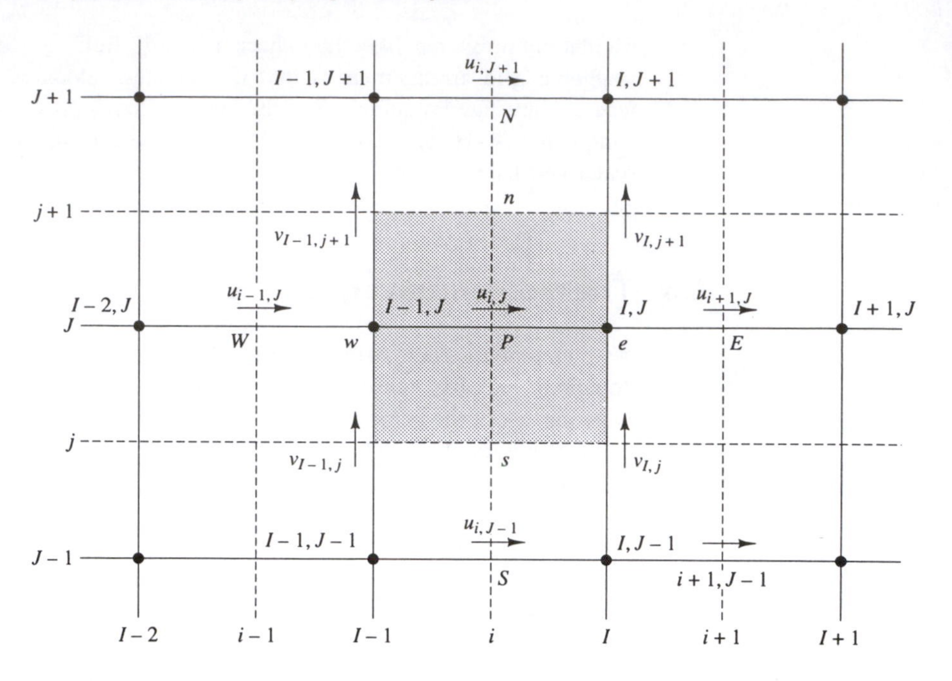

Fig. 6.3 A u-control volume and its neighbouring velocity components

system we give the values of F and D for each of the faces e, w, n and s of the u-control volume:

$$F_w = (\rho u)_w = \frac{F_{i,J} + F_{i-1,J}}{2}$$
$$= \frac{1}{2}\left[\left(\frac{\rho_{I,J} + \rho_{I-1,J}}{2}\right)u_{i,J} + \left(\frac{\rho_{I-1,J} + \rho_{I-2,J}}{2}\right)u_{i-1,J}\right] \quad (6.9a)$$

$$F_e = (\rho u)_e = \frac{F_{i+1,J} + F_{i,J}}{2}$$
$$= \frac{1}{2}\left[\left(\frac{\rho_{I+1,J} + \rho_{I,J}}{2}\right)u_{i+1,J} + \left(\frac{\rho_{I,J} + \rho_{I-1,J}}{2}\right)u_{i,J}\right] \quad (6.9b)$$

$$F_s = (\rho v)_s = \frac{F_{I,j} + F_{I-1,j}}{2}$$
$$= \frac{1}{2}\left[\left(\frac{\rho_{I,J} + \rho_{I,J-1}}{2}\right)v_{I,j} + \left(\frac{\rho_{I-1,J} + \rho_{I-1,J-1}}{2}\right)v_{I-1,j}\right] \quad (6.9c)$$

$$F_n = (\rho v)_n = \frac{F_{I,j+1} + F_{I-1,j+1}}{2}$$
$$= \frac{1}{2}\left[\left(\frac{\rho_{I,J+1} + \rho_{I,J}}{2}\right)v_{I,j+1} + \left(\frac{\rho_{I-1,J+1} + \rho_{I-1,J}}{2}\right)v_{I-1,j+1}\right] \quad (6.9d)$$

$$D_w = \frac{\Gamma_{I-1,J}}{x_i - x_{i-1}} \quad (6.9e)$$

$$D_e = \frac{\Gamma_{I,J}}{x_{i+1} - x_i} \quad (6.9f)$$

$$D_s = \frac{\Gamma_{I-1,J} + \Gamma_{I,J} + \Gamma_{I-1,J-1} + \Gamma_{I,J-1}}{4(y_J - y_{J-1})} \quad (6.9g)$$

$$D_n = \frac{\Gamma_{I-1,J+1} + \Gamma_{I,J+1} + \Gamma_{I-1,J} + \Gamma_{I,J}}{4(y_{J+1} - y_J)} \quad (6.9h)$$

The formulae (6.9) show that where scalar variables or velocity components are not available at a u-control volume cell face a suitable two- or four-point average is formed over the nearest points where values are available. During each iteration the u- and v-velocity components used to evaluate the above expressions are those obtained as the outcome of the previous iteration (or the initial guess in the first iteration). It should be noted that these **known** u- and v-values contribute to the coefficients a in equation (6.8). These are distinct from $u_{i,J}$ and u_{nb} in this equation which denote the **unknown** scalars.

By analogy the v-momentum equation becomes

$$a_{I,j}v_{I,j} = \sum a_{nb}v_{nb} + (p_{I,J-1} - p_{I,J})A_{I,j} + b_{I,j} \tag{6.10}$$

The neighbours involved in the summation $\sum a_{nb}v_{nb}$ and prevailing velocities are as shown in Figure 6.4.

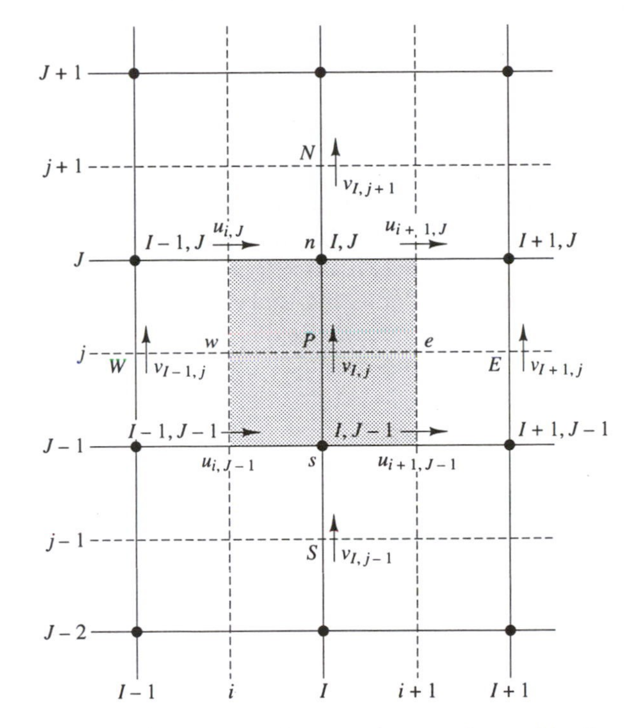

Fig. 6.4 A v-control volume and its neighbouring velocity components

Coefficients $a_{I,j}$ and a_{nb} again contain combinations of the convective flux per unit mass F and the diffusive conductance D at v-control volume cell faces. Their values are obtained by the same averaging procedure adopted for the u-control volume and are given below:

$$F_w = (\rho u)_w = \frac{F_{i,J} + F_{i,J-1}}{2}$$
$$= \frac{1}{2}\left[\left(\frac{\rho_{I,J} + \rho_{I-1,J}}{2}\right)u_{i,J} + \left(\frac{\rho_{I-1,J-1} + \rho_{I,J-1}}{2}\right)u_{i,J-1}\right] \tag{6.11a}$$

$$F_e = (\rho u)_e = \frac{F_{i+1,J} + F_{i+1,J-1}}{2}$$
$$= \frac{1}{2}\left[\left(\frac{\rho_{I+1,J} + \rho_{I,J}}{2}\right)u_{i+1,J} + \left(\frac{\rho_{I,J-1} + \rho_{I+1,J-1}}{2}\right)u_{i+1,J-1}\right] \tag{6.11b}$$

$$F_s = (\rho v)_s = \frac{F_{I,j-1} + F_{I,j}}{2}$$
$$= \frac{1}{2}\left[\left(\frac{\rho_{I,J-1} + \rho_{I,J-2}}{2}\right)v_{I,j-1} + \left(\frac{\rho_{I,J} + \rho_{I,J-1}}{2}\right)v_{I,J}\right] \quad (6.11c)$$

$$F_n = (\rho v)_n = \frac{F_{I,j} + F_{I,j+1}}{2}$$
$$= \frac{1}{2}\left[\left(\frac{\rho_{I,J} + \rho_{I,J-1}}{2}\right)v_{I,j} + \left(\frac{\rho_{I,J+1} + \rho_{I,J}}{2}\right)v_{I,j+1}\right] \quad (6.11d)$$

$$D_w = \frac{\Gamma_{I-1,J-1} + \Gamma_{I,J-1} + \Gamma_{I-1,J} + \Gamma_{I,J}}{4(x_I - x_{I-1})} \quad (6.11e)$$

$$D_e = \frac{\Gamma_{I,J-1} + \Gamma_{I+1,J-1} + \Gamma_{I,J} + \Gamma_{I+1,J}}{4(x_{I+1} - x_I)} \quad (6.11f)$$

$$D_s = \frac{\Gamma_{I,J-1}}{y_j - y_{j-1}} \quad (6.11g)$$

$$D_n = \frac{\Gamma_{I,J}}{y_{j+1} - y_j} \quad (6.11h)$$

Again at each iteration level the values of F are computed using the u- and v-velocity components resulting from the previous iteration.

Given a pressure field p, discretised momentum equations of the form (6.8) and (6.10) can be written for each u- and v-control volume and then solved to obtain the velocity fields. If the pressure field is correct the resulting velocity field will satisfy continuity. As the pressure field is unknown, we need a method for calculating pressure.

6.4 The SIMPLE algorithm

The acronym SIMPLE stands for Semi-Implicit Method for Pressure-Linked Equations. The algorithm was originally put forward by Patankar and Spalding (1972) and is essentially a guess-and-correct procedure for the calculation of pressure on the staggered grid arrangement introduced above. The method is illustrated by considering the two-dimensional laminar steady flow equations in Cartesian co-ordinates.

To initiate the SIMPLE calculation process a pressure field p^* is guessed. **Discretised momentum equations** (6.8) and (6.10) **are solved** using the guessed pressure field to yield velocity components u^* and v^* as follows:

$$a_{i,J} u^*_{i,J} = \sum a_{nb} u^*_{nb} + (p^*_{I-1,J} - p^*_{I,J})A_{i,J} + b_{i,J} \quad (6.12)$$

$$a_{I,j} v^*_{I,j} = \sum a_{nb} v^*_{nb} + (p^*_{I,J-1} - p^*_{I,J})A_{I,j} + b_{I,j} \quad (6.13)$$

Now we define the correction p' as the difference between the correct pressure field

p and the guessed pressure field p^*, so that

$$p = p^* + p' \tag{6.14}$$

Similarly we define velocity corrections u' and v' to relate the correct velocities u and v to the guessed velocities u^* and v^*

$$u = u^* + u' \tag{6.15}$$

$$v = v^* + v' \tag{6.16}$$

Substitution of the correct pressure field p into the momentum equations yields the correct velocity field (u, v). Discretised equations (6.8) and (6.10) link the correct velocity fields with the correct pressure field.

Subtraction of equations (6.12) and (6.13) from (6.8) and (6.10), respectively, gives

$$a_{i,J}(u_{i,J} - u^*_{i,J}) = \sum a_{nb}\left(u_{nb} - u^*_{nb}\right) + \left[\left(p_{I-1,J} - p^*_{I-1,J}\right) - \left(p_{I,J} - p^*_{I,J}\right)\right]A_{i,J} \tag{6.17}$$

$$a_{I,j}\left(v_{I,j} - v^*_{I,j}\right) = \sum a_{nb}\left(v_{nb} - v^*_{nb}\right) + \left[\left(p_{I,J-1} - p^*_{I,J-1}\right) - \left(p_{I,J} - p^*_{I,J}\right)\right]A_{I,j} \tag{6.18}$$

Using correction formulae (6.14–6.16) the equations (6.17–6.18) may be rewritten as follows:

$$a_{i,J}u'_{i,J} = \sum a_{nb}u'_{nb} + \left(p'_{I-1,J} - p'_{I,J}\right)A_{i,J} \tag{6.19}$$

$$a_{I,j}v'_{I,j} = \sum a_{nb}v'_{nb} + \left(p'_{I,J-1} - p'_{I,J}\right)A_{I,j} \tag{6.20}$$

At this point an approximation is introduced: $\sum a_{nb}u'_{nb}$ and $\sum a_{nb}v'_{nb}$ are dropped to simplify equations (6.19) and (6.20) for the velocity corrections. **Omission of these terms is the main approximation** of the SIMPLE algorithm. We obtain

$$u'_{i,J} = d_{i,J}\left(p'_{I-1,J} - p'_{I,J}\right) \tag{6.21}$$

$$v'_{I,j} = d_{I,j}\left(p'_{I,J-1} - p'_{I,J}\right) \tag{6.22}$$

$$\text{where} \quad d_{i,J} = \frac{A_{i,J}}{a_{i,J}} \text{ and } d_{I,j} = \frac{A_{I,j}}{a_{I,j}} \tag{6.23}$$

Equations (6.21) and (6.22) describe the corrections to be applied to velocities through formulae (6.15) and (6.16), which gives

$$u_{i,J} = u^*_{i,J} + d_{i,J}\left(p'_{I-1,J} - p'_{I,J}\right) \tag{6.24}$$

$$v_{I,j} = v^*_{I,j} + d_{I,j}\left(p'_{I,J-1} - p'_{I,J}\right) \tag{6.25}$$

Similar expressions exist for $u_{i+1,J}$ and $v_{I,j+1}$:

$$u_{i+1,J} = u^*_{i+1,J} + d_{i+1,J}\left(p'_{I,J} - p'_{I+1,J}\right) \tag{6.26}$$

$$v_{I,j+1} = v^*_{I,j+1} + d_{I,j+1}\left(p'_{I,J} - p'_{I,J+1}\right) \tag{6.27}$$

$$\text{where } d_{i+1,J} = \frac{A_{i+1,J}}{a_{i+1,J}} \text{ and } d_{I,j+1} = \frac{A_{I,j+1}}{a_{I,j+1}} \tag{6.28}$$

Thus far we have only considered the momentum equations but, as mentioned earlier, the velocity field is also subject to the constraint that it should satisfy continuity equation (6.3). Continuity is satisfied in discretised form for the scalar control volume shown in Figure 6.5:

$$\left[(\rho uA)_{i+1,J} - (\rho uA)_{i,J}\right] + \left[(\rho vA)_{I,j+1} - (\rho vA)_{I,j}\right] = 0 \tag{6.29}$$

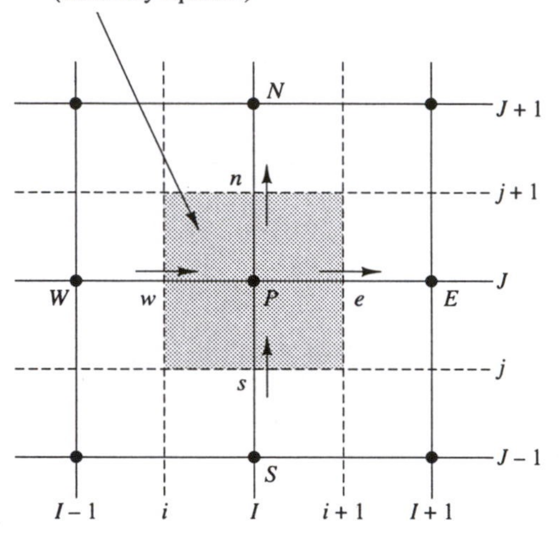

Fig. 6.5 The scalar control volume used for the discretisation of the continuity equation

Substitution of the corrected velocities of equations (6.24–6.27) into discretised continuity equation (6.29) gives

$$\begin{aligned}&\Big[\rho_{i+1,J}A_{i+1,J}\Big(u^*_{i+1,J} + d_{i+1,J}\big(p'_{I,J} - p'_{I+1,J}\big)\Big)\\ &\quad -\rho_{i,J}A_{i,J}\Big(u^*_{i,J} + d_{i,J}\big(p'_{I-1,J} - p'_{I,J}\big)\Big)\Big]\\ &\quad +\Big[\rho_{I,j+1}A_{I,j+1}\Big(v^*_{I,j+1} + d_{I,j+1}\big(p'_{I,J} - p'_{I,J+1}\big)\Big)\\ &\quad -\rho_{I,j}A^*_{I,j}\Big(v^*_{I,j} + d_{I,j}\big(p'_{I,J-1} - p'_{I,J}\big)\Big)\Big] = 0\end{aligned} \tag{6.30}$$

This may be re-arranged to give

$$\begin{aligned}&\left[(\rho dA)_{i+1,J} + (\rho dA)_{i,J} + (\rho dA)_{I,j+1} + (\rho dA)_{I,j}\right]p'_{I,J}\\ &\quad = (\rho dA)_{i+1,J}p'_{I+1,J} + (\rho dA)_{i,J}p'_{I-1,J} + (\rho dA)_{I,j+1}p'_{I,J+1}\\ &\qquad + (\rho dA)_{I,j}p'_{I,J-1}\\ &\qquad + \left[(\rho u^*A)_{i,J} - (\rho u^*A)_{i+1,J} + (\rho v^*A)_{I,j} - (\rho v^*A)_{I,j+1}\right]\end{aligned} \tag{6.31}$$

Identifying the coefficients of p' this may be written as

$$a_{I,J}p'_{I,J} = a_{I+1,J}p'_{I+1,J} + a_{I-1,J}p'_{I-1,J} + a_{I,J+1}p'_{I,J+1} + a_{I,J-1}p'_{I,J-1} + b'_{I,J} \tag{6.32}$$

where $a_{I,J} = a_{I+1,J} + a_{I-1,J} + a_{I,J+1} + a_{I,J-1}$ and the coefficients are given below:

$a_{I+1,J}$	$a_{I-1,J}$	$a_{I,J+1}$	$a_{i,J-1}$	$b'_{I,J}$
$(\rho dA)_{i+1,J}$	$(\rho dA)_{i,J}$	$(\rho dA)_{I,j+1}$	$(\rho dA)_{I,j}$	$(\rho u^* A)_{i,J} - (\rho u^* A)_{i+1,J} + (\rho v^* A)_{I,j} - (\rho v^* A)_{I,j+1}$

Equation (6.32) represents the discretised continuity equation as an **equation for pressure correction p'**. The source term b' in the equation is the continuity imbalance arising from the incorrect velocity field u^*, v^*. By solving equation (6.32), the pressure correction field p' can be obtained at all points. Once the pressure correction field is known, the correct pressure field may be obtained using formula (6.14) and velocity components through correction formulae (6.24–6.27). The omission of terms such as $\sum a_{nb}u'_{nb}$ in the derivation does not affect the final solution because the pressure correction and velocity corrections will all be zero in a converged solution giving $p^* = p$, $u^* = u$ and $v^* = v$.

The pressure correction equation is susceptible to divergence unless some **under-relaxation** is used during the iterative process and new, improved, pressures p^{new} are obtained with

$$p^{new} = p^* + \alpha_p p' \tag{6.33}$$

where α_p is the pressure under-relaxation factor. If we select α_p equal to 1 the guessed pressure field p^* is corrected by p'. However, the corrections p', in particular when the guessed field p^* is far away from the final solution, is often too large for stable computations. A value of α_p equal to zero would apply no correction at all, which is also undesirable. Taking α_p between 0 and 1 allows us to add to guessed field p^* a fraction of the correction field p' that is large enough to move the iterative improvement process forward, but small enough to ensure stable computation.

The velocities are also under-relaxed. The iteratively improved velocity components u^{new} and v^{new} are obtained from

$$u^{new} = \alpha_u u + (1 - \alpha_u)u^{(n-1)} \tag{6.34}$$

$$v^{new} = \alpha_v v + (1 - \alpha_v)v^{(n-1)} \tag{6.35}$$

where α_u and α_v are the u- and v-velocity under-relaxation factors with values between 0 and 1, u and v are the corrected velocity components without relaxation and $u^{(n-1)}$ and $v^{(n-1)}$ represent their values obtained in the previous iteration. After some algebra it can be shown that with under-relaxation the discretised u-momentum equation takes the form

$$\frac{a_{i,J}}{\alpha_u}u_{i,J} = \sum a_{nb}u_{nb} + (p_{I-1,J} - p_{I,J})A_{i,J} + b_{i,J} + \left[(1 - \alpha_u)\frac{a_{i,J}}{\alpha_u}\right]u_{i,J}^{(n-1)} \tag{6.36}$$

and the discretised v-momentum equation

$$\frac{a_{I,j}}{\alpha_v} v_{I,j} = \sum a_{nb} v_{nb} + (p_{I,J-1} - p_{I,J}) A_{I,j} + b_{I,j} + \left[(1 - \alpha_v) \frac{a_{I,j}}{\alpha_v} \right] v_{I,j}^{(n-1)} \tag{6.37}$$

The pressure correction equation is also affected by velocity under-relaxation and it can be shown that d-terms of the pressure correction equation become

$$d_{i,J} = \frac{A_{i,J}\alpha_u}{a_{i,J}}, \quad d_{i+1,J} = \frac{A_{i+1,J}\alpha_u}{a_{i+1,J}}, \quad d_{I,j} = \frac{A_{I,j}\alpha_v}{a_{I,j}}$$

and

$$d_{I,j+1} = \frac{A_{I,j+1}\alpha_v}{a_{I,j+1}}$$

Note that in these formulae $a_{i,J}, a_{i+1,J}, a_{I,j}$ and $a_{I,j+1}$ are the central coefficients of discretised velocity equations at positions (i,J), $(i+1,J)$, (I,j) and $(I,j+1)$ of a scalar cell centred around P.

A correct choice of under-relaxation factors α is essential for cost-effective simulations. Too large a value of α may lead to oscillatory or even divergent iterative solutions and a value which is too small will cause extremely slow convergence. Unfortunately, the optimum values of under-relaxation factors are flow dependent and must be sought on a case-by-case basis. The use of under-relaxation will be discussed further in Chapter 8.

6.5 Assembly of a complete method

The SIMPLE algorithm gives a method of calculating pressure and velocities. The method is iterative and when other scalars are coupled to the momentum equations, the calculation needs to be done sequentially. The sequence of operations in a CFD procedure which employs the SIMPLE algorithm is given in Figure 6.6.

6.6 The SIMPLER algorithm

The SIMPLER (SIMPLE Revised) algorithm of Patankar (1980) is an improved version of SIMPLE. In this algorithm the discretised continuity equation (6.29) is used to derive a **discretised equation for pressure**, instead of a pressure correction equation as in SIMPLE. Thus the intermediate pressure field is obtained directly without the use of a correction. Velocities are, however, still obtained through the velocity corrections (6.24–6.27) of SIMPLE.

The discretised momentum equations (6.12–6.13) are re-arranged as

$$u_{i,J} = \frac{\sum a_{nb} u_{nb} + b_{i,J}}{a_{i,J}} + \frac{A_{i,J}}{a_{i,J}} (p_{I-1,J} - p_{I,J}) \tag{6.38}$$

$$v_{I,j} = \frac{\sum a_{nb} v_{nb} + b_{I,j}}{a_{I,j}} + \frac{A_{I,j}}{a_{I,j}} (p_{I,J-1} - p_{I,J}) \tag{6.39}$$

Fig. 6.6 The SIMPLE algorithm

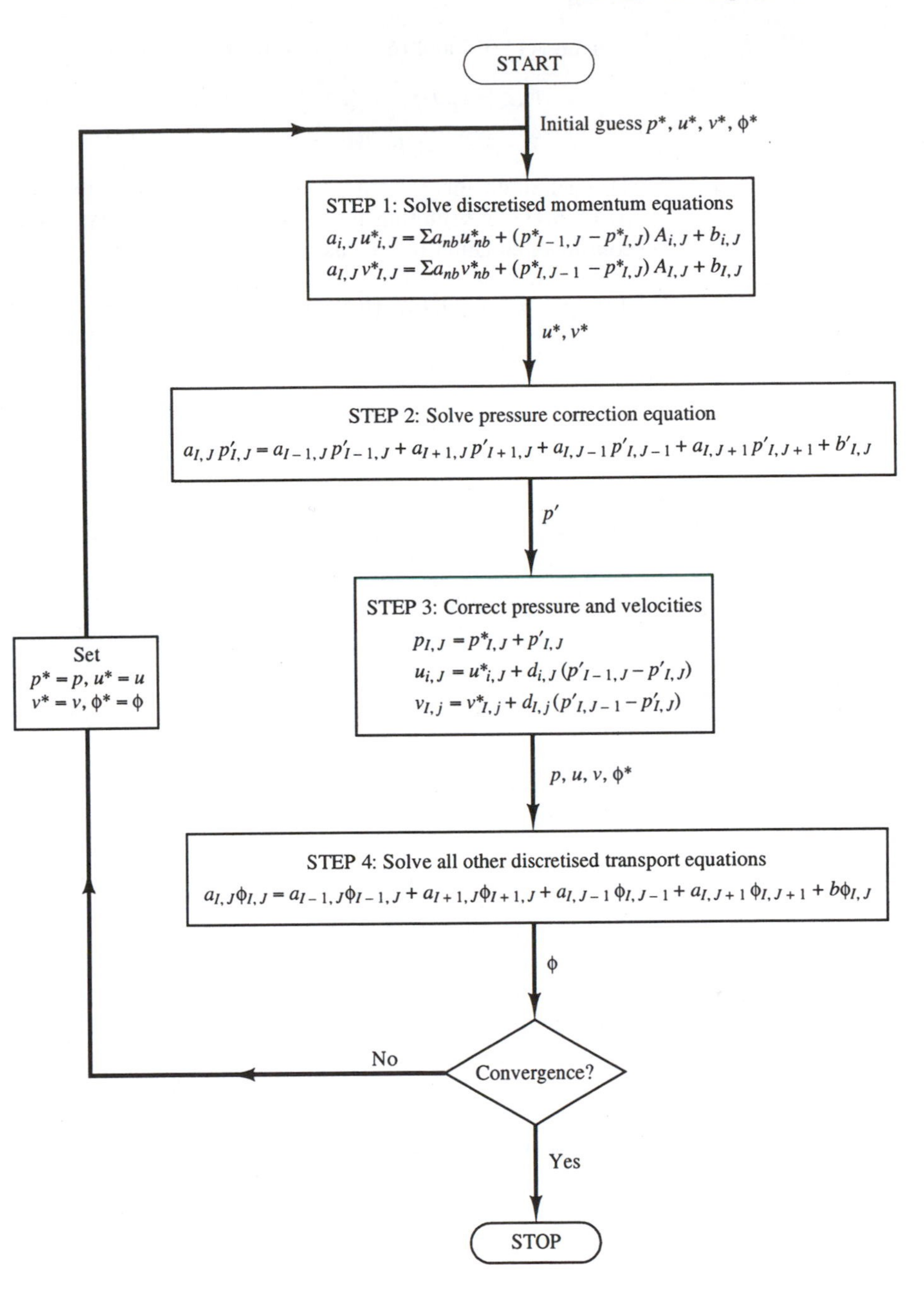

In the SIMPLER algorithm pseudo-velocities $\hat{u}$ and $\hat{v}$ are now defined as follows:

$$\hat{u}_{i,J} = \frac{\sum a_{nb}u_{nb} + b_{i,J}}{a_{i,J}} \tag{6.40}$$

$$\hat{v}_{I,j} = \frac{\sum a_{nb}v_{nb} + b_{I,j}}{a_{I,j}} \tag{6.41}$$

Equations (6.38) and (6.39) can now be written as

$$u_{i,J} = \hat{u}_{i,J} + d_{i,J}(p_{I-1,J} - p_{I,J}) \tag{6.42}$$

$$v_{I,j} = \hat{v}_{I,j} + d_{I,j}(p_{I,J-1} - p_{I,J}) \tag{6.43}$$

The definition for d, introduced in the developments of section 6.4, is applied in (6.42–6.43). Substituting for $u_{i,J}$ and $v_{I,j}$ from these equations into the discretised continuity equation (6.29), using similar forms for $u_{i+1,J}$ and $v_{I,j+1}$, results in

$$\begin{aligned}&\Big[\rho_{i+1,J}A_{i+1,J}\big(\hat{u}_{i+1,J} + d_{i+1,J}(p_{I,J} - p_{I+1,J})\big)\\ &\quad -\rho_{i,J}A_{i,J}\big(\hat{u}_{i,J} + d_{i,J}(p_{I-1,J} - p_{I,J})\big)\Big]\\ &\quad +\Big[\rho_{I,j+1}A_{I,j+1}\big(\hat{v}_{I,j+1} + d_{I,j+1}(p_{I,J} - p_{I,J+1})\big)\\ &\quad -\rho_{I,j}A_{I,j}\big(\hat{v}_{I,j} + d_{I,j}(p_{I,J-1} - p_{I,J})\big)\Big] = 0\end{aligned} \tag{6.44}$$

Equation (6.44) may be re-arranged to give a discretised pressure equation

$$\begin{aligned}a_{I,J}p_{I,J} = {}& a_{I+1,J}p_{I+1,J} + a_{I-1,J}p_{I-1,J} + a_{I,J+1}p_{I,J+1}\\ &+ a_{I,J-1}p_{I,J-1} + b_{I,J}\end{aligned} \tag{6.45}$$

where $a_{I,J} = a_{I+1,J} + a_{I-1,J} + a_{I,J+1} + a_{I,J-1}$ and the coefficients are given below:

$a_{I+1,J}$	$a_{I-1,J}$	$a_{I,J+1}$	$a_{I,J-1}$	$b_{I,J}$
$(\rho dA)_{i+1,J}$	$(\rho dA)_{i,J}$	$(\rho dA)_{I,j+1}$	$(\rho dA)_{I,j}$	$(\rho\hat{u}A)_{i,J} - (\rho\hat{u}A)_{i+1,J} + (\rho\hat{v}A)_{I,j} - (\rho\hat{v}A)_{I,j+1}$

Note that the coefficients of equation (6.45) are the same as those in the discretised pressure correction equation (6.32), with the difference that the source term b is evaluated using the pseudo-velocities. Subsequently, the discretised momentum equations (6.12–6.13) are solved using the pressure field obtained above. This yields the velocity components u^* and v^*. The velocity correction equations (6.24–6.27) are used in the SIMPLER algorithm to obtain corrected velocities. Therefore, the p'-equation (6.32) must also be solved to obtain the pressure corrections needed for the velocity corrections. The full sequence of operations is described in Figure 6.7.

6.7 The SIMPLEC algorithm

The SIMPLEC (SIMPLE-Consistent) algorithm of Van Doormal and Raithby (1984) follows the same steps as the SIMPLE algorithm, with the difference that the momentum equations are manipulated so that the SIMPLEC velocity correction equations omit terms that are less significant than those omitted in SIMPLE.

The u-velocity correction equation of SIMPLEC is given by

$$u'_{i,J} = d_{i,J}\left(p'_{I-1,J} - p'_{I,J}\right) \tag{6.46}$$

where

$$d_{i,J} = \frac{A_{i,J}}{a_{i,J} - \sum a_{nb}} \tag{6.47}$$

Fig. 6.7

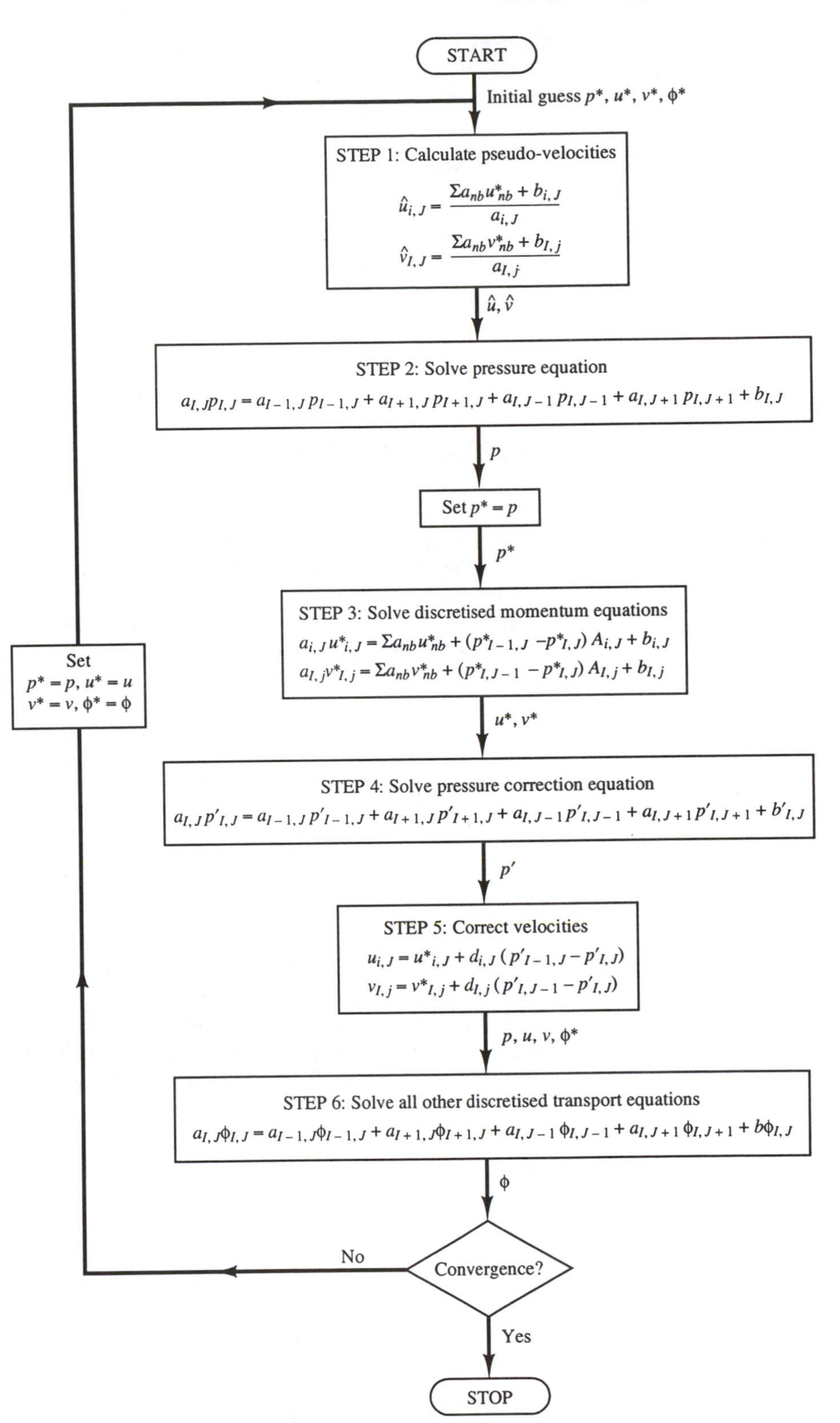
START
Initial guess p^*, u^*, v^*, ϕ^*
STEP 1: Calculate pseudo-velocities
$\hat{u}_{i,J} = \dfrac{\Sigma a_{nb}u^*_{nb} + b_{i,J}}{a_{i,J}}$
$\hat{v}_{I,J} = \dfrac{\Sigma a_{nb}v^*_{nb} + b_{I,j}}{a_{I,j}}$
$\hat{u}, \hat{v}$
STEP 2: Solve pressure equation
$a_{I,J}p_{I,J} = a_{I-1,J}\,p_{I-1,J} + a_{I+1,J}\,p_{I+1,J} + a_{I,J-1}\,p_{I,J-1} + a_{I,J+1}\,p_{I,J+1} + b_{I,J}$
p
Set $p^* = p$
p^*
STEP 3: Solve discretised momentum equations
$a_{i,J}\,u^*_{i,J} = \Sigma a_{nb}u^*_{nb} + (p^*_{I-1,J} - p^*_{I,J})\,A_{i,J} + b_{i,J}$
$a_{I,j}v^*_{I,j} = \Sigma a_{nb}v^*_{nb} + (p^*_{I,J-1} - p^*_{I,J})\,A_{I,j} + b_{I,j}$
u^*, v^*
STEP 4: Solve pressure correction equation
$a_{I,J}\,p'_{I,J} = a_{I-1,J}\,p'_{I-1,J} + a_{I+1,J}\,p'_{I+1,J} + a_{I,J-1}\,p'_{I,J-1} + a_{I,J+1}\,p'_{I,J+1} + b'_{I,J}$
p'
STEP 5: Correct velocities
$u_{i,J} = u^*_{i,J} + d_{i,J}\,(p'_{I-1,J} - p'_{I,J})$
$v_{I,j} = v^*_{I,j} + d_{I,j}\,(p'_{I,J-1} - p'_{I,J})$
p, u, v, ϕ^*
STEP 6: Solve all other discretised transport equations
$a_{I,J}\phi_{I,J} = a_{I-1,J}\phi_{I-1,J} + a_{I+1,J}\phi_{I+1,J} + a_{I,J-1}\,\phi_{I,J-1} + a_{I,J+1}\,\phi_{I,J+1} + b\phi_{I,J}$
ϕ
Convergence?
No
Set
$p^* = p, u^* = u$
$v^* = v, \phi^* = \phi$
Yes
STOP

Similarly the modified v-velocity correction equation is

$$v'_{I,j} = d_{I,j}\left(p'_{I,J-1} - p'_{I,J}\right) \tag{6.48}$$

$$\text{where} \quad d_{I,j} = \frac{A_{I,j}}{a_{I,j} - \sum a_{nb}} \tag{6.49}$$

The discretised pressure correction equation is the same as in SIMPLE, except that the d-terms are calculated from equations (6.47) and (6.49). The sequence of operations of the SIMPLEC algorithm is identical to that of SIMPLE (see section 6.5).

6.8 The PISO algorithm

The PISO algorithm, which stands for Pressure Implicit with Splitting of Operators, of Issa (1986) is a pressure–velocity calculation procedure developed originally for the non-iterative computation of unsteady compressible flows. It has been adapted successfully for the iterative solution of steady state problems. PISO involves one predictor step and two corrector steps and may be seen as an extension of SIMPLE, with a **further corrector step** to enhance it.

Predictor step

Discretised momentum equations (6.12–6.13) are solved with a guessed or intermediate pressure field p^* to give velocity components u^* and v^* using the same method as the SIMPLE algorithm.

Corrector step 1

The fields u^* and v^* will not satisfy continuity unless the pressure field p^* is correct. The first corrector step of SIMPLE is introduced to give a velocity field (u^{**}, v^{**}) which satisfies the discretised continuity equation. The resulting equations are the same as the velocity correction equations (6.21–6.22) of SIMPLE but, since there is a further correction step in the PISO algorithm, we use a slightly different notation:

$$p^{**} = p^* + p'$$
$$u^{**} = u^* + u'$$
$$v^{**} = v^* + v'$$

These formulae are used to define corrected velocities u^{**} and v^{**}:

$$u^{**}_{i,J} = u^{*}_{i,J} + d_{i,J}\left(p'_{I-1,J} - p'_{I,J}\right) \tag{6.50}$$

$$v^{**}_{I,j} = v^{*}_{I,j} + d_{I,j}\left(p'_{I,J-1} - p'_{I,J}\right) \tag{6.51}$$

As in the SIMPLE algorithm equations (6.50–6.51) are substituted into the discretised continuity equation (6.29) to yield the pressure correction equation (6.32) with its coefficients and source term. In the context of the PISO method equation (6.32) is called the first pressure correction equation. It is solved to yield the first pressure correction field p'. Once the pressure corrections are known, the velocity components u^{**} and v^{**} can be obtained through equations (6.50–6.51).

Corrector step 2

To enhance the SIMPLE procedure PISO performs a second corrector step. The discretised momentum equations for u^{**} and v^{**} are

$$a_{i,J} u^{**}_{i,J} = \sum a_{nb} u^{*}_{nb} + (p^{**}_{I-1,J} - p^{**}_{I,J}) A_{i,J} + b_{i,J} \tag{6.12}$$

$$a_{I,j} v^{**}_{I,j} = \sum a_{nb} v^{*}_{nb} + (p^{**}_{I,J-1} - p^{**}_{I,J}) A_{I,j} + b_{I,j} \tag{6.13}$$

A twice-corrected velocity field (u^{***}, v^{***}) may be obtained by solving the momentum equations once more:

$$a_{i,J} u^{***}_{i,J} = \sum a_{nb} u^{**}_{nb} + (p^{***}_{I-1,J} - p^{***}_{I,J}) A_{i,J} + b_{i,J} \tag{6.52}$$

$$a_{I,j} v^{***}_{I,j} = \sum a_{nb} v^{**}_{nb} + (p^{***}_{I,J-1} - p^{***}_{I,J}) A_{I,j} + b_{I,j} \tag{6.53}$$

Note that the summation terms are evaluated using the velocities u^{**} and v^{**} calculated in the previous corrector step.

Subtraction of equation (6.12) from (6.52) and (6.13) from (6.53) gives

$$u^{***}_{i,J} = u^{**}_{i,J} + \frac{\sum a_{nb} \left(u^{**}_{nb} - u^{*}_{nb} \right)}{a_{i,J}} + d_{i,J} \left(p''_{I-1,J} - p''_{I,J} \right) \tag{6.54}$$

$$v^{***}_{I,j} = v^{**}_{I,j} + \frac{\sum a_{nb} \left(v^{**}_{nb} - v^{*}_{nb} \right)}{a_{I,j}} + d_{I,j} \left(p''_{I,J-1} - p''_{I,J} \right) \tag{6.55}$$

where p'' is the second pressure correction so that p^{***} may be obtained by

$$p^{***} = p^{**} + p'' \tag{6.56}$$

Substitution of u^{***} and v^{***} in the discretised continuity equation (6.29) yields a second pressure correction equation

$$\begin{aligned} a_{I,J} p''_{I,J} = {} & a_{I+1,J} p''_{I+1,J} + a_{I-1,J} p''_{I-1,J} + a_{I,J+1} p''_{I,J+1} \\ & + a_{I,J-1} p''_{I,J-1} + b''_{I,J} \end{aligned} \tag{6.57}$$

with $a_{I,J} = a_{I+1,J} + a_{I-1,J} + a_{I},\ J+1 + a_{I,J-1}$ and the coefficients are as follows:

$a_{I+1,J}$	$a_{I-1,J}$	$a_{I,J+1}$	$a_{I,J-1}$	$b''_{I,J}$
$(\rho dA)_{i+1,J}$	$(\rho dA)_{i,J}$	$(\rho dA)_{I,j+1}$	$(\rho dA)_{I,j}$	$\left[\left(\frac{\rho A}{a}\right)_{i,J}\sum a_{nb}(u^{**}_{nb}-u^{*}_{nb}) - \left(\frac{\rho A}{a}\right)_{i+1,J}\sum a_{nb}(u^{**}_{nb}-u^{*}_{nb}) + \left(\frac{\rho A}{a}\right)_{I,j}\sum a_{nb}(v^{**}_{nb}-v^{*}_{nb}) - \left(\frac{\rho A}{a}\right)_{I,j+1}\sum a_{nb}(v^{**}_{nb}-v^{*}_{nb})\right]$

In the derivation of (6.57) the source term

$$\left[(\rho A u^{**})_{i,J} - (\rho A u^{**})_{i+1,J} + (\rho A v^{**})_{I,j} - (\rho A v^{**})_{I,j+1}\right]$$

is zero since the velocity components u^{**} and v^{**} satisfy continuity.

Equation (6.57) is solved to obtain the second pressure correction field p'' and a twice-corrected pressure field is obtained from

$$\boxed{p^{***} = p^{**} + p'' = p^{*} + p' + p''} \qquad (6.58)$$

Finally the twice-corrected velocity field is obtained from equations (6.54–6.55).

In the non-iterative calculation of unsteady flows the pressure field p^{***} and the velocity fields u^{***} and v^{***} are considered to be the correct u, v and p. The sequence of operations for an iterative steady state PISO calculation is given in Figure 6.8.

The PISO algorithm solves the pressure correction equation twice so the method requires additional storage for calculating the source term of the second pressure correction equation. As before under-relaxation is required with the above procedure to stabilise the calculation process. Although this method implies a considerable increase in computational effort it has been found to be efficient and fast. For example, for a benchmark, laminar, backward-facing step problem Issa *et al* (1986) report a reduction of CPU time by a factor of 2 compared to standard SIMPLE.

The PISO algorithm presented above is the adapted, steady state version of an algorithm that was originally developed for non-iterative time-dependent calculations. The transient algorithm can also be applied to steady state calculations by starting with guessed initial conditions and solving as a transient problem for a long period of time until the steady state is achieved. This will be discussed in Chapter 8.

6.9 General comments on SIMPLE, SIMPLER, SIMPLEC and PISO

The SIMPLE algorithm is relatively straightforward and has been successfully implemented in numerous CFD procedures. The other variations of SIMPLE can

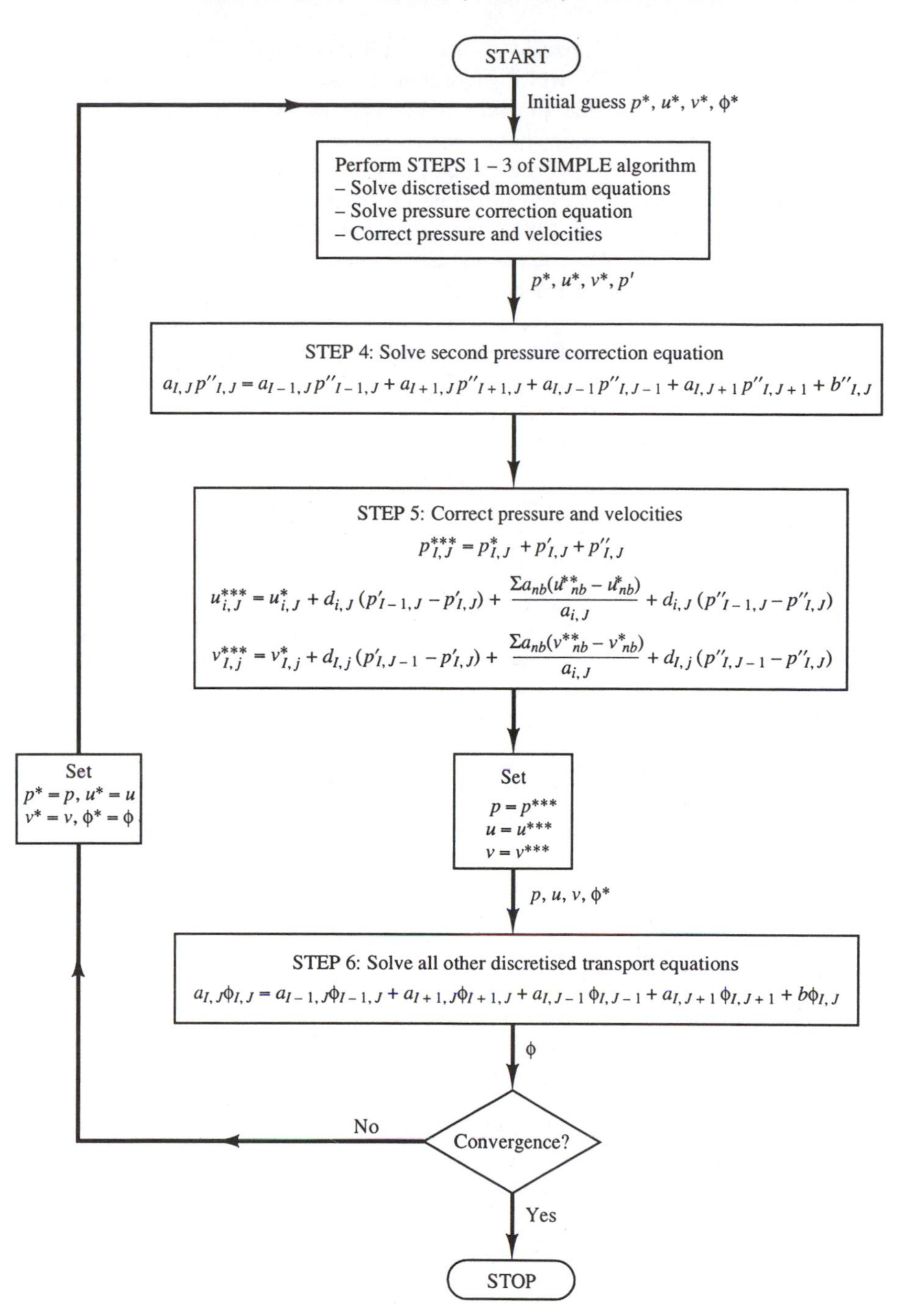

Fig. 6.8 The PISO algorithm

produce savings in computational effort due to improved convergence. In SIMPLE, the pressure correction p' is satisfactory for correcting velocities but not so good for correcting pressure. Hence the improved procedure SIMPLER uses the pressure corrections to obtain velocity corrections only. A separate, more effective, pressure equation is solved to yield the correct pressure field. Since no terms are omitted to derive the discretised pressure equation in SIMPLER the resulting pressure field corresponds to the velocity field. Therefore, in SIMPLER the application of the correct velocity field results in the correct pressure field, whereas it does not in the SIMPLE algorithm. Consequently, the method is highly effective in calculating the pressure field correctly. This has significant advantages when solving the momentum equations. Although the number of calculations involved in the SIMPLER algorithm is about 30% larger than that for SIMPLE, the fast convergence rate reportedly reduces the computer time by 30–50% (Anderson *et al*, 1984). The SIMPLER algorithm is therefore often used as the default procedure in commercial CFD codes. Further details of SIMPLE and its variants may be found in Patankar (1980).

SIMPLEC and PISO have proved to be as efficient as SIMPLER in certain types of flows but it is not clear whether it can be categorically stated that they are better than SIMPLER. Comparisons have shown that the performance of each algorithm depends on the flow conditions, the degree of coupling between the momentum equation and scalar equations – in combusting flows, for example, due to the dependence of the local density on concentration and temperature – and on the amount of under-relaxation used, and sometimes even on the details of the numerical technique used for solving the algebraic equations. A comprehensive comparison of PISO, SIMPLER and SIMPLEC methods for a variety of steady flow problems by Jang *et al* (1986) showed that, for problems in which momentum equations are not coupled to a scalar variable, the PISO algorithm showed robust convergence behaviour and required less computational effort than SIMPLER and SIMPLEC. It was also observed that when the scalar variables were closely linked to velocities, PISO had no significant advantage over the other methods. Iterative methods using SIMPLER and SIMPLEC have robust convergence characteristics in strongly coupled problems, but it could not be ascertained which of SIMPLER or SIMPLEC was superior.

6.10 Summary

The most popular solution algorithms for pressure and velocity calculations with the finite volume method have been discussed. They all possess the following common characteristics:

- The problems associated with the non-linearity of the momentum equations and the coupling between transport equations are tackled by adopting an iterative solution strategy.
- Velocity components are defined on staggered grids to avoid problems associated with pressure field oscillations of high spatial frequency.
- In the staggered grid arrangement velocities are stored at the cell faces of scalar control volumes. The discretised momentum equations are solved on staggered control volumes whose cell faces contain the pressure nodes.
- The SIMPLE algorithm is an iterative procedure for the calculation of pressure and velocity fields. Starting from an initial pressure field p^* its principal steps are:
  - solve the discretised momentum equation to yield the intermediate velocity field (u^*, v^*).
  - solve the continuity equation in the form of an equation for pressure correction p'.

– correct pressure and velocity by means of

$$p_{I,J} = p^*_{I,J} + p'_{I,J}$$
$$u_{i,J} = u^*_{i,J} + d_{i,J}(p'_{I-1,J} - p'_{I,J})$$
$$v_{I,j} = v^*_{I,j} + d_{I,j}(p'_{I,J-1} - p'_{I,J})$$

– solve all other discretised transport equations for scalars ϕ.
– repeat until the fields p, u, v and ϕ have all converged.

- Refinements to SIMPLE have produced more economical and stable iteration methods.
- The steady state PISO algorithm adds an extra correction step to SIMPLE to enhance its performance per iteration.
- The SIMPLER method is currently used as the default algorithm in many commercial CFD codes. It is still unclear which of the SIMPLE variants is the best for general purpose computation.
- Under-relaxation is required in all methods to ensure stability of the iteration process.

7

Solution of Discretised Equations

7.1 Introduction

In the previous chapters we have discussed methods of discretising the governing equations of fluid flow and heat transfer. This process results in a system of linear algebraic equations which needs to be solved. The complexity and size of the set of equations depends on the dimensionality of the problem, the number of grid nodes and the discretisation practice. Although any valid procedure can be used to solve the algebraic equations, the available computer resources set a powerful constraint. There are two families of solution techniques for linear algebraic equations: **direct methods** and **indirect or iterative methods**. Simple examples of direct methods are Cramer's rule matrix inversion and Gaussian elimination. The number of operations to the solution of a system of N equations with N unknowns by means of a direct method can be determined beforehand and is on the order of N^3. The simultaneous storage of all N^2 coefficients of the set of equations in core memory is required.

Iterative methods are based on the repeated application of a relatively simple algorithm leading to eventual convergence after a – sometimes large – number of repetitions. Well-known examples are the Jacobi and Gauss–Seidel point-by-point iteration methods. The total number of operations, typically on the order of N per iteration cycle, cannot be predicted in advance. Stronger still, it is not possible to guarantee convergence unless the system of equations satisfies fairly exacting criteria. The main advantage of iterative solution methods is that only non-zero coefficients of the equations need to be stored in core memory.

The one-dimensional conduction example in Chapter 4, section 4.3.1, led to a tri-diagonal system – a system with only three non-zero coefficients per equation. When QUICK differencing is applied to a convection–diffusion problem this gives rise to a penta-diagonal system that has five non-zero coefficients, which is somewhat more complex to deal with. Nevertheless, the finite volume method usually yields systems of equations each of which has a vast majority of zero entries. Since the systems are often very large – up to 100000 or 1 million equations – we find that iterative methods are generally much more economical than direct methods.

Jacobi and Gauss–Seidel iterative methods are easy to implement in simple computer programs, but they can be slow to converge when the system of equations is large. Hence they are not considered suitable for general CFD procedures. Thomas (1949) developed a technique for rapidly solving tri-diagonal systems that is now called the Thomas algorithm or the tri-diagonal matrix algorithm (TDMA). The TDMA is actually a direct method for one-dimensional situations, but it can be applied iteratively, in a line-by-line fashion, to solve multi-dimensional problems and is widely used in CFD programs. It is computationally inexpensive and has the advantage that it requires a minimum amount of storage. In this chapter the TDMA method is explained in detail together with alternative methods which will be discussed briefly.

7.2 The tri-diagonal matrix algorithm

Consider a system of equations that has a tri-diagonal form

$$\phi_1 = C_1 \quad (7.1a)$$
$$-\beta_2\phi_1 + D_2\phi_2 - \alpha_2\phi_3 = C_2 \quad (7.1b)$$
$$-\beta_3\phi_2 + D_3\phi_3 - \alpha_3\phi_4 = C_3 \quad (7.1c)$$
$$-\beta_4\phi_3 + D_4\phi_4 - \alpha_3\phi_4 = C_4$$
$$. \quad . \quad . \quad . \quad = \quad . \quad .$$
$$-\beta_n\phi_{n-1} + D_n\phi_n - \alpha_n\phi_{n+1} = C_n \quad (7.1n)$$
$$\phi_{n+1} = C_{n+1} \quad (7.1n+1)$$

In the above set of equations ϕ_1 and ϕ_{n+1} are known boundary values. The general form of any single equation is

$$-\beta_j\phi_{j-1} + D_j\phi_j - \alpha_j\phi_{j+1} = C_j \quad (7.2)$$

Equations (7.1b–n) of the above set can be rewritten as

$$\phi_2 = \frac{\alpha_2}{D_2}\phi_3 + \frac{\beta_2}{D_2}\phi_1 + \frac{C_2}{D_2} \quad (7.3a)$$

$$\phi_3 = \frac{\alpha_3}{D_3}\phi_4 + \frac{\beta_3}{D_3}\phi_2 + \frac{C_3}{D_3} \quad (7.3b)$$

$$\phi_4 = \frac{\alpha_4}{D_4}\phi_5 + \frac{\beta_4}{D_4}\phi_3 + \frac{C_4}{D_4} \quad (7.3c)$$

. . . .

$$\phi_n = \frac{\alpha_n}{D_n}\phi_{n+1} + \frac{\beta_n}{D_n}\phi_{n-1} + \frac{C_n}{D_n}$$

These equations can be solved by forward elimination and back-substitution. The **forward elimination** process starts by removing ϕ_2 from equation (7.3b) by substitution from equation (7.3a) to give

$$\phi_3 = \left(\frac{\alpha_3}{D_3 - \beta_3\dfrac{\alpha_2}{D_2}}\right)\phi_4 + \left(\frac{\beta_3\left(\dfrac{\beta_2}{D_2}\phi_1 + \dfrac{C_2}{D_2}\right) + C_3}{D_3 - \beta_3\dfrac{\alpha_2}{D_2}}\right) \quad (7.4a)$$

If we adopt the notation

$$A_2 = \frac{\alpha_2}{D_2} \text{ and } C'_2 = \frac{\beta_2}{D_2}\phi_1 + \frac{C_2}{D_2} \tag{7.4b}$$

equation (7.4a) can be written as

$$\phi_3 = \left(\frac{\alpha_3}{D_3 - \beta_3 A_2}\right)\phi_4 + \left(\frac{\beta_3 C'_2 + C_3}{D_3 - \beta_3 A_2}\right) \tag{7.4c}$$

If we let

$$A_3 = \frac{\alpha_3}{D_3 - \beta_3 A_2} \quad \text{and} \quad C'_3 = \frac{\beta_3 C'_2 + C_3}{D_3 - \beta_3 A_2}$$

equation (7.4c) can be re-cast as

$$\phi_3 = A_3\phi_4 + C'_3 \tag{7.5}$$

Formula (7.5) can now be used to eliminate ϕ_3 from (7.3c) and the procedure can be repeated up to the last equation of the set. This constitutes the forward elimination process.

For the **back-substitution** we use the general form of recurrence relationship (7.5):

$$\phi_j = A_j\phi_{j+1} + C'_j \tag{7.6a}$$

where

$$A_j = \frac{\alpha_j}{D_j - \beta_j A_{j-1}} \tag{7.6b}$$

$$C'_j = \frac{\beta_j C'_{j-1} + C_j}{D_j - \beta_j A_{j-1}} \tag{7.6c}$$

The formulae can be made to apply at the boundary points $j = 1$ and $j = n + 1$ by setting the following values for A and C':

$$A_1 = 0 \quad \text{and} \quad C'_1 = \phi_1$$

$$A_{n+1} = 0 \quad \text{and} \quad C'_{n+1} = \phi_{n+1}$$

In order to solve a system of equations it is first arranged in the form of equation (7.2) and α_j, β_j, D_j and C'_j are identified. The values of A_j and C'_j are subsequently calculated starting at $j = 2$ and going up to $j = n$ using (7.6b–c). Since the value of ϕ is known at boundary location $(n + 1)$ the values for ϕ_j can be obtained in reverse order $(\phi_n, \phi_{n-1}, \phi_{n-2}, \ldots, \phi_2)$ by means of the recurrence formula (7.6a). The method is simple and easy to incorporate into CFD programs. A FORTRAN subroutine for the TDMA method is given in Anderson *et al* (1984).

In the above derivation of the TDMA method we assumed that boundary values ϕ_1 and ϕ_{n+1} were given. To implement a fixed gradient (or flux) boundary condition, for example at $j = 1$, the coefficient β_2 in equation (7.1b) is set to zero and the flux across the boundary is incorporated in source term C_2. The actual value of the variable at the boundary is now not directly used in the formulation. The absence of the first or the last value does not pose a problem in applying the TDMA method as will be illustrated in examples below.

7.3 Application of TDMA to two-dimensional problems

The TDMA can be applied iteratively to solve a system of equations for two-dimensional problems. Consider the grid in Figure 7.1 and a general two-dimensional discretised transport equation of the form

$$a_P\phi_P = a_W\phi_W + a_E\phi_E + a_S\phi_S + a_N\phi_N + b \tag{7.7}$$

To solve the system TDMA is applied along chosen, for example north–south ($n-s$), lines. The discretised equation is re-arranged in the form

$$-a_S\phi_S + a_P\phi_P - a_N\phi_N = a_W\phi_W + a_E\phi_E + b \tag{7.8}$$

The right hand side of (7.8) is assumed to be temporarily known. Equation (7.8) is in the form of equation (7.2) where $\alpha_j \equiv a_N, \beta_j \equiv a_S, D_j \equiv a_P$ and $C_j \equiv a_W\phi_W + a_E\phi_E + b$. Now we can solve along the n–s direction of the chosen line for values $j = 2, 3, 4, \ldots, n$ as shown in Figure 7.1.

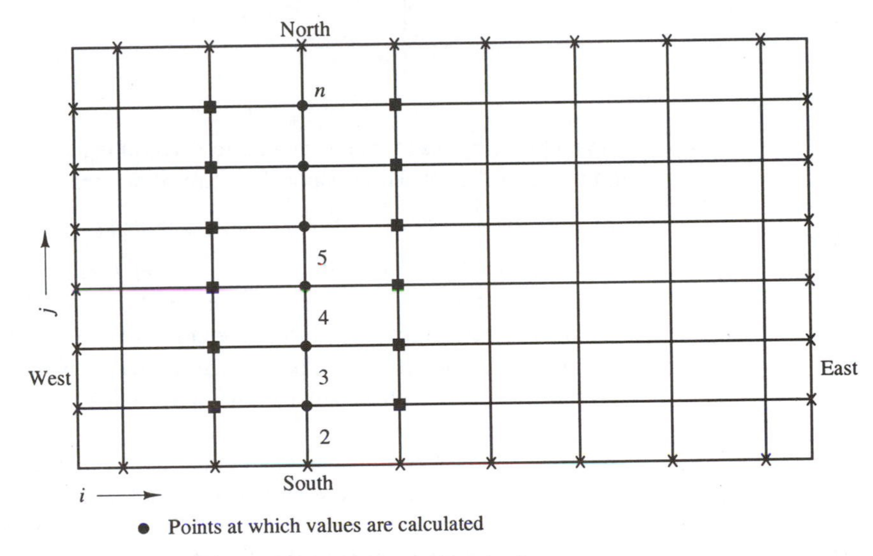

Fig. 7.1 Line-by-line application of the TDMA method

Subsequently the calculation is moved to the next north–south line. The sequence in which lines are chosen is known as the sweep direction. If we sweep from west to east the values of ϕ_W to the west of point P are known from the calculations on the previous line. Values of ϕ_E to its east, however, are unknown so the solution process must be iterative. At each iteration cycle ϕ_E is taken to have its value at the end of the previous iteration or a given initial value (e.g. zero) at the first iteration. The line-by-line calculation procedure is repeated several times until a converged solution is obtained.

7.4 Application of the TDMA method to three-dimensional problems

For three-dimensional problems the TDMA method is applied line by line on a selected plane and then the calculation is moved to the next plane, scanning the

Fig. 7.2 Application of the TDMA method in a 3D geometry

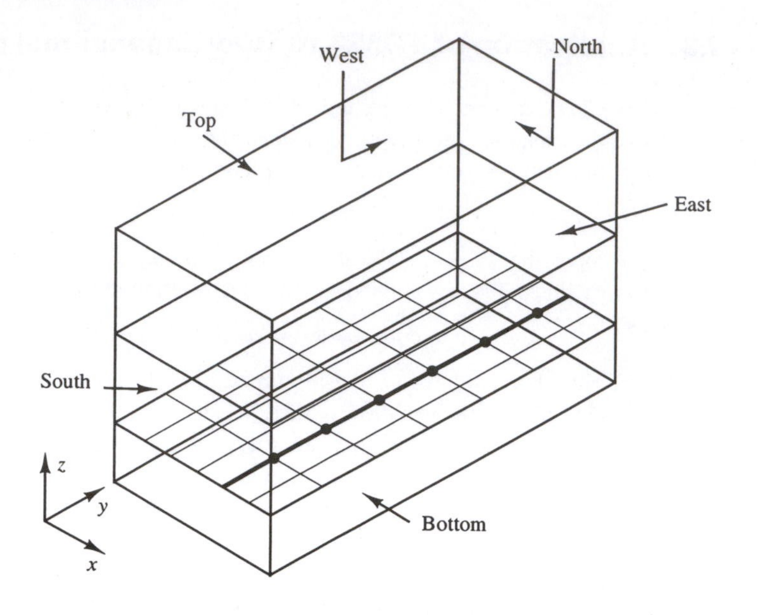

domain plane by plane. For example, to solve along an n–s line in the x–y plane of Figure 7.2, a discretised transport equation is written as

$$-a_S\phi_S + a_P\phi_P - a_N\phi_N = a_W\phi_W + a_E\phi_E + a_B\phi_B + a_T\phi_T + b \qquad (7.9)$$

The values at W and E as well as those at B and T on the right hand side of equation (7.9) are considered to be temporarily known. Using the TDMA procedure values of ϕ along a selected north–south line are computed. The calculation is moved to the next line and subsequently swept through the whole plane until all unknown values on each line have been calculated. After completion of one plane the process is moved on to the next plane.

In two- and three-dimensional computations the convergence can often be accelerated by **alternating the sweep direction** so that all boundary information is fed into the calculation more effectively. To solve along an east–west line in the present three-dimensional case the discretised equation is re-arranged as follows:

$$-a_W\phi_W + a_P\phi_P - a_E\phi_E = a_S\phi_S + a_N\phi_N + a_B\phi_B + a_T\phi_T + b \qquad (7.10)$$

7.5 Examples

Example 7.1 *An illustration of TDMA in one dimension.* We consider the one-dimensional steady state conductive/convective heat transfer from a bar of material first discussed in Example 4.3 of section 4.3.3. The geometry is shown in Figure 7.3. The temperature on the left hand boundary is taken to be 100 °C and the right hand boundary is insulated so the heat flux across it is zero. Heat is lost to the surroundings by convective heat transfer. Solve the matrix equation (3.52) for this problem using TDMA.

Fig. 7.3 The grid for Example 7.1

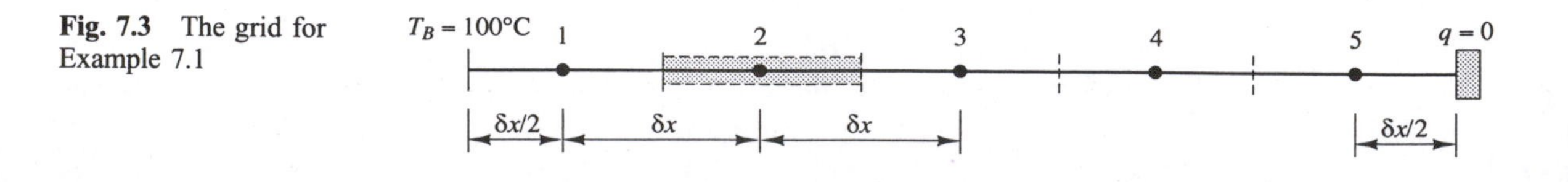

Solution The matrix equation found in section 4.3.3 was

$$\begin{bmatrix} 20 & -5 & 0 & 0 & 0 \\ -5 & 15 & -5 & 0 & 0 \\ 0 & -5 & 15 & -5 & 0 \\ 0 & 0 & -5 & 15 & -5 \\ 0 & 0 & 0 & -5 & 10 \end{bmatrix} \begin{bmatrix} \phi_1 \\ \phi_2 \\ \phi_3 \\ \phi_4 \\ \phi_5 \end{bmatrix} = \begin{bmatrix} 1100 \\ 100 \\ 100 \\ 100 \\ 100 \end{bmatrix} \tag{4.49}$$

The general form of the equation used in the TDMA method is

$$-\beta_j\phi_{j-1} + D_j\phi_j - \alpha_j\phi_{j+1} = C_j \tag{7.2}$$

Nodes 1 and 5 are boundary nodes so we set $\beta_1 = 0$ and $\alpha_5 = 0$. The ϕ at the boundaries is not used; the boundary conditions enter into the calculation through the source terms C_j.

To show the results most clearly the values of α, β, D and C are tabulated for each point in Table 7.1 and A_j and C_j', calculated using the recurrence formulae (7.6b) and (7.6c), are given in Table 7.2.

Table 7.1

Point	β_j	D_j	α_j	C_j	A_j	C_j'
1	0	20	5	1100	0.25	55
2	5	15	5	100	0.3636	27.2737
3	5	15	5	100	0.3793	17.9311
4	5	15	5	100	0.3816	14.4736
5	5	10	0	100	0.00	22.1852

Table 7.2 Specimen calculation

$$A_j = \frac{\alpha_j}{D_j - \beta_j A_{j-1}} \qquad C_j' = \frac{\beta_j C_{j-1}' + C_j}{D_j - \beta_j A_{j-1}}$$

$$A_1 = \frac{5}{(20-0)} = 0.25 \qquad C_1' = \frac{0+1100}{(20-0)} = 55$$

$$A_2 = \frac{5}{(15 - 5 \times 0.25)} = 0.3636 \qquad C_2' = \frac{5 \times 55 + 100}{(15 - 5 \times 0.25)} = 27.2727$$

$$A_3 = \frac{5}{(15 - 5 \times 0.3636)} = 0.3793 \qquad C_3' = \frac{5 \times 27.2727 + 100}{(15 - 5 \times 0.3636)} = 17.9308$$

$$A_4 = \frac{5}{(15 - 5 \times 0.3793)} = 0.3816 \qquad C_4' = \frac{5 \times 17.9308 + 100}{(15 - 5 \times 0.3793)} = 14.4735$$

$$A_5 = 0 \qquad C_5' = \frac{5 \times 14.4735 + 100}{(10 - 5 \times 0.3816)} = 21.30$$

Solution with the back-substitution formula (7.6a), $\phi_j = A_j\phi_{j+1} + C_j'$, gives

$$\begin{aligned} \phi_5 &= 0 + 21.30 \\ &= 21.30 \\ \phi_4 &= 0.3816 \times 21.30 + 14.4735 \\ &= 22.60 \\ \phi_3 &= 0.3793 \times 22.60 + 17.9308 \\ &= 26.50 \\ \phi_2 &= 0.3636 \times 26.50 + 27.2727 \\ &= 36.91 \\ \phi_1 &= 0.25 \times 36.91 + 55 \\ &= 64.22 \end{aligned}$$

Example 7.2 *A two-dimensional line-by-line application of TDMA.* In Figure 7.4 a two-dimensional plate of thickness 1 cm is shown. The thermal conductivity of the plate material is $k = 1000$ W/m/K. The west boundary receives a steady heat flux of 500 kW/m^2 and the south and east boundaries are insulated. If the north boundary is maintained at a temperature of 100 °C, use a uniform grid with $\Delta x = \Delta y = 0.1$ m to calculate the steady state temperature distribution at nodes 1, 2, 3, 4, ... etc.

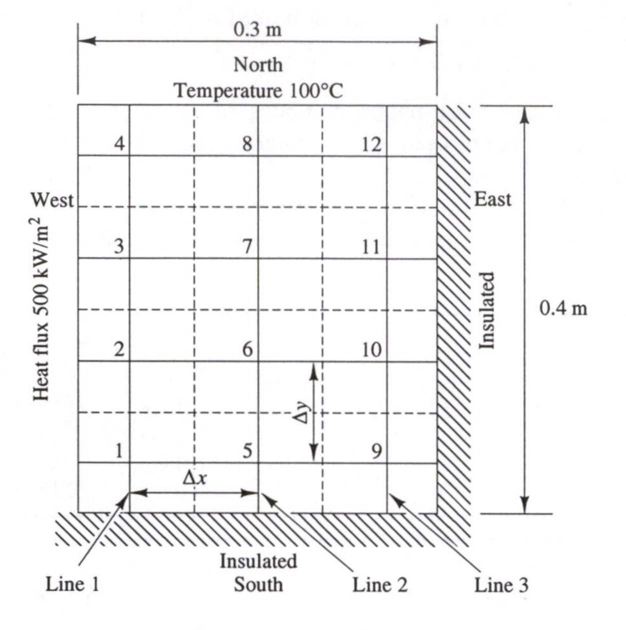

Fig. 7.4 Boundary conditions for the 2D heat transfer problem described in Example 7.2

Solution The two-dimensional steady state heat transfer in the plate is governed by

$$\frac{\partial}{\partial x}\left(k\frac{\partial T}{\partial x}\right) + \frac{\partial}{\partial y}\left(k\frac{\partial T}{\partial y}\right) = 0 \tag{7.11}$$

This can be written in discretised form as

$$a_P T_P = a_W T_W + a_E T_E + a_S T_S + a_N T_N \tag{7.12a}$$

where

$$a_W = \frac{k}{\Delta x}A_w; \quad a_E = \frac{k}{\Delta x}A_e; \quad a_S = \frac{k}{\Delta y}A_s; \quad a_N = \frac{k}{\Delta y}A_n \tag{7.12b}$$

$$a_P = a_W + a_E + a_S + a_N \tag{7.12c}$$

In this case, the values of all neighbour coefficients are equal:

$$a_W = a_E = a_N = a_S = \frac{1000}{0.1} \times (0.1 \times 0.01) = 10$$

At interior points 6 and 7

$$a_P = a_W + a_E + a_S + a_N = 40$$

So the discretised equation at node 6 is

$$40T_6 = 10T_2 + 10T_{10} + 10T_5 + 10T_7$$

All nodes except 6 and 7 are adjacent to boundaries.

At a boundary node the discretised equation takes the form

$$a_P T_P = a_W T_W + a_E T_E + a_S T_S + a_N T_N + S_u$$

$$a_P = a_W + a_E + a_S + a_N - S_p$$

The boundary conditions are incorporated into the discretised equations by setting the relevant coefficient to zero and by the inclusion of source terms through S_u and S_p. Otherwise, the procedure is the same as in the one-dimensional example 7.1. We demonstrate the approach by forming the discretised equations for boundary nodes 1 and 4.

At node 1

West is a constant flux boundary; let b_W be the contribution to the source term from the west:

$$a_W = 0$$

$$b_W = q_w \,.\, A_w = 500 \times 10^3 \times (0.1 \times 0.01) = 500$$

South is an insulated boundary; no flux enters the control volume through the south boundary:

$$a_S = 0$$

$$b_S = 0$$

Total source

$$S_u = b_W + b_S = 500$$

$$S_p = 0$$

The discretised equation at node 1 is

$$20T_1 = 10T_2 + 10T_5 + 500$$

At node 4

West is a constant flux boundary

$$a_W = 0$$

$$b_W = 500 \times 10^3 \times (0.1 \times 0.01) = 500$$

North is a constant temperature boundary

$$a_N = 0$$

$$b_N = \frac{2k}{\Delta y} A_n \times 100 = 2000$$

$$S_{P_N} = -\frac{2k}{\Delta y} A_n = -20$$

Total source

$$S_u = b_W + b_N = 500 + 2000$$

$$S_p = -20$$

Now we have

$$a_P = a_S + a_E - S_P = 10 + 10 + 20 = 40$$
$$S_u = 2500$$

The discretised equation at node 4 is

$$40T_4 = 10T_3 + 10T_8 + 2500$$

The coefficients and the source term of the discretisation equation for all points are summarised in Table 7.3.

Table 7.3

Point	a_N	a_S	a_W	a_E	a_P	S_u
1	10	0	0	10	20	500
2	10	10	0	10	30	500
3	10	10	0	10	30	500
4	0	10	0	10	40	2500
5	10	0	10	10	30	0
6	10	10	10	10	40	0
7	10	10	10	10	40	0
8	0	10	10	10	50	2000
9	10	0	10	0	20	0
10	10	10	10	0	30	0
11	10	10	10	0	30	0
12	0	10	10	0	40	2000

Let us apply TDMA along north–south lines, sweeping from west to east. The discretisation equation is given by

$$-a_S T_S + a_P T_P - a_N T_N = a_W T_W + a_E T_E + b \tag{7.13}$$

For convenience the line in Figure 7.4 containing points 1 to 4 is referred to as line 1, the one containing points 5 to 8 as line 2, and the one with points 9 to 12 as line 3.

Table 7.4

Point	β_j	D_j	α_j	C_j	A_j	C'_j
1	0	20	10	500	0.5	25
2	10	30	10	500	0.4	30
3	10	30	10	500	0.385	30.769
4	10	40	0	2500	0	77.667

All west coefficients are zero at points 1, 2, 3 and 4, and hence the values to the west of line 1 do not enter into the calculation. East values (points 5, 6, 7 and 8) are required for the evaluation of C. They are unknown at this stage and are assumed to be zero as an initial guess. The values of α, β, D and C can be calculated using the equations (7.2) and (7.13). Now we have $\alpha_j = -a_N, \beta_j = -a_S, D_j = a_P$ and $C_j = a_W T_W + a_E T_E + S_u$. The values of α, β, D and C and A_j and C'_j for line 1 are summarised in Table 7.4 and the calculations for A_j and C'_j in Table 7.5.

Table 7.5

$A_j = \dfrac{\alpha_j}{D_j - \beta_j A_{j-1}}$	$C'_j = \dfrac{\beta_j C'_{j-1} + C_j}{D_j - \beta_j A_{j-1}}$
$A_1 = \dfrac{10}{(20-0)} = 0.5$	$C'_1 = \dfrac{0+500}{(20-0)} = 25$
$A_2 = \dfrac{10}{(30 - 10 \times 0.5)} = 0.4$	$C'_2 = \dfrac{10 \times 25 + 500}{30 - 10 \times 0.5} = 30$
$A_3 = \dfrac{10}{(30 - 10 \times 0.4)} = 0.385$	$C'_3 = \dfrac{10 \times 30 + 500}{(30 - 10 \times 0.4)} = 30.769$
$A_4 = 0$	$C'_4 = \dfrac{10 \times 30.769 + 2500}{(40 - 10 \times 0.385)} = 77.66$

Solution by back-substitution gives

$$
\begin{aligned}
T_4 &= 0 + 77.667 \\
&= 77.66 \\
T_3 &= 0.385 \times 77.667 + 30.769 \\
&= 60.67 \\
T_2 &= 0.4 \times 60.67 + 30 \\
&= 54.26 \\
T_1 &= 0.5 \times 54.268 + 25 \\
&= 52.13
\end{aligned}
$$

The TDMA calculation procedure for line 2 is similar to line 1. Here the values to the west are known from the calculations given above and values to the east are assumed to be zero. We leave the detailed calculations as an exercise for the reader. The values of α, β, D and C for points 5, 6, 7 and 8 are summarised in Table 7.6.

Table 7.6

Point	β_j	D_j	α_j	C_j
5	0	30	10	521.3
6	10	40	10	542.6
7	10	40	10	606.5
8	10	50	0	2777.6

The TDMA solution for line 2 is $T_5 = 27.38$, $T_6 = 30.03$, $T_7 = 38.47$ and $T_8 = 63.23$. We can now proceed to the third line containing points 9, 10, 11 and 12. The values of α, β, D and C for points 9, 10, 11 and 12 are summarised in Table 7.7.

At the end of the first iteration we have the values shown in Table 7.8 for the entire field.

Table 7.7

Point	β_j	D_j	α_j	C_j
9	0	20	10	273.8
10	10	30	10	303.3
11	10	30	10	384.7
12	10	40	0	2632.3

Table 7.8 Values at the end of the first iteration

Point	*1*	*2*	*3*	*4*	*5*	*6*	*7*	*8*	*9*	*10*	*11*	*12*
T	52.13	54.26	60.65	77.66	27.38	30.03	38.47	63.23	32.79	38.21	51.82	78.76

Table 7.9 The converged solution after 37 iterations

Point	*1*	*2*	*3*	*4*	*5*	*6*	*7*	*8*	*9*	*10*	*11*	*12*
T	260.0	242.2	205.6	146.3	222.7	211.1	178.1	129.7	212.1	196.5	166.2	124.0

The entire procedure is now repeated until a converged solution is obtained. In this case after 37 iterations we obtain the converged solution (total error less than 1.0) shown in Table 7.9.

7.6 Other solution techniques used in CFD

For two- and three-dimensional problems, the TDMA method can be applied only in a line-by-line fashion and therefore the spread of boundary information into the calculation domain is slow. In CFD calculations the convergence rate depends on the direction of flow, with sweeping from upstream to downstream producing much faster convergence than sweeping against the flow or parallel to the flow direction. Although convergence problems can be alleviated by alternating the sweep directions, in three-dimensional recirculating flows, where the dominant flow direction is unknown in advance, convergence can be slow. When overall stability considerations require coupling between the values over the whole calculation domain the TDMA can be unsatisfactory for the solution of discretised equations.

Higher order schemes for the discretisation process will link each discretisation equation to nodes other than the immediate neighbours. Here, the TDMA method can only be applied by incorporating a large number of neighbouring contributions in the source term. This may be undesirable in terms of stability, can impair the effectiveness of the higher order scheme and may hinder, if applicable, the implicit nature of the scheme.

When the system of equations to be solved has the form of a penta-diagonal matrix, as may be the case in QUICK and other higher order discretisation schemes, a generalised version of TDMA, known as the penta-diagonal matrix algorithm, is available. Basically a sequence of operations is carried out on the original matrix to reduce it to upper-triangular form and back-substitution is performed to obtain the solution. Details of the method can be found in Fletcher (1991). The method is, of course, not nearly as economical as the TDMA.

In advanced CFD procedures that use body-fitted co-ordinate systems, the discretised equations normally contain a large number of contributions from surrounding nodes and therefore the TDMA method may prove awkward to incorporate. The strongly implicit procedure (SIP) due to Stone (1968), in particular with the improvements suggested by Schneider and Zedan (1981), is more suitable in this case. Details are not presented here in the interest of brevity and the interested reader is referred to Anderson *et al* (1984). Another solution procedure which is being used in CFD calculations is the conjugate gradient method (CGM) of Hestenes and Stiefel (1952). This method is based on matrix factorisation techniques. Improvements by Reid (1971), Concus *et al* (1976) and Kershaw (1978) ensure accelerated convergence in the CFD calculations. The CGM, nevertheless, requires greater storage than the other iterative methods described earlier. Further details of the method can also be found in Press *et al* (1992).

7.7 Summary

We have discussed the solution of systems of equations with the TDMA method. This algorithm is highly economical for tri-diagonal systems. It consists of a forward elimination and a back-substitution stage:

- *Forward elimination*
  - arrange system of equations in the form of (7.2): $-\beta_j\phi_{j-1}+D_j\phi_j - \alpha_j\phi_{j+1} = C_j$
  - calculate coefficients α_j, β_j, D_j and C_j
  - starting at $j = 2$ calculate A_j and C'_j using (7.6b–c): $A_j = \alpha_j(D_j - \beta_j A_{j-1})^{-1}$ and $C'_j = (\beta_j C'_{j-1} + C_j)(D_j - \beta_j A_{j-1})^{-1}$
  - repeat for $j = 3$ to $j = n$
- *Back-substitution*
  - starting at $j = n$ obtain ϕ_n by evaluating (7.6a): $\phi_j = A_j\phi_{j+1} + C'_j$
  - repeat for $j = n - 1$ to $j = 2$ giving ϕ_{n-1} to ϕ_2 in reverse order

The TDMA can be applied, in an iterative fashion, to two- and three-dimensional computations and is the standard algorithm for the solution of the flow equations in Cartesian co-ordinates. Problems arise if discretisation schemes are used that incorporate influences from locations other than the immediate neighbours, or if body-fitted co-ordinate systems are chosen. In these cases it may be necessary to resort to alternative techniques such as the penta-diagonal matrix algorithm or Stone's strongly implicit procedure and the conjugate gradient method.

8

The Finite Volume Method for Unsteady Flows

8.1 Introduction

Having finished the task of developing the finite volume method for steady flows we are now in a position to consider the more complex category of time-dependent problems. The conservation law for the transport of a scalar in an unsteady flow has the general form

$$\frac{\partial}{\partial t}(\rho\phi) + div(\rho\mathbf{u}\phi) = div(\Gamma\ grad\ \phi) + S_\phi \tag{8.1}$$

The first term of the equation represents the rate of change term and is zero for steady flows. To predict transient problems we must retain this term in the discretisation process. The finite volume integration of equation (8.1) over a control volume (CV) must be augmented with a further integration over a finite time step Δt. By replacing the volume integrals of the convective and diffusive terms with surface integrals as before (see section 2.5) and changing the order of integration in the rate of change term we obtain

$$\int_{CV}\left(\int_t^{t+\Delta t}\frac{\partial}{\partial t}(\rho\phi)dt\right)dV + \int_t^{t+\Delta t}\left(\int_A n\,.\,(\rho\mathbf{u}\phi)dA\right)dt$$
$$= \int_t^{t+\Delta t}\left(\int_A n\,.\,(\Gamma\ grad\ \phi)dA\right)dt + \int_t^{t+\Delta t}\int_{CV} S_\phi\ dV\ dt \tag{8.2}$$

So far we have made no approximations but to make progress we need techniques for evaluating the integrals. The control volume integration is essentially the same as in steady flows and the measures explained in Chapters 4 and 5 are again adopted to ensure successful treatment of convection, diffusion and source terms. Here we focus our attention on methods necessary for the time integration. The process is illustrated below using the one-dimensional unsteady diffusion (heat transfer) equation and is later extended to multi-dimensional unsteady diffusion and convection-diffusion problems.

8.2 One-dimensional unsteady heat conduction

Unsteady one-dimensional heat conduction is governed by the equation

$$\rho c \frac{\partial T}{\partial t} = \frac{\partial}{\partial x}\left(k \frac{\partial T}{\partial x}\right) + S \tag{8.3}$$

In addition to the usual variables we have c, the specific heat of the material (J/kg/K).

Fig. 8.1

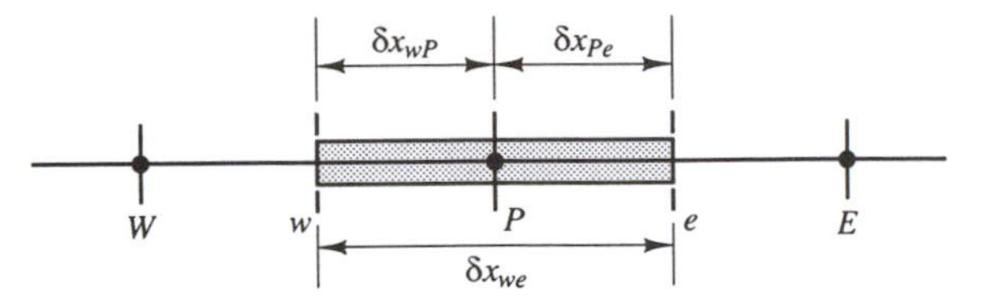

Consider the one-dimensional control volume in Figure 8.1. Integration of equation (8.3) over the control volume and over a time interval from t to $t + \Delta t$ gives

$$\int_t^{t+\Delta t} \int_{CV} \rho c \frac{\partial T}{\partial t} dV\, dt = \int_t^{t+\Delta t} \int_{CV} \frac{\partial}{\partial x}\left(k \frac{\partial T}{\partial x}\right) dV\, dt + \int_t^{t+\Delta t} \int_{CV} s dV\, dt \tag{8.4}$$

This may be written as

$$\int_w^e \left[\int_t^{t+\Delta t} \rho c \frac{\partial T}{\partial t} dt\right] dV = \int_t^{t+\Delta t} \left[\left(kA \frac{\partial T}{\partial x}\right)_e - \left(kA \frac{\partial T}{\partial x}\right)_w\right] dt + \int_t^{t+\Delta t} \bar{S} \Delta V\, dt \tag{8.5}$$

In equation (8.5), A is the face area of the control volume, ΔV is its volume, which is equal to $A\Delta x$ where Δx is the width of the control volume, and $\bar{S}$ is the average source strength. If the temperature at a node is assumed to prevail over the whole control volume, the left hand side can be written as

$$\int_{CV} \left[\int_t^{t+\Delta t} \rho c \frac{\partial T}{\partial t} dt\right] dV = \rho c (T_P - T_P^o) \Delta V \tag{8.6}$$

In equation (8.6) superscript 'o' refers to temperatures at time t; temperatures at time level $t + \Delta t$ are not superscripted. The same result as (8.6) would be obtained by substituting $(T_p - T_p^o)/\Delta t$ for $\partial T/\partial t$ so this term has been discretised using a first-order (backward) differencing scheme. Higher order schemes, which may be used to discretise this term, will be discussed briefly later in this chapter. If we apply central differencing to the diffusion terms on the right hand side equation (8.5) may be written as

$$\rho c (T_P - T_P^o) \Delta V = \int_t^{t+\Delta t} \left[\left(k_e A \frac{T_E - T_P}{\delta x_{PE}}\right) - \left(k_w A \frac{T_P - T_W}{\delta x_{WP}}\right)\right] dt + \int_t^{t+\Delta t} \bar{S} \Delta V\, dt \tag{8.7}$$

To evaluate the right hand side of this equation we need to make an assumption about the variation of T_P, T_E and T_W with time. We could use temperatures at time t or at time $t + \Delta t$ to calculate the time integral or, alternatively, a combination of temperatures at time t and $t + \Delta t$. We may generalise the approach by means of a weighting parameter θ between 0 and 1 and write the integral I_T of temperature T_P with respect to time as

$$I_T = \int_t^{t+\Delta t} T_P \, dt = \left[\theta T_P + (1-\theta)T_P^o\right]\Delta t \tag{8.8}$$

Hence

θ	0	1/2	1
I_T	$T_P^o \Delta t$	$\frac{1}{2}(T_P + T_P^o)\Delta t$	$T_P \Delta t$

We have highlighted the following values of integral I_T: if $\theta = 0$ the temperature at (old) time level t is used; if $\theta = 1$ the temperature at new time level $t + \Delta t$ is used; and finally if $\theta = 1/2$, the temperatures at t and $t + \Delta t$ are equally weighted.

Using formula (8.8) for T_W and T_E in equation (8.7), and dividing by $A\Delta t$ throughout, we have

$$\rho c\left(\frac{T_P - T_P^o}{\Delta t}\right)\Delta x = \theta\left[\frac{k_e(T_E - T_P)}{\delta x_{PE}} - \frac{k_w(T_P - T_W)}{\delta x_{WP}}\right] + (1-\theta)\left[\frac{k_e\left(T_E^o - T_P^o\right)}{\delta x_{PE}} - \frac{k_w\left(T_P^o - T_W^o\right)}{\delta x_{WP}}\right] + \bar{S}\Delta x \tag{8.9}$$

which may be re-arranged to give

$$\left[\rho c\frac{\Delta x}{\Delta t} + \theta\left(\frac{k_e}{\delta x_{PE}} + \frac{k_w}{\delta x_{WP}}\right)\right]T_P = \frac{k_e}{\delta x_{PE}}\left[\theta T_E + (1-\theta)T_E^o\right] + \frac{k_w}{\delta x_{WP}}\left[\theta T_W + (1-\theta)T_W^o\right] + \left[\rho c\frac{\Delta x}{\Delta t} - (1-\theta)\frac{k_e}{\delta x_{PE}} - (1-\theta)\frac{k_w}{\delta x_{WP}}\right]T_P^o + \bar{S}\Delta x \tag{8.10}$$

Now we identify the coefficients of T_W and T_E as a_W and a_E and write equation (8.10) in the familiar standard form:

$$a_P T_P = a_W\left[\theta T_W + (1-\theta)T_W^o\right] + a_E\left[\theta T_E + (1-\theta)T_E^o\right] + \left[a_P^o - (1-\theta)a_W - (1-\theta)a_E\right]T_P^o + b \tag{8.11}$$

where $$a_P = \theta(a_W + a_E) + a_P^o$$

and $$a_P^o = \rho c\frac{\Delta x}{\Delta t}$$

with

a_W	a_E	b
$\dfrac{k_w}{\delta x_{WP}}$	$\dfrac{k_e}{\delta x_{PE}}$	$\bar{S}\Delta x$

The exact form of the final discretised equation depends on the value of θ. When θ is zero, we only use temperatures T_P^o, T_W^o and T_E^o at the old time level t on the right hand side of equation (8.11) to evaluate T_P at the new time; the resulting scheme is called **explicit**. When $0 < \theta \leq 1$ temperatures at the new time level are used on both sides of the equation; the resulting schemes are called **implicit**. The extreme case of $\theta = 1$ is termed **fully implicit** and the case corresponding to $\theta = 1/2$ is called **the Crank–Nicolson** scheme (Crank and Nicolson, 1947).

8.2.1 Explicit scheme

In the explicit scheme the source term is linearised as $b = S_u + S_p T_p^o$. Now the substitution of $\theta = 0$ into (8.11) gives the **explicit discretisation** of the unsteady conductive heat transfer equation:

$$a_P T_P = a_W T_W^o + a_E T_E^o + \left[a_P^o - \left(a_W + a_E - S_p\right)\right] T_P^o + S_u \tag{8.12}$$

where $a_P = a_P^o$

and $a_P^o = \rho c \dfrac{\Delta x}{\Delta t}$

a_W	a_E
$\dfrac{k_w}{\delta x_{WP}}$	$\dfrac{k_e}{\delta x_{PE}}$

The right hand side of equation (8.12) only contains values at the old time step so the left hand side can be calculated by forward marching in time. The scheme is based on backward differencing and its Taylor series truncation error accuracy is first-order with respect to time. As explained in Chapter 5 all coefficients need to be positive in the discretised equation. The coefficient of T_P^o may be viewed as the neighbour coefficient connecting the values at the old time level to those at the new time level. For this coefficient to be positive we must have $a_p^o - a_W - a_E > 0$. For constant k and uniform grid spacing, $\delta x_{PE} = \delta x_{WP} = \Delta x$, this condition may be written as

$$\rho c \frac{\Delta x}{\Delta t} > \frac{2k}{\Delta x} \tag{8.13a}$$

or

$$\Delta t < \rho c \frac{(\Delta x)^2}{2k} \tag{8.13b}$$

This inequality sets a stringent maximum limit to the time step size and represents a serious limitation for the explicit scheme. It becomes very expensive to improve spatial accuracy because the maximum possible time step needs to be reduced as the square of Δx. Consequently, this method is not recommended for general transient problems. Explicit schemes with greater formal accuracy than the above one have been designed. Examples are the Richardson and the DuFort–Frankel method which use temperatures at more than two time levels. These methods also have fewer stability restrictions than the ordinary explicit method. Details of such schemes can be found in Abbot and Basco (1990), Anderson *et al* (1984) and Fletcher (1991). Nevertheless, provided that the time step size is chosen with care, the explicit scheme described above is efficient for simple conduction calculations. This will be illustrated through a further example in section 8.3.

8.2.2 Crank–Nicolson scheme

The Crank–Nicolson method results from setting $\theta = 1/2$ in equation (8.11). Now the discretised unsteady heat conduction equation is

$$a_P T_P = a_E \left[\frac{T_E + T_E^o}{2}\right] + a_W \left[\frac{T_W + T_W^o}{2}\right] + \left[a_P^o - \frac{a_E}{2} - \frac{a_W}{2}\right] T_P^o + b \tag{8.14}$$

where $$a_P = \frac{1}{2}(a_W + a_E) + a_P^o - \frac{1}{2} S_p$$

and $$a_P^o = \rho c \frac{\Delta x}{\Delta t}$$

a_W	a_E	b
$\dfrac{k_w}{\delta x_{WP}}$	$\dfrac{k_e}{\delta x_{PE}}$	$S_u + \frac{1}{2} S_p T_P^o$

Since more than one unknown value of T at the new time level is present in equation (8.14) the method is implicit and simultaneous equations for all node points need to be solved at each time step. Although schemes with $1/2 \leq \theta \leq 1$, including the Crank–Nicolson scheme, are unconditionally stable for all values of the time step (Fletcher, 1991) it is more important to ensure that all coefficients are positive for physically realistic and bounded results. This is the case if the coefficient of T_p^o

satisfies the following condition:

$$a_P^o > \left[\frac{a_E + a_W}{2}\right]$$

which leads to

$$\boxed{\Delta t < \rho c \frac{\Delta x^2}{k}} \tag{8.15}$$

This time step limitation is only slightly less restrictive than (8.13) associated with the explicit method. The Crank–Nicolson method is based on central differencing and hence it is second-order accurate in time. With sufficiently small time steps it is possible to achieve considerably greater accuracy than with the explicit method. The overall accuracy of a computation depends also on the spatial differencing practice, so the Crank–Nicolson scheme is normally used in conjunction with spatial central differencing.

8.2.3 The fully implicit scheme

When the value of θ is set equal to 1 we obtain the fully implicit scheme. The discretised equation is

$$\boxed{a_P T_P = a_W T_W + a_E T_E + a_P^o T_P^o + S_u} \tag{8.16}$$

where $\boxed{a_P = a_P^o + a_W + a_E - S_p}$

and $\boxed{a_P^o = \rho c \frac{\Delta x}{\Delta t}}$

with

a_W	a_E
$\dfrac{k_w}{\delta x_{WP}}$	$\dfrac{k_e}{\delta x_{PE}}$

Both sides of the equation contain temperatures at the new time step, and a system of algebraic equations must be solved at each time level (see Example 8.2). The time marching procedure starts with a given initial field of temperatures T^o. The system of equations (8.16) is solved after selecting time step Δt. Next the solution T is assigned to T^o and the procedure is repeated to progress the solution by a further time step.

It can be seen that all coefficients are positive, which makes the implicit scheme unconditionally stable for any size of time step. Since the accuracy of the scheme is only first-order in time, small time steps are needed to ensure the accuracy of results. The implicit method is recommended for general purpose transient calculations because of its robustness and unconditional stability.

8.3 Illustrative examples

We now demonstrate the properties of the explicit and implicit discretisation schemes by means of a comparison of numerical results for a one-dimensional unsteady conduction example with analytical solutions to assess the accuracy of the methods.

Example 8.1 A thin plate is initially at a uniform temperature of 200 °C. At a certain time $t = 0$ the temperature of the east side of the plate is suddenly reduced to 0 °C. The other surface is insulated. Use the explicit finite volume method in conjunction with a suitable time step size to calculate the transient temperature distribution of the slab and compare it with the analytical solution at time (i) $t = 40$ s, (ii) $t = 80$ s and (iii) $t = 120$ s. Recalculate the numerical solution using a time step size equal to the limit given by (8.13) for $t = 40$ s and compare the results with the analytical solution. The data are: plate thickness $L = 2$ cm, thermal conductivity $k = 10$ W/m/K and $\rho c = 10 \times 10^6$ J/m^3/K.

Solution The one-dimensional transient heat conduction equation is

$$\rho c \frac{\partial T}{\partial t} = \frac{\partial}{\partial x}\left(k \frac{\partial T}{\partial x}\right) \tag{8.17}$$

and the initial conditions are

$$T = 200 \quad \text{at} \quad t = 0$$

and the boundary conditions are

$$\frac{\partial T}{\partial x} = 0 \quad \text{at} \quad x = 0,\ t > 0$$

$$T = 0 \quad \text{at} \quad x = L,\ t > 0$$

The analytical solution is given in Ozisik (1985) as

$$\frac{T(x,\ t)}{200} = \frac{4}{\pi} \sum_{n=1}^{\infty} \frac{(-1)^{n+1}}{2n-1} \exp\left(-\alpha \lambda_n^2 t\right) \cos(\lambda_n x) \tag{8.18}$$

$$\text{where } \lambda_n = \frac{(2n-1)\pi}{2L} \text{ and } \alpha = k/\rho c$$

The numerical solution with the explicit method is generated by dividing the domain width L into five equal control volumes with $\Delta x = 0.004$ m. The resulting one-dimensional grid is shown in Figure 8.2.

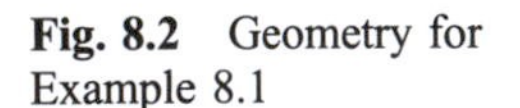
Fig. 8.2 Geometry for Example 8.1

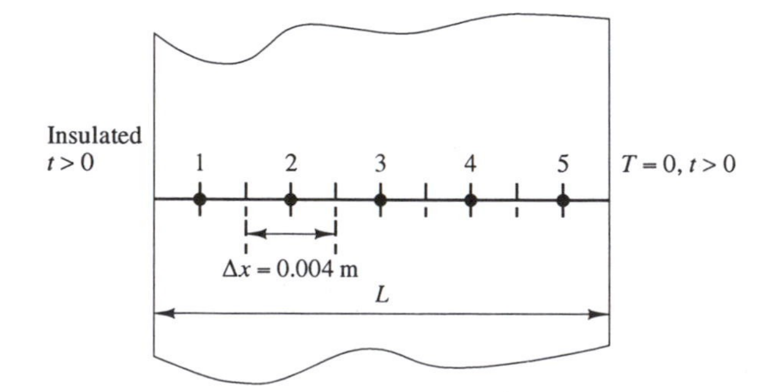

The discretised form of governing equation (8.17) for an internal control volume using the explicit method is given by (8.12). Control volumes 1 and 5 adjoin boundaries, so the links are cut in the direction of the boundary and the boundary fluxes are included in the source terms. At the control volume 1, the west boundary is insulated; hence the flux across that boundary is zero. We modify the equation (8.9) where the physics can be most easily discerned. The discretised equation at node 1 becomes

$$\rho c \frac{(T_P - T_P^o)}{\Delta t} \Delta x = \left[\frac{k}{\Delta x}(T_E^o - T_P^o)\right] - 0 \tag{8.19}$$

For time $t > 0$, the temperature of the east boundary of control volume 5 is constant (say T_B). The discretised equation at node 5 becomes

$$\rho c \frac{(T_P - T_P^o)}{\Delta t} \Delta x = \left[\frac{k}{\Delta x/2}(T_B - T_P^o)\right] - \left[\frac{k}{\Delta x}(T_P^o - T_W^o)\right] \tag{8.20}$$

All discretised equations can now be written in standard form:

$$a_P T_P = a_W T_W^o + a_E T_E^o + \left[a_P^o - (a_W + a_E)\right] T_P^o + S_u \tag{8.21}$$

where $a_P = a_P^o = \rho c \dfrac{\Delta x}{\Delta t}$

and

Node	a_W	a_E	S_u
1	0	$k/\Delta x$	0
2, 3, 4	$k/\Delta x$	$k/\Delta x$	0
5	$k/\Delta x$	0	$\frac{2k}{\Delta x}(T_B - T_P^o)$

The time step for the explicit method is subject to the condition that

$$\Delta t < \frac{\rho c(\Delta x)^2}{2k}$$

$$\Delta t < \frac{10 \times 10^6 (0.004)^2}{2 \times 10}$$

$$\Delta t < 8\ \text{s}$$

Let us select $\Delta t = 2$ s. Substituting numerical values we have

$$\frac{k}{\Delta x} = \frac{10}{0.004} = 2500$$

$$\rho c \frac{\Delta x}{\Delta t} = 10 \times 10^6 \times \frac{0.004}{2} = 20000$$

After substitution of numerical values and some simplification the discretisation equations for the various nodes are:

$$\begin{aligned} &\text{Node 1:} && 200T_P = 25T_E^o + 175T_P^o \\ &\text{Nodes 2–4:} && 200T_P = 25T_W^o + 25T_E^o + 150T_P^o \\ &\text{Node 5:} && 200T_P = 25T_W^o + 125T_P^o \end{aligned} \tag{8.22}$$

Starting with the initial condition where all the nodes are at a temperature of 200 °C, the solution at each time step is obtained using equations (8.22). Although the calculations are not complicated, their number is large and they are most effectively carried out by a computer program. Table 8.1 gives a sample of the calculations for the first two time steps.

Table 8.2 shows the results for 10 consecutive time steps and Table 8.3 shows the numerical and analytical results at times 40, 80 and 120 s. As can be seen from the

Table 8.1 Specimen calculations for the explicit method

Time	*Node 1*	*Node 2*	*Node 3*	*Node 4*	*Node 5*
$t = 0\text{s}$	$T_1^0 = 200$	$T_2^0 = 200$	$T_3^0 = 200$	$T_4^0 = 200$	$T_5^0 = 200$
1	$200T_1^1 = 25 \times 200$ $+175 \times 200$	$200T_2^1 = 25 \times 200$ $+25 \times 200$	$200T_3^1 = 25 \times 200$ $+25 \times 200$	$200T_4^1 = 25 \times 200$ $+25 \times 200$	$200T_5^1 = 25 \times 200$ $+125 \times 200$
$t = 2\text{s}$	$+150 \times 200$ $T_1^1 = 200$	$+150 \times 200$ $T_2^1 = 200$	$+150 \times 200$ $T_3^1 = 200$	$T_4^1 = 200$	$T_5^1 = 150$
2	$200T_1^2 = 25 \times 200$ $+175 \times 200$	$200T_2^2 = 25 \times 200$ $+25 \times 200$ $+150 \times 200$	$200T_3^2 = 25 \times 200$ $+25 \times 200$ $+150 \times 200$	$200T_4^2 = 25 \times 200$ $+25 \times 150$ $+150 \times 200$	$200T_5^2 = 25 \times 200$ $+125 \times 150$
$t = 4\text{s}$	$T_1^2 = 200$	$T_2^2 = 200$	$T_3^2 = 200$	$T_4^2 = 193.75$	$T_5^2 = 118.75$

Note: Subscripts denote the node number, superscripts denote the time step

Table 8.2 Results for Example 8.1 (explicit method)

Time step	*Time (s)*	*Node number*						
			1	*2*	*3*	*4*	*5*	
		$x = 0.0$	$x = 0.002$	$x = 0.006$	$x = 0.01$	$x = 0.014$	$x = 0.016$	$x = 0.018$
0	0	200	200	200	200	200	200	200
1	2	200	200	200	200	200	150	0
2	4	200	200	200	200	193.75	118.75	0
3	6	200	200	200	199.21	185.16	98.43	0
4	8	200	200	199.9	197.55	176.07	84.66	0
5	10	199.98	199.98	199.62	195.16	167.33	74.92	0
6	12	199.94	199.94	199.11	192.24	159.26	67.74	0
7	14	199.83	199.83	198.35	188.98	151.94	62.24	0
8	16	199.65	199.65	197.36	185.52	145.36	57.89	0
9	18	199.37	199.37	196.17	181.98	139.45	54.35	0
10	20	198.97	198.97	194.79	178.44	134.12	51.40	0

Table 8.3

Point	*Time = 40 s*			*Time = 80 s*			*Time = 120 s*		
	Numerical	Analytical	% error	Numerical	Analytical	% error	Numerical	Analytical	% error
1	188.64	188.39	−0.13	153.33	152.65	−0.43	120.53	119.87	−0.55
2	176.41	175.76	−0.36	139.05	138.36	−0.50	108.82	108.21	−0.56
3	148.29	147.13	−0.79	111.29	110.63	−0.59	86.47	85.96	−0.58
4	100.76	99.50	−1.26	72.06	71.56	−0.69	55.58	55.25	−0.60
5	35.94	35.38	−1.57	24.96	24.77	−0.75	19.16	19.05	−0.59

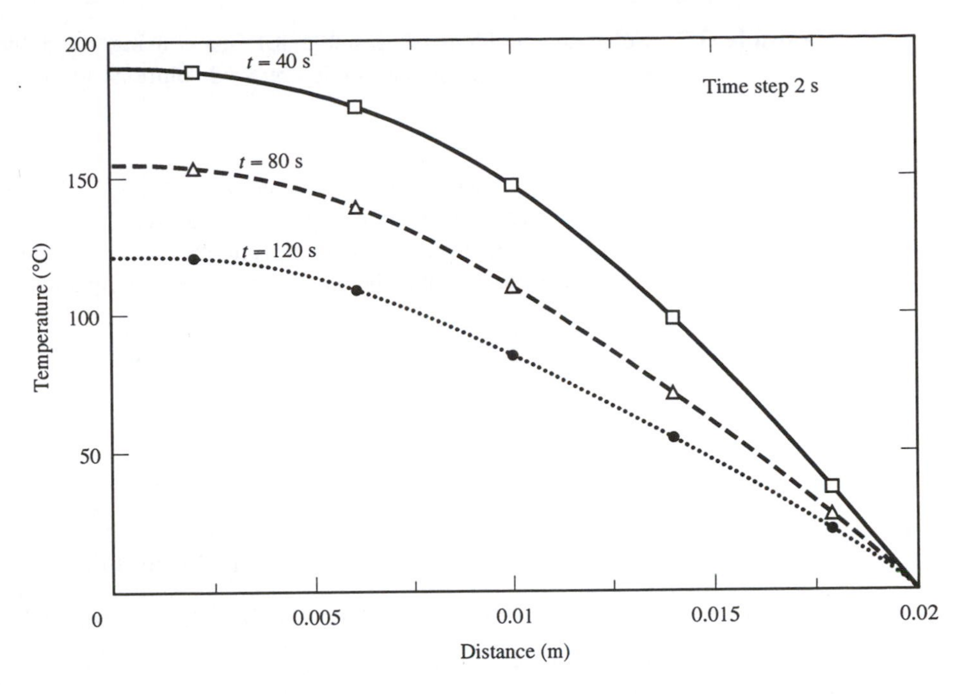

Fig. 8.3 Comparison of numerical and analytical solutions at different times

error analysis the results are in good agreement with the analytical solution. Figure 8.3 shows the comparison in a graphical form.

Figure 8.4 shows the solution for time $t = 40$ s with a time step of 8 s. The previous result with a step size of 2 s and the exact solution are also shown for comparison. We conclude that a time step equal to the limiting value of 8 s gives a very inaccurate and unrealistic numerical solution that oscillates about the exact solution.

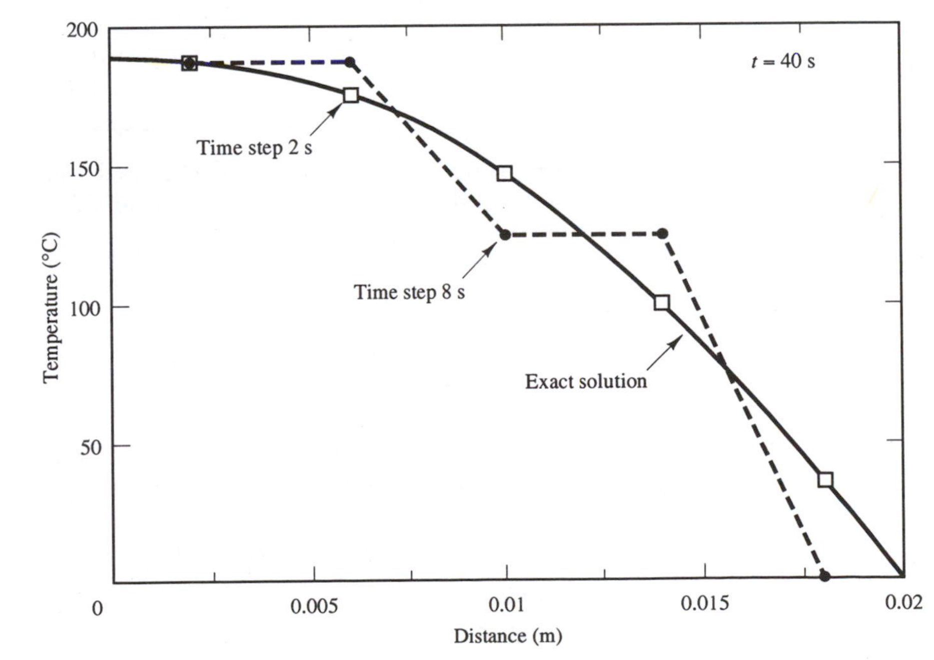

Fig. 8.4 Comparison of results obtained using different time step values

Example 8.2 Solve the problem of Example 8.1 again using the fully implicit method and compare the explicit and implicit method solutions for a time step of 8 s.

Solution Let us use the same grid arrangement as in Figure 8.2. The fully implicit method describes events at internal control volumes 2, 3 and 4 by means of discretised equation (8.16). Boundary control volumes 1 and 5 again need special treatment. Upon incorporating the boundary conditions into equation (8.9) we get for node 1:

$$\rho c \frac{(T_P - T_P^o)}{\Delta t} \Delta x = \left[\frac{k}{\Delta x}(T_E - T_P)\right] - 0 \tag{8.23}$$

and for node 5:

$$\rho c \frac{(T_P - T_P^o)}{\Delta t} \Delta x = \left[\frac{k}{\Delta x/2}(T_B - T_P)\right] - \left[\frac{k}{\Delta x}(T_P - T_W)\right] \tag{8.24}$$

The discretised equations are written in standard form:

$$a_P T_P = a_W T_W + a_E T_E + a_P^o T_P^o + S_u \tag{8.25}$$

where $a_P = a_W + a_E + a_P^o - S_p$

and $a_P^o = \rho c \dfrac{\Delta x}{\Delta t}$

and

Node	a_W	a_E	S_p	S_u
1	0	$k/\Delta x$	0	0
2, 3, 4	$k/\Delta x$	$k/\Delta x$	0	0
5	$k/\Delta x$	0	$-\dfrac{2k}{\Delta x}$	$\dfrac{2k}{\Delta x} T_B$

Although the implicit method permits large values for the time step Δt, we will use reasonably small time steps of 2 s to ensure good accuracy. The grid spacing and other data are as before so again we have

$$\frac{k}{\Delta x} = \frac{10}{0.004} = 2500$$

$$\rho c \frac{\Delta x}{\Delta t} = 10 \times 10^6 \times \frac{0.004}{2} = 20000$$

After substitution of numerical values and the necessary simplification the discretised equations for the various nodes are:

$$\begin{aligned} &\text{Node 1}: && 225 T_P = 25 T_E + 200 T_P^o \\ &\text{Nodes 2–4}: && 250 T_P = 25 T_W + 25 T_E + 200 T_P^o \\ &\text{Node 5}: && 275 T_P = 25 T_W + 200 T_P^o + 50 T_B \end{aligned}$$

Noting that $T_B = 0$, the set of equations to be solved at each time step is

$$\begin{bmatrix} 225 & -25 & 0 & 0 & 0 \\ -25 & 250 & -25 & 0 & 0 \\ 0 & -25 & 250 & -25 & 0 \\ 0 & 0 & -25 & 250 & -25 \\ 0 & 0 & 0 & -25 & 275 \end{bmatrix} \begin{bmatrix} T_1 \\ T_2 \\ T_3 \\ T_4 \\ T_5 \end{bmatrix} = \begin{bmatrix} 200T_1^o \\ 200T_2^o \\ 200T_3^o \\ 200T_4^o \\ 200T_5^o \end{bmatrix} \tag{8.26}$$

The matrix form emphasises that the equations for each point contain unknown neighbouring temperatures. The explicit scheme involves a straightforward evaluation of a single algebraic equation to find each new nodal temperature, but the fully implicit method requires the (more expensive) solution of a system (8.26) at each time level. The values of temperature at the previous time level are used to calculate the right hand side. Table 8.4 and Figure 8.5 show that the numerical results again compare favourably with the analytical solution.

In Figure 8.6 we give the solution at $t = 40$ s obtained using the implicit and explicit methods with a time step of 8 s along with the analytical solution. Whereas the explicit method gives unrealistic oscillations at this step size, the implicit method

Table 8.4

Point	*Time* = 40 *s*			*Time* = 80 *s*			*Time* = 120 *s*		
	Numerical	Analytical	% error	Numerical	Analytical	% error	Numerical	Analytical	% error
1	187.38	188.38	0.51	153.72	152.65	−0.70	121.52	119.87	−1.42
2	176.28	175.76	−0.29	139.79	138.36	−1.03	109.78	108.21	−1.24
3	150.04	147.13	−1.97	112.38	110.63	−1.57	87.33	85.96	−1.59
4	103.69	99.50	−4.20	73.09	71.56	−2.13	56.20	55.25	−1.71
5	37.51	35.38	−6.02	25.38	24.77	−2.46	19.39	19.05	−1.78

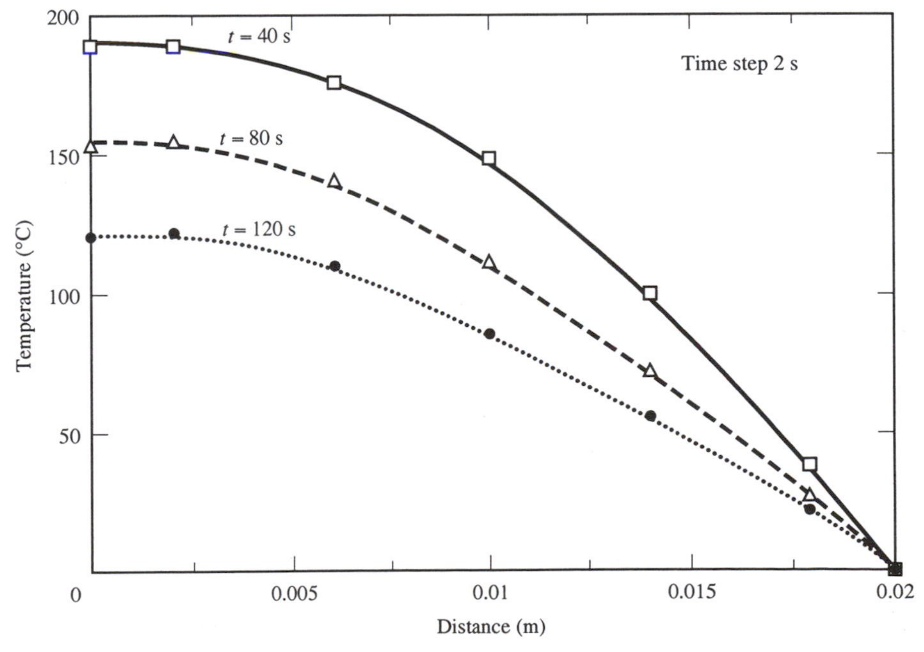

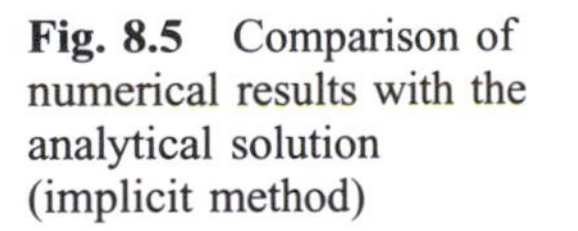
Fig. 8.5 Comparison of numerical results with the analytical solution (implicit method)

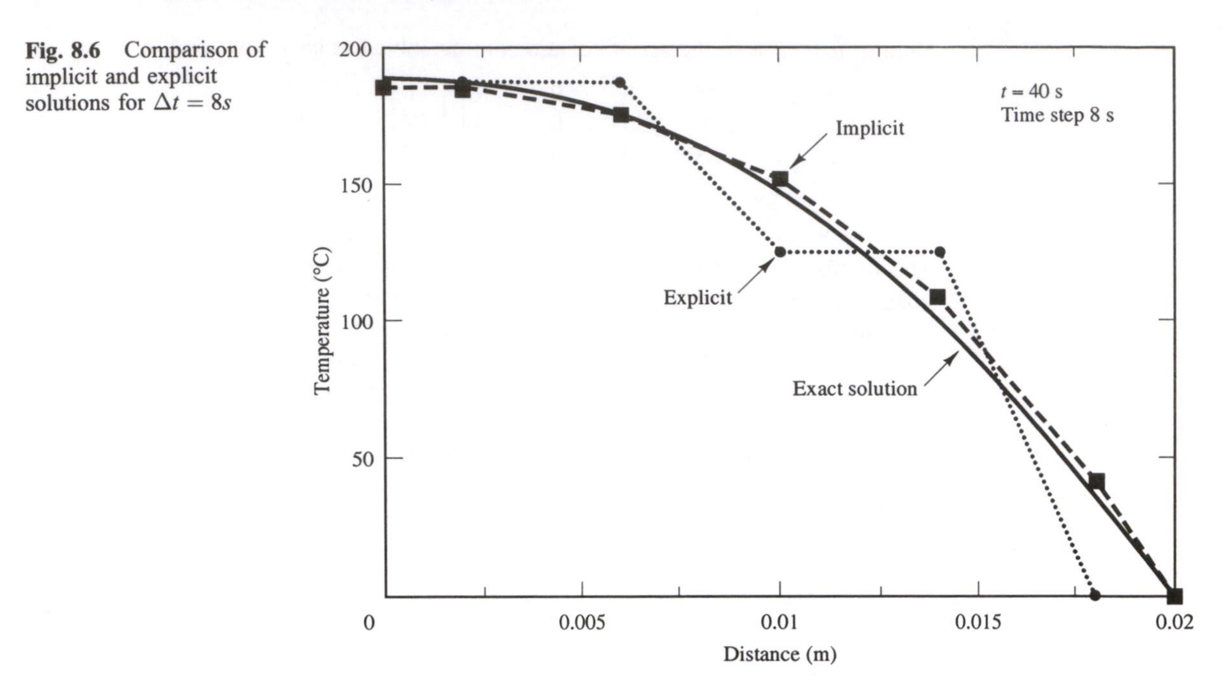

Fig. 8.6 Comparison of implicit and explicit solutions for $\Delta t = 8s$

gives results that are in reasonable agreement with the exact solution. This clearly illustrates a key advantage of the implicit method, which tolerates much larger time steps. However, we stress that good solution accuracy can, of course, only be achieved with small time steps.

8.4 Implicit method for two- and three-dimensional problems

The fully implicit method is recommended for general purpose CFD computations on the grounds of its superior stability. We now quote its extension to calculations in two and three space dimensions. Transient diffusion in three dimensions is governed by

$$\rho c \frac{\partial \phi}{\partial t} = \frac{\partial}{\partial x}\left(k \frac{\partial \phi}{\partial x}\right) + \frac{\partial}{\partial y}\left(k \frac{\partial \phi}{\partial y}\right) + \frac{\partial}{\partial z}\left(k \frac{\partial \phi}{\partial z}\right) + S \tag{8.27}$$

A three-dimensional control volume is considered for the discretisation. The resulting equation is

$$\boxed{\begin{aligned} a_P\phi_P = {} & a_W\phi_W + a_E\phi_E + a_S\phi_S + a_N\phi_N + a_B\phi_B + a_T\phi_T \\ & + a_P^o\phi_P^o + S_u \end{aligned}} \tag{8.28}$$

where $$\boxed{a_P = a_W + a_E + a_S + a_N + a_B + a_T + a_P^o - S_p}$$

$$\boxed{a_P^o = \rho c \frac{\Delta V}{\Delta t}}$$

The neighbouring coefficients are a_W, a_E in one-dimensional problems, and a_W, a_E, a_S, a_N in two and $a_W, a_E, a_S, a_N, a_B, a_T$ in three dimensions; $b = (S_u + S_p\phi_P)$ is the linearised source. A summary of the relevant neighbour coefficients is given below:

	a_W	a_E	a_S	a_N	a_B	a_T
1D	$\frac{\Gamma_w A_w}{\delta x_{WP}}$	$\frac{\Gamma_e A_e}{\delta x_{PE}}$	–	–	–	–
2D	$\frac{\Gamma_w A_w}{\delta x_{WP}}$	$\frac{\Gamma_e A_e}{\delta x_{PE}}$	$\frac{\Gamma_s A_s}{\delta y_{SP}}$	$\frac{\Gamma_n A_n}{\delta y_{PN}}$	–	–
3D	$\frac{\Gamma_w A_w}{\delta x_{WP}}$	$\frac{\Gamma_e A_e}{\delta x_{PE}}$	$\frac{\Gamma_s A_s}{\delta y_{SP}}$	$\frac{\Gamma_n A_n}{\delta y_{PN}}$	$\frac{\Gamma_b A_b}{\delta z_{BP}}$	$\frac{\Gamma_t A_t}{\delta z_{PT}}$

The following values for the volume and cell face areas apply in the three cases:

	1D	2D	3D
ΔV	Δx	$\Delta x \Delta y$	$\Delta x \Delta y \Delta z$
$A_w = A_e$	1	Δy	$\Delta y \Delta z$
$A_n = A_s$	–	Δx	$\Delta x \Delta z$
$A_b = A_t$	–	–	$\Delta x \Delta y$

8.5 Discretisation of transient convection–diffusion equation

In the fully implicit discretisation approach outlined above for multi-dimensional diffusion problems, the term arising from temporal discretisation appears as (i) the contribution of a_p^o to the central coefficient a_P and (ii) the contribution of $a_p^o \phi_p^o$ as an additional source term on the right hand side. The other coefficients are unaltered and are the same as in the discretised equations for steady state problems. Using this as a basis the discretised equations for transient convection–diffusion equations are also simple to obtain. The unsteady transport of a property ϕ is given by

$$\frac{\partial}{\partial t}(\rho\phi) + div(\rho \mathbf{u}\phi) = div(\Gamma\ grad\ \phi) + S_\phi \tag{8.29}$$

The hybrid differencing scheme was recommended in Chapter 5 on the grounds of its stability as the preferred method for the treatment of convection terms, so here we quote the implicit/hybrid difference form of the transient convection–diffusion equations.

Transient three-dimensional convection–diffusion of a general property ϕ in a velocity field **u** is governed by

$$\frac{\partial(\rho\phi)}{\partial t} + \frac{\partial(\rho u\phi)}{\partial x} + \frac{\partial(\rho v\phi)}{\partial y} + \frac{\partial(\rho w\phi)}{\partial z} = \frac{\partial}{\partial x}\left(\Gamma\frac{\partial\phi}{\partial x}\right) + \frac{\partial}{\partial y}\left(\Gamma\frac{\partial\phi}{\partial y}\right) + \frac{\partial}{\partial z}\left(\Gamma\frac{\partial\phi}{\partial z}\right) + S \tag{8.30}$$

The fully implicit discretisation equation is

$$a_P\phi_P = a_W\phi_W + a_E\phi_E + a_S\phi_S + a_N\phi_N + a_B\phi_B + a_T\phi_T + a_P^o\phi_P^o + S_u \tag{8.31}$$

where $$a_P = a_W + a_E + a_S + a_N + a_B + a_T + a_P^o + \Delta F - S_p$$

with $$a_P^o = \frac{\rho_P^o \Delta V}{\Delta t}$$

and $$\bar{S}\Delta V = S_u + S_p\phi_P$$

The neighbour coefficients of this equation for the hybrid differencing scheme are as follows:

	One-dimensional flow	Two-dimensional flow	Three-dimensional flow
a_W	$\max\left[F_w, \left(D_w + \frac{F_w}{2}\right), 0\right]$	$\max\left[F_w, \left(D_w + \frac{F_w}{2}\right), 0\right]$	$\max\left[F_w, \left(D_w + \frac{F_w}{2}\right), 0\right]$
a_E	$\max\left[-F_e, \left(D_e + \frac{F_e}{2}\right), 0\right]$	$\max\left[-F_e, \left(D_e - \frac{F_e}{2}\right), 0\right]$	$\max\left[-F_e, \left(D_e - \frac{F_e}{2}\right), 0\right]$
a_S	–	$\max\left[F_s, \left(D_s + \frac{F_s}{2}\right), 0\right]$	$\max\left[F_s, \left(D_s + \frac{F_s}{2}\right), 0\right]$
a_N	–	$\max\left[-F_n, \left(D_n - \frac{F_n}{2}\right), 0\right]$	$\max\left[-F_n, \left(D_n - \frac{F_n}{2}\right), 0\right]$
a_B	–	–	$\max\left[F_b, \left(D_b + \frac{F_b}{2}\right), 0\right]$
a_T	–	–	$\max\left[-F_t, \left(D_t - \frac{F_t}{2}\right), 0\right]$
ΔF	$F_e - F_w$	$F_e - F_w + F_n - F_s$	$F_e - F_w + F_n - F_s + F_t - F_b$

In the above expressions the values of F and D are calculated with the following formulae:

Face	w	e	s	n	b	t
F	$(\rho u)_w A_w$	$(\rho u)_e A_e$	$(\rho v)_s A_s$	$(\rho v)_n A_n$	$(\rho w)_b A_b$	$(\rho w)_t A_t$
D	$\frac{\Gamma_w}{\delta x_{WP}} A_w$	$\frac{\Gamma_e}{\delta x_{PE}} A_e$	$\frac{\Gamma_s}{\delta y_{SP}} A_s$	$\frac{\Gamma_n}{\delta y_{PN}} A_n$	$\frac{\Gamma_b}{\delta z_{BP}} A_b$	$\frac{\Gamma_t}{\delta z_{PT}} A_t$

The volumes and cell face areas given in section 8.4 apply here as well.

Other schemes such as linear upwind or QUICK may be incorporated into these equations by substituting the appropriate expressions for the coefficients as will be demonstrated in the following example.

8.6 Worked example of transient convection–diffusion using QUICK differencing

Example 8.3 Consider convection and diffusion in the one-dimensional domain sketched in Figure 8.7. Calculate the transient temperature field if the initial temperature is zero everywhere and the boundary conditions are $\phi = 0$ at $x = 0$ and $\partial\phi/\partial x = 0$ at $x = L$. The data are $L = 1.5$ m, $u = 2$ m/s, $\rho = 1.0$ kg/m^3 and $\Gamma = 0.03$ kg/m/s. The

Fig. 8.7

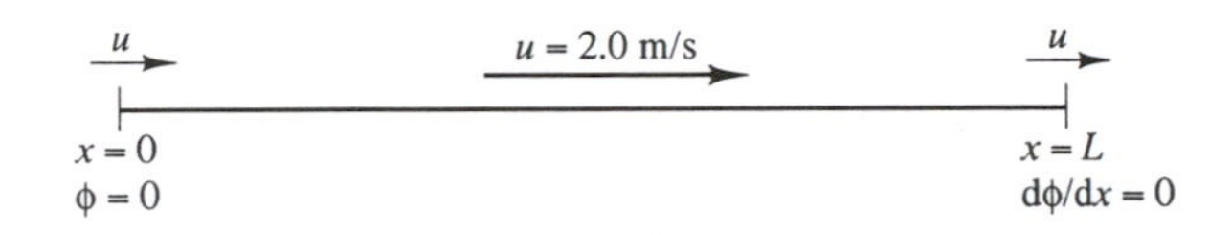

source distribution defined by Figure 8.8 applies at times $t > 0$ with $a = -200$, $b = 100$, $x_1 = 0.6$ m, $x_2 = 0.2$ m. Write a computer program to calculate the transient temperature distribution until it reaches a steady state using the implicit method for time integration and the Hayase *et al* variant of the QUICK scheme for the convective and diffusive terms and compare this result with the analytical steady state solution.

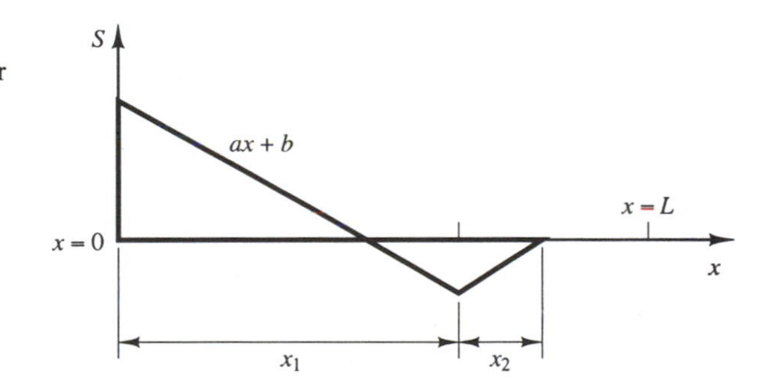

Fig. 8.8 Geometry and the source distribution for Example 8.3

Solution Transient convection–diffusion of a property ϕ subjected to a distributed source term is governed by

$$\frac{\partial(\rho\phi)}{\partial t} + \frac{\partial(\rho u\phi)}{\partial x} = \frac{\partial}{\partial x}\left(\Gamma\frac{\partial\phi}{\partial x}\right) + S \tag{8.32}$$

We use a 45 point grid to subdivide the domain and perform all calculations with a computer program. It is convenient to use the Hayase *et al* formulation of QUICK (see Section 5.9.3) since it gives a tri-diagonal system of equations which can be solved iteratively with the TDMA (see section 7.2).

The velocity is $u = 2.0$ m/s and the cell width is $\Delta x = 0.0333$ so $F = \rho u = 2.0$ and $D = \Gamma/\delta x = 0.9$ everywhere. The Hayase *et al* formulation gives ϕ at cell faces by means of the following formulae:

$$\phi_e = \phi_P + \frac{1}{8}(3\phi_E - 2\phi_P - \phi_W) \tag{8.33}$$

$$\phi_w = \phi_W + \frac{1}{8}(3\phi_P - 2\phi_W - \phi_{WW}) \tag{8.34}$$

The implicit discretisation equation at a general node with Hayase's *et al* QUICK scheme is given by

$$\begin{aligned}\frac{\rho(\phi_P - \phi_P^o)\Delta x}{\Delta t} &+ F_e\left[\phi_P + \frac{1}{8}(3\phi_E - 2\phi_P - \phi_W)\right] \\ &- F_w\left[\phi_W + \frac{1}{8}(3\phi_P - 2\phi_W - \phi_{WW})\right] \\ &= D_e(\phi_E - \phi_P) - D_w(\phi_P - \phi_W)\end{aligned} \tag{8.35}$$

The first and last nodes need to be treated separately. At control volume 1 the mirror node approach, introduced in section 5.9.1, can be used to create a west (W) node beyond the boundary at $x = 0$. Since $\phi_A = 0$ at this boundary (A) the linearly extrapolated value at the mirror node is given by

$$\phi_0 = -\phi_P \tag{8.36}$$

and the diffusive flux at the boundary by

$$\Gamma\frac{\partial\phi}{\partial x}\bigg|_A = \frac{D_A}{3}(9\phi_P - 8\phi_A - \phi_E) \tag{8.37}$$

The discretisation equation at node 1 may be written as

$$\begin{aligned}\frac{\rho(\phi_P - \phi_P^o)\Delta x}{\Delta t} &+ F_e\left[\phi_P + \frac{1}{8}(3\phi_E - \phi_P)\right] - F_A\phi_A \\ &= D_e(\phi_E - \phi_P) - \frac{D_A}{3}(9\phi_P - 8\phi_A - \phi_E)\end{aligned} \tag{8.38}$$

At the last control volume, the zero gradient boundary condition applies so the diffusive flux through the boundary B equals zero and the value ϕ at the boundary is equal to the upstream nodal value, i.e. $\phi_B = \phi_P$. The discretisation equation for control volume 45 becomes

$$\begin{aligned}\frac{\rho(\phi_P - \phi_P^o)\Delta x}{\Delta t} &+ F_B\phi_P - F_w\left[\phi_W + \frac{1}{8}(3\phi_P - 2\phi_W - \phi_{WW})\right] \\ &= 0 - D_w(\phi_P - \phi_W)\end{aligned}$$

These discretisation equations (8.35), (8.37) and (8.40) are now cast in standard form:

$$a_P\phi_P = a_W\phi_W + a_E\phi_E + a_P^o\phi_P^o + S_u \tag{8.40}$$

with $\quad a_P = a_W + a_E + a_P^o + (F_e - F_w) - S_p$

$$a_P^o = \frac{\rho\Delta x}{\Delta t}$$

and

Node	a_W	a_E	S_p	S_u
1	0	$D_e + \frac{D_A}{3}$	$-\left(\frac{8}{3}D_A + F_A\right)$	$\left(\frac{8}{3}D_A + F_A\right)\phi_A + \frac{1}{8}F_e(\phi_P - 3\phi_E)$
2	$D_w + F_w$	D_e	0	$\frac{1}{8}F_w(3\phi_P - \phi_W) + \frac{1}{8}F_e(\phi_W + 2\phi_P - 3\phi_E)$
3–44	$D_w + F_w$	D_e	0	$\frac{1}{8}F_w(3\phi_P - 2\phi_W - \phi_{WW}) + \frac{1}{8}F_e(\phi_W + 2\phi_P - 3\phi_E)$
45	$D_w + F_w$	0	0	$\frac{1}{8}F_w(3\phi_P - 2\phi_W - \phi_{WW})$

The discretisation equation for control volume 2 has been adjusted to take into account the special expression that was used to evaluate the convective flux through the cell face it has in common with control volume 1.

A time step $\Delta t = 0.01$ s is selected, which is well within the stability limit for explicit schemes so we can look forward to reasonably accurate and stable results with the implicit method. At any given time level substitution of numerical values gives the coefficients summarised in Table 8.5.

Table 8.5

Node	a_W	a_E	a_P^o	*Total source*	S_p	a_P
1	0	1.2	3.33	$4.4\phi_A + 0.25(\phi_P - 3\phi_E) + 3.33\phi_P^o$	−4.4	8.93
2	2.9	0.9	3.33	$0.025(5\phi_P - 3\phi_E) + 3.33\phi_P^o$	0	7.13
3–44	2.9	0.9	3.33	$0.25(5\phi_P - \phi_W - \phi_{WW} - 3\phi_E) + 3.33\phi_P^o$	0	7.13
45	2.9	0	3.33	$0.25(3\phi_P - 2\phi_W - \phi_{WW}) + 3.33\phi_P^o$	0	6.23

Starting with an initial field of $\phi_P^o = 0$ at all nodes, the set of equations defined by the coefficients and source contributions in Table 8.5 is solved iteratively until a converged solution ϕ_P is obtained. Subsequently, the ϕ_P-values at the current time level are assigned to ϕ_P^o and the solution proceeds to the next time level. To monitor whether the steady state has been reached we track the difference between old and new ϕ_P-values. When this attains a magnitude less than a prescribed small tolerance (say 10^{-9}) the solution is regarded as having reached the steady state.

The Analytical Solution

To find the exact steady state solution of (8.32) its time derivative is set to zero and the resulting ordinary differential equation is integrated twice with respect to x. The even periodic extension of the source distribution on an interval $(-L, L)$ is represented by means of a Fourier cosine series, which gives the forcing function in the differential equation. Under the given boundary conditions the solution to the problem is as follows:

$$\phi(x) = C_1 + C_2 e^{Px} - \frac{a_0}{P^2}(Px+1) - \sum_{n=1}^{\infty} a_n \left(\frac{L}{n\pi}\right)\left[P \sin\left(\frac{n\pi x}{L}\right) + \left(\frac{n\pi}{L}\right)\cos\left(\frac{n\pi x}{L}\right)\right] \Big/ \left[P^2 + \left(\frac{n\pi}{L}\right)^2\right] \quad (8.41)$$

with $P = \dfrac{\rho u}{\Gamma}$; $C_2 = \dfrac{a_0}{P^2 e^{PL}} + \sum_{n=1}^{\infty} \dfrac{a_n}{e^{PL}} \cos(n\pi) \Big/ \left[P^2 + \left(\dfrac{n\pi}{L}\right)^2\right]$

and $C_1 = -C_2 + \dfrac{a_0}{P^2} + \sum_{n=1}^{\infty} a_n \Big/ \left[P^2 + \left(\dfrac{n\pi}{L}\right)^2\right]$

and

$$a_0 = \frac{(x_1+x_2)(ax_1+b)+bx_1}{2L}$$

$$a_n = \frac{2L}{n^2\pi^2}\left\{\left(\frac{a(x_1+x_2)+b}{x_2}\right)\cos\left(\frac{n\pi x_1}{L}\right) - \left[a + \left(\frac{ax_1+b}{x_2}\right)\cos\left(\frac{n\pi(x_1+x_2)}{L}\right)\right]\right\}$$

The analytical and numerical steady state solutions are compared in Figure 8.9. As can be seen the use of the QUICK scheme and a fine grid for spatial discretisation ensure near-perfect agreement.

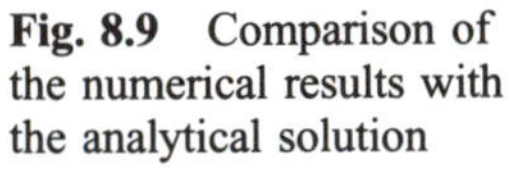

Fig. 8.9 Comparison of the numerical results with the analytical solution

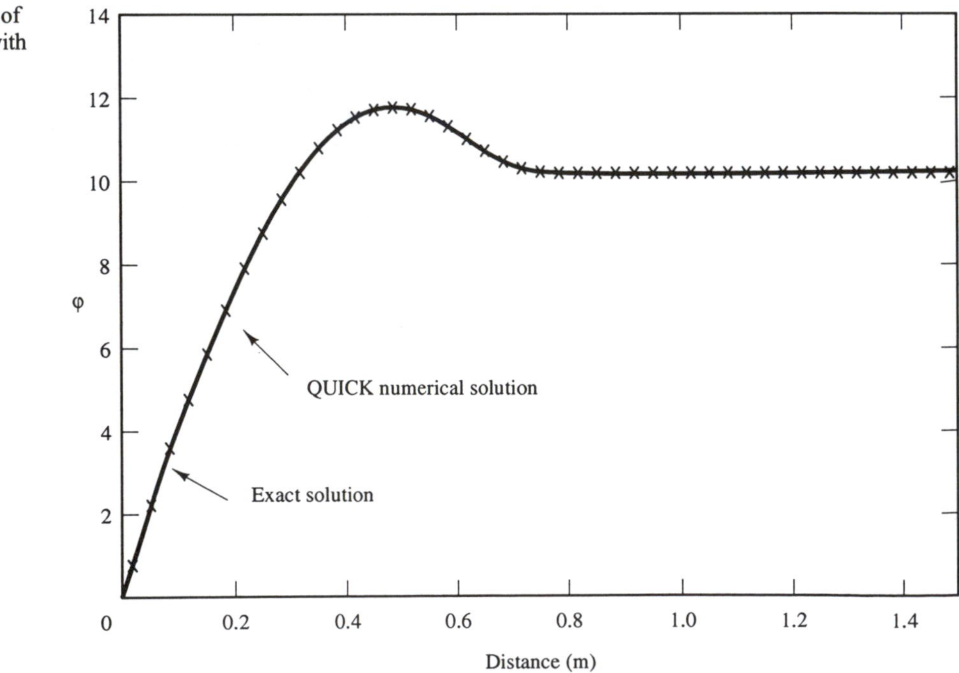

8.7 Solution procedures for unsteady flow calculations

8.7.1 Transient SIMPLE

Algorithms such as SIMPLE, described in Chapter 6 for the calculation of steady flows, may be extended to transient calculations. The discretised momentum equations will now include transient terms formulated with the procedure described in section 8.5. An additional term is also required in the pressure correction equation. The continuity equation in a transient two-dimensional flow is given by

$$\frac{\partial \rho}{\partial t} + \frac{\partial(\rho u)}{\partial x} + \frac{\partial(\rho v)}{\partial y} = 0 \tag{8.42}$$

The integrated form of this equation over a two-dimensional scalar control volume becomes

$$\frac{(\rho_P - \rho_P^o)}{\Delta t}\Delta V + \left[(\rho u A)_e - (\rho u A)_w\right] + \left[(\rho u A)_n - (\rho u A)_s\right] = 0 \tag{8.43}$$

The pressure correction equation is derived from the continuity equation and should therefore contain terms representing its transient behaviour. For example, the equivalent of pressure correction equation (5.32) for a two-dimensional transient flow will take the form

$$\begin{aligned} a_{I,J}p'_{I,J} &= a_{I+1,J}p'_{I+1,J} + a_{I-1,J}p'_{I-1,J} + a_{I,J+1}p'_{I,J+1} \\ &\quad + a_{I,J-1}p'_{I,J-1} + b'_{I,J} \end{aligned} \tag{8.44}$$

where $a_{I,J} = a_{I+1,J} + a_{I-1,J} + a_{I,J+1} + a_{I,J-1}$

and $b'_{I,J} = (\rho u^* A)_{i,J} - (\rho u^* A)_{i+1,J} + (\rho v^* A)_{I,j} - (\rho v^* A)_{I,j+1} + \dfrac{(\rho_P^o - \rho_P)\Delta V}{\Delta t}$

with neighbour coefficients

$a_{I-1,J}$	$a_{I+1,J}$	$a_{I,J-1}$	$a_{I,J+1}$
$(\rho d A)_{i,J}$	$(\rho d A)_{i+1,J}$	$(\rho d A)_{I,j}$	$(\rho d A)_{I,j+1}$

The extension to three-dimensional flows includes the same extra term in the source.

In transient flow calculations with the implicit formulation, the iterative procedures described for steady state calculations employing SIMPLE, SIMPLER or SIMPLEC are applied at each time level until convergence is achieved. Figure 8.10 shows the algorithm structure.

8.7.2 The transient PISO algorithm

The PISO algorithm is a non-iterative transient calculation procedure. It relies on the temporal accuracy gained by the discretisation practice, in particular the operator splitting technique (Issa, 1986). In the transient algorithm all time-dependent terms are retained in the momentum and continuity equations. This gives the following

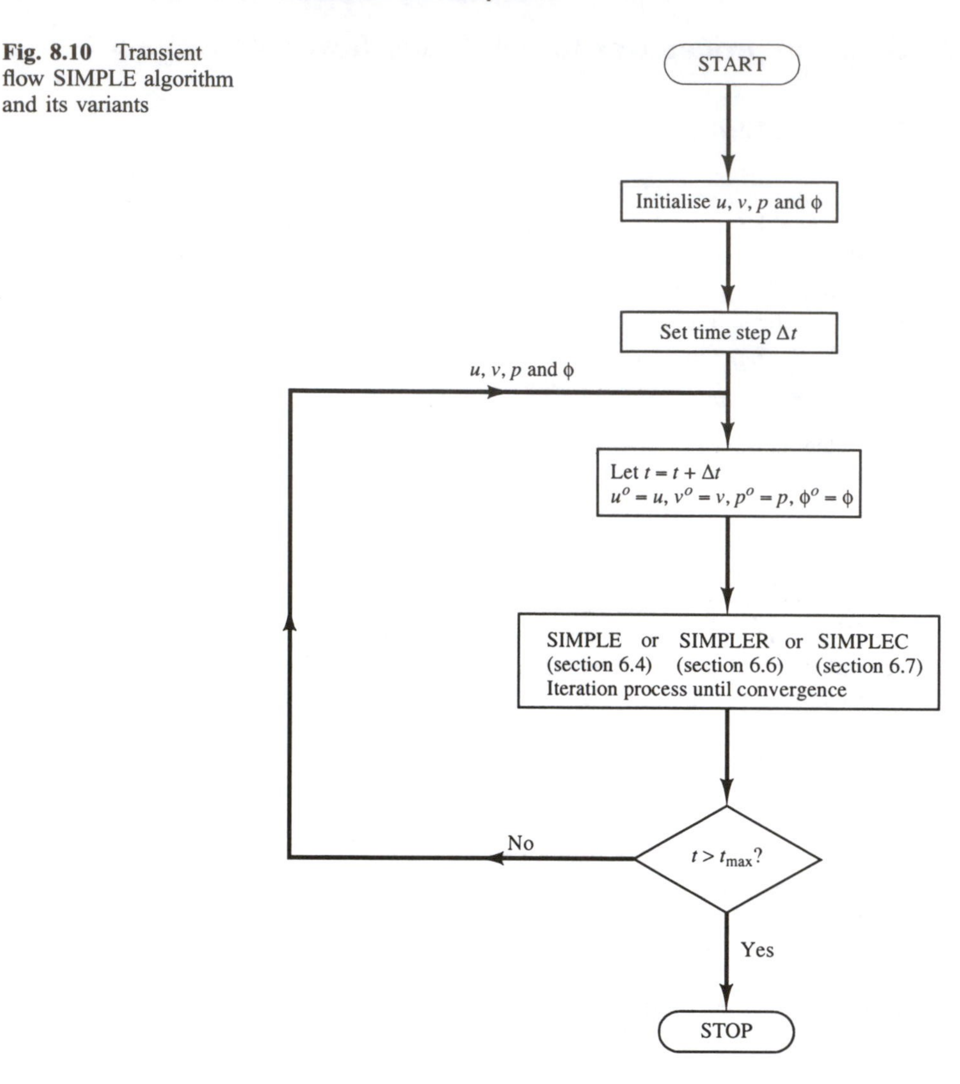

Fig. 8.10 Transient flow SIMPLE algorithm and its variants

additional contributions to the momentum and pressure correction equations in the transient form of PISO:

- add $a_P^o = \rho_P^o \Delta V / \Delta t$ to the central coefficients of the discretised u- and v-momentum equations (5.12–5.13) and (5.52–5.53) respectively
- add $a_P^o u_P^o$ and $a_P^o v_P^o$ to the source terms of the u- and v-momentum equations
- add $(\rho_P^o - \rho_P)\Delta V / \Delta t$ to the source term of both the first and second discretised pressure correction equations.

Otherwise the basic equations and steps involved in the transient version of the PISO algorithm are the same as those set out in section 6.8. The PISO procedure explained there is carried out at each time level to calculate the velocity and pressure fields. Issa (1986) showed that the temporal accuracy achieved by the predictor–corrector process for pressure and momentum is of order 3 (Δt^3) and 4 (Δt^4) respectively. Therefore, the pressure and velocity fields obtained at the end of the PISO process with a suitably small time step are considered to be accurate enough to proceed to the next time step immediately and the algorithm is non-iterative.

Since the algorithm relies on the higher order temporal accuracy gained by the splitting technique, small time steps are recommended to ensure accurate results. If necessary a higher order temporal differencing scheme may be incorporated in the algorithm for improved performance, such as a second-order implicit scheme that uses three time levels $n+1, n, n-1$ at intervals of Δt. We may use the gradient at time level n of the quadratic profile passing through T^{n+1}, T^n and T^{n-1} to evaluate $\partial T/\partial t$. The resulting time discretisation with second-order accuracy is

$$\frac{\partial T}{\partial t} = \frac{1}{2\Delta t}\left(3T^{n+1} - 4T^n + T^{n-1}\right) \tag{8.45}$$

Incorporation of the scheme to formulate discretised equations is relatively straightforward. The values at time level n and $n-1$ known from previous time steps are treated as source terms and are placed on the right hand side of the equation.

The PISO method has yielded accurate results with sufficiently small time steps (see, for example, Issa *et al*, 1986; Kim and Benson, 1992). Since the PISO method does not require iterations within a time level it is less expensive than the implicit SIMPLE algorithm. CFD simulation of flow and heat transfer in internal combustion engines requires transient calculations that are inevitably time consuming and expensive especially with three-dimensional geometries. Ahmadi-Befrui *et al* (1990) have presented a version of PISO known as EPISO suitable for predicting engine flows.

8.8 Steady state calculations using the pseudo-transient approach

It was mentioned in Chapter 6 that under-relaxation is necessary to stabilise the iterative process of obtaining steady state solutions. The under-relaxed form of the two-dimensional u-momentum equation, for example, takes the form

$$\frac{a_{i,J}}{\alpha_u}u_{i,J} = \sum a_{nb}u_{nb} + (p_{I-1,J} - p_{I,J})A_{i,J} + b_{i,J} + \left[(1-\alpha_u)\frac{a_{i,J}}{\alpha_u}\right]u_{i,J}^{(n-1)} \tag{8.46}$$

Compare this with the transient (implicit) u-momentum equation

$$\left(a_{i,J} + \frac{\rho_{i,J}^o \Delta V}{\Delta t}\right)u_{i,J} = \sum a_{nb}u_{nb} + (p_{I-1,J} - p_{I,J})A_{i,J} + b_{i,J} + \frac{\rho_{i,J}^o \Delta V}{\Delta t}u_{i,J}^o \tag{8.47}$$

In equation (8.46) the superscript $n-1$ indicates the previous iteration and in equation (8.47) superscript o represents the previous time level. We immediately note a clear analogy between transient calculations and under-relaxation in steady state calculations. It can be easily deduced that

$$(1-\alpha_u)\frac{a_{i,J}}{\alpha_u} = \frac{\rho_{i,J}^o \Delta V}{\Delta t} \tag{8.48}$$

This formula shows that it is possible to achieve the effects of under-relaxed iterative steady state calculations from a given initial field by means of a pseudo-transient computation starting from the same initial field by taking a step size that satisfies

(8.48). Alternatively steady state calculations may be interpreted as pseudo-transient solutions with spatially varying time steps. The pseudo-transient approach is useful for situations in which governing equations gives rise to stability problems, e.g. buoyant flows, highly swirling flows and compressible flows with shocks.

8.9 A brief note on other transient schemes

Other transient flow calculation procedures such as MAC (Harlow and Welch, 1965), SMAC (Amsden and Harlow, 1970), ICE (Harlow and Amsden, 1971) and ICED-ALE (Hirt *et al*, 1974) are available to the user. The calculation methodology of this class of schemes includes the direct solution of a Poisson equation for the pressure as a central feature of the algorithm. The overall calculation process is, therefore, substantially different from the techniques explained here and the interested reader is referred to the cited references for more details. In the well-known engine prediction code KIVA-II the ICED-ALE method is used as the core solution procedure. The method has been shown to be reliable for predicting practical internal combustion engine flows and is widely used for internal combustion engine research (see Amsden *et al*, 1985, 1989; Zellat *et al*, 1990; Blunsdon *et al*, 1992, 1993). Kim and Benson (1992) compared the PISO method with SMAC algorithms for the prediction of unsteady flows and reported that SMAC was more efficient, faster and more accurate than PISO. The MAC/ICE class of methods are, however, mathematically complex and not widely used in general purpose CFD procedures.

8.10 Summary

Techniques for the solution of transient flow problems were developed by considering the unsteady diffusion and convection–diffusion equations. We distinguish between the following time-stepping algorithms for the computation of a variable ϕ at a new time level:

- explicit – uses only ϕ from the previous time level
- Crank–Nicolson – uses a mixture of ϕ from the previous time level and ϕ at a new time level
- implicit – uses mainly surrounding ϕ-values at the new time level

The stability and accuracy properties of each of the schemes are given in Table 8.6 and described further below.

Table 8.6

Scheme	*Stability*	*Accuracy*	*Positive coefficient criterion*
Explicit	Conditionally stable	First order	$\Delta t < \rho(\delta x)^2/2\Gamma$
Crank–Nicolson	Unconditionally stable	Second order	$\Delta t < \rho(\Delta x)^2/\Gamma$
Implicit	Unconditionally stable	First order	Always positive

- For robust general purpose transient CFD calculations the implicit scheme is recommended. The unconditional stability of this and the Crank–Nicolson scheme is, however, bought at the price of having to solve a system of equations at each time level. In two- and three-dimensional calculations this requires intermediate iterative stages.
- The (fully implicit) transient discretisation equations for diffusion and convection–diffusion are practically the same as those of steady problems apart from minor changes to the central coefficient a_P and the source term b_P:

$$a_P^{(t)} = a_P^{(s)} + a_P^o \quad \text{and} \quad b_P^{(t)} = b_P^{(s)} + a_P^o \phi_P^o \quad \text{with} \quad a_P^o = \rho_P^o \Delta V / \Delta t$$

  The superscript (t) refers to the transient form and (s) to the steady form.
- In addition to the above modifications to the momentum equations in SIMPLE its pressure correction equation also requires an addition of $(\rho_p^o - \rho_p)\Delta V/\Delta t$ to the source term b_P. The time stepping procedure creates an extra loop outside the main iteration cycles of SIMPLE.
- The time accuracy of the second corrector step of PISO makes it very attractive for non-iterative transient calculations.
- The similarity between the under-relaxed iterative solution and the pseudo-transient was highlighted. The pseudo-transient strategy has been widely used to combat stability problems in flows with complex physics.

9

Implementation of Boundary Conditions

9.1 Introduction

All CFD problems are defined in terms of initial and boundary conditions. It is important that the user specifies these correctly and understands their role in the numerical algorithm. In transient problems the initial values of all the flow variables need to be specified at all solution points in the flow domain. Since this involves no special measures other than initialising the appropriate data arrays in the CFD code we do not need to discuss this topic further. The present chapter describes the implementation of the following most common boundary conditions in the discretised equations of the finite volume method:

- inlet
- outlet
- wall
- prescribed pressure
- symmetry
- periodicity (or cyclic boundary condition)

In constructing a staggered grid arrangement we set up additional nodes surrounding the physical boundary, as illustrated in Figure 9.1. The calculations are performed at internal nodes only ($I = 2$ and $J = 2$ onwards). Two notable features of the arrangement are (i) the physical boundaries coincide with the scalar control volume boundaries and (ii) the nodes just outside the inlet of the domain (along $I = 1$ in Figure 9.1) are available to store the inlet conditions. This enables the introduction of boundary conditions to be achieved with small modifications to the discretised equations for near-boundary internal nodes.

In Chapters 4 and 5 we have seen that boundary conditions enter the discretised equations by suppression of the link to the boundary side and modification of the source terms. The appropriate coefficient of the discretised equation is set to zero and the boundary side flux – exact or linearly approximated – is introduced through source terms S_u and S_p. We shall frequently make use of this device to fix the flux of

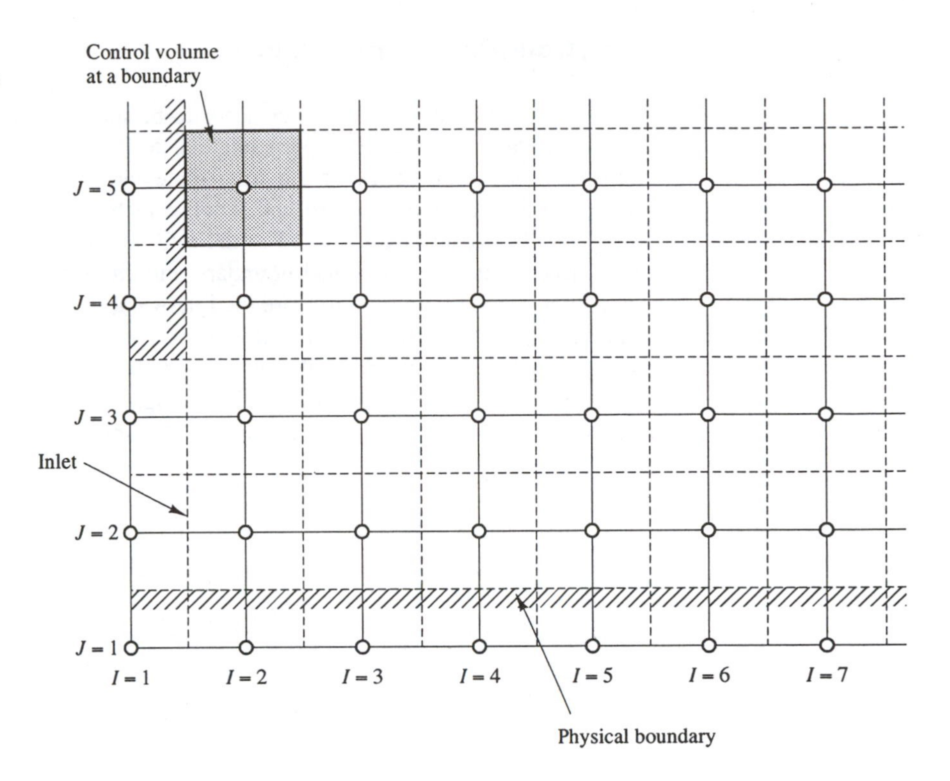

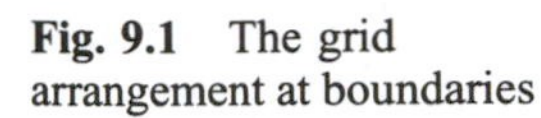
Fig. 9.1 The grid arrangement at boundaries

a variable at a cell face, but we also need a technique to cope with situations where we need to set the value of a variable at a node. This can be done by introducing two overwhelmingly large source terms into the relevant discretised equation. For example, to set the variable ϕ at node P to a value ϕ_{fix} the following source term modification is used in its discretised equation:

$$S_p = -10^{30} \quad \text{and} \quad S_u = 10^{30}\phi_{fix} \tag{9.1}$$

With these sources added to the discretised equation we have

$$\left(a_P + 10^{30}\right)\phi_P = \sum a_{nb}\phi_{nb} + 10^{30}\phi_{fix} \tag{9.2}$$

The actual magnitude of the number 10^{30} is arbitrary as long as it is very large compared with all the coefficients in the original discretised equation. Thus if a_P and a_{nb} are all negligible the discretised equation effectively states that

$$\phi_P = \phi_{fix} \tag{9.3}$$

which fixes the value of ϕ at P.

In addition to setting the value of a variable at internal nodes this treatment is also useful for dealing with solid obstacles within a domain by taking $\phi_{fix} = 0$ (or any other desired value) at nodes within a solid region. The system of discretised flow equations can be solved as normal without having to deal with the obstacles separately.

Details of the modifications needed to implement the most common boundary conditions will be further explained in the text to follow. We make the following assumptions: (i) the flow is always subsonic ($M < 1$), (ii) k–ε turbulence modelling is used, (iii) the hybrid differencing method is used for discretisation and (iv) the SIMPLE solution algorithm is applied.

9.2 Inlet boundary conditions

The distribution of all flow variables needs to be specified at inlet boundaries. Here we discuss the case of an inlet perpendicular to the x-direction. Figures 9.2 to 9.5 show the grid arrangement in the immediate vicinity of an inlet for u- and v-momentum, scalar and pressure correction equation cells. The flow direction is assumed to be broadly from the left to the right in the diagrams. As mentioned, the grid extends outside the physical boundary and the nodes along the line $I = 1$ (or $i = 2$ for u-velocity) are used to store the inlet values of flow variables (indicated by u_{in}, v_{in} and p'_{in}). Just downstream of this extra node we start to solve the discretised equation for the first internal cell, which is shaded.

The diagrams also show the 'active' neighbours and cell faces which are represented in the discretised equation for the shaded cell assuming that hybrid

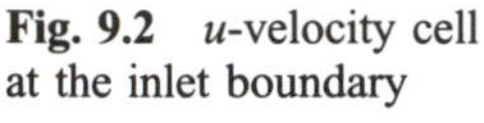

Fig. 9.2 u-velocity cell at the inlet boundary

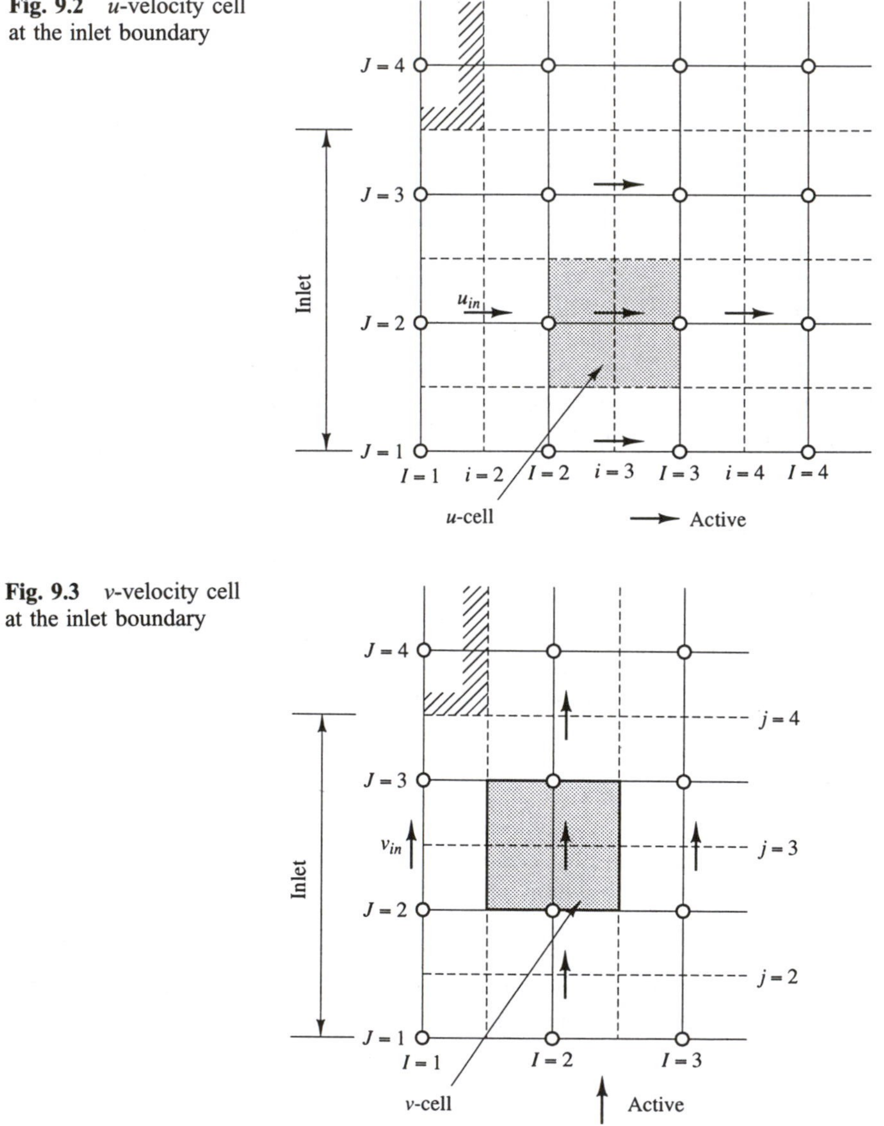

Fig. 9.3 v-velocity cell at the inlet boundary

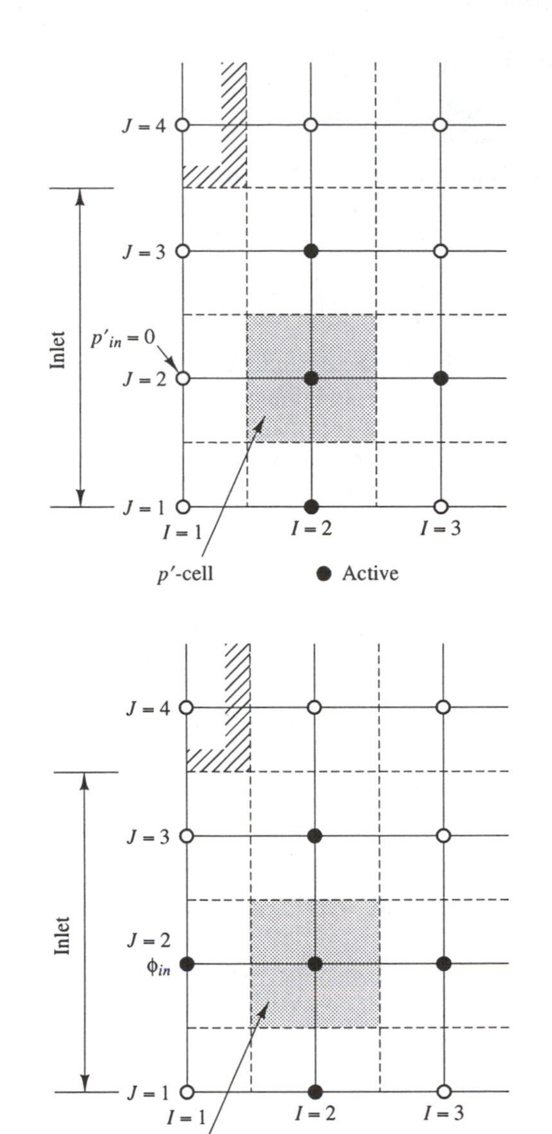

Fig. 9.4 Pressure correction cell at inlet boundary

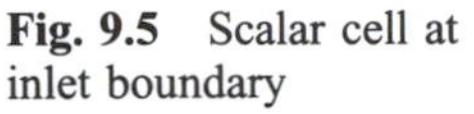

Fig. 9.5 Scalar cell at inlet boundary

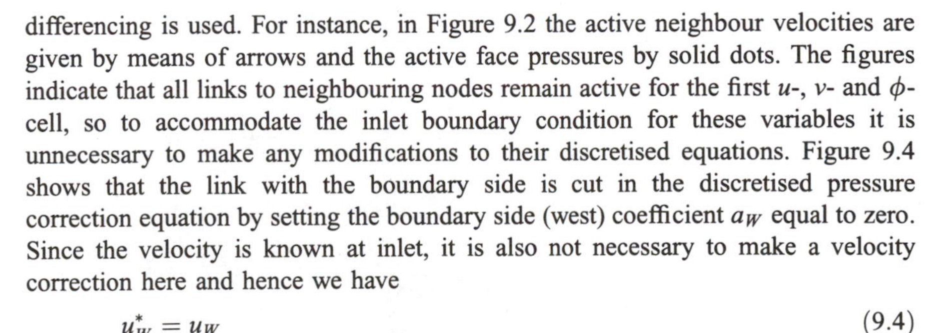

differencing is used. For instance, in Figure 9.2 the active neighbour velocities are given by means of arrows and the active face pressures by solid dots. The figures indicate that all links to neighbouring nodes remain active for the first u-, v- and ϕ-cell, so to accommodate the inlet boundary condition for these variables it is unnecessary to make any modifications to their discretised equations. Figure 9.4 shows that the link with the boundary side is cut in the discretised pressure correction equation by setting the boundary side (west) coefficient a_W equal to zero. Since the velocity is known at inlet, it is also not necessary to make a velocity correction here and hence we have

$$u_W^* = u_W \tag{9.4}$$

in the source associated with discretised pressure correction (6.32).

Reference pressure

The pressure field obtained by solving the pressure correction equation does not give absolute pressures (Patankar, 1980). It is common practice to fix the absolute pressure at one inlet node and set the pressure correction to zero at that node. Having specified a reference value the absolute pressure field inside the domain can now be obtained.

Estimation of k and ε at inlet boundaries

The most accurate simulations can only be achieved by supplying measured values of turbulent kinetic energy k and dissipation rate ε. However, if we perform outline design calculation such data are often not available. In this case commercial CFD codes often estimate k and ε with the approximate formulae described in section 3.5.2, based on a turbulence intensity – typically between 1 and 6% – and a length scale.

Inlet boundaries perpendicular to the y-direction

The above procedure is, of course, not restricted to an inlet boundary perpendicular to the x-direction. When we have an inlet perpendicular to the y-direction the velocity component v, for which inlet value v_{in} is available at $j = 2$, takes the place of velocity component u and the calculations start at $j = 3$. The inlet values of the remaining variables are stored at $J = 1$ and solution starts at $J = 2$. They are otherwise treated as above.

9.3 Outlet boundary conditions

Outlet boundary conditions may be used in conjunction with the inlet boundary conditions of section 9.2. If the location of the outlet is selected far away from

Fig. 9.6 u-control volume at an outlet boundary

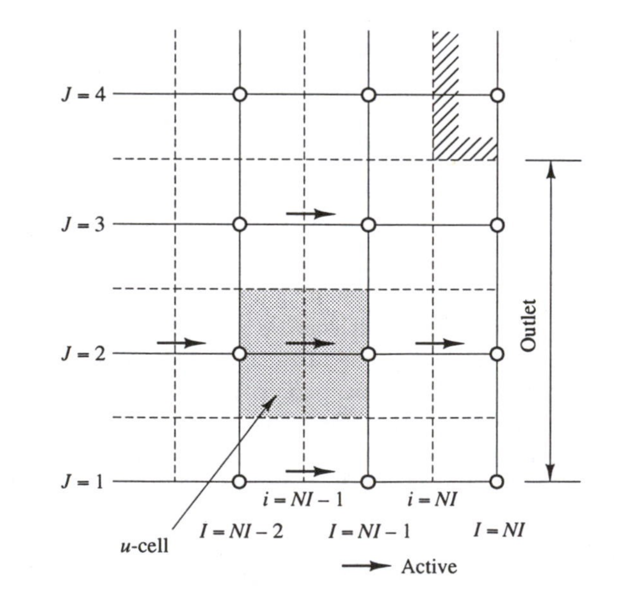

geometrical disturbances the flow often reaches a fully developed state where no change occurs in the flow direction. In such a region we can place an outlet surface and state that the gradients of all variables (except pressure) are zero in the flow direction. It is normally possible to make a reasonably accurate prediction of the flow direction far away from obstacles. This gives us the opportunity to locate the outlet surface perpendicular to the flow direction and take gradients in the direction normal to the outlet surface equal to zero.

Figures 9.6 to 9.9 show the grid arrangements near such an outlet boundary. We have shaded the last cells upstream of the outlet, for which a discretised equation is solved, and, as before, highlighted the active neighbours and faces.

If NI is the total number of nodes in the x-direction, the equations are solved for cells up to I (or i) $= NI - 1$. Before the relevant equations are solved the values of flow variables at the next node (NI), just outside the domain, are determined by extrapolation from the interior on the assumption of zero gradient at the outlet plane.

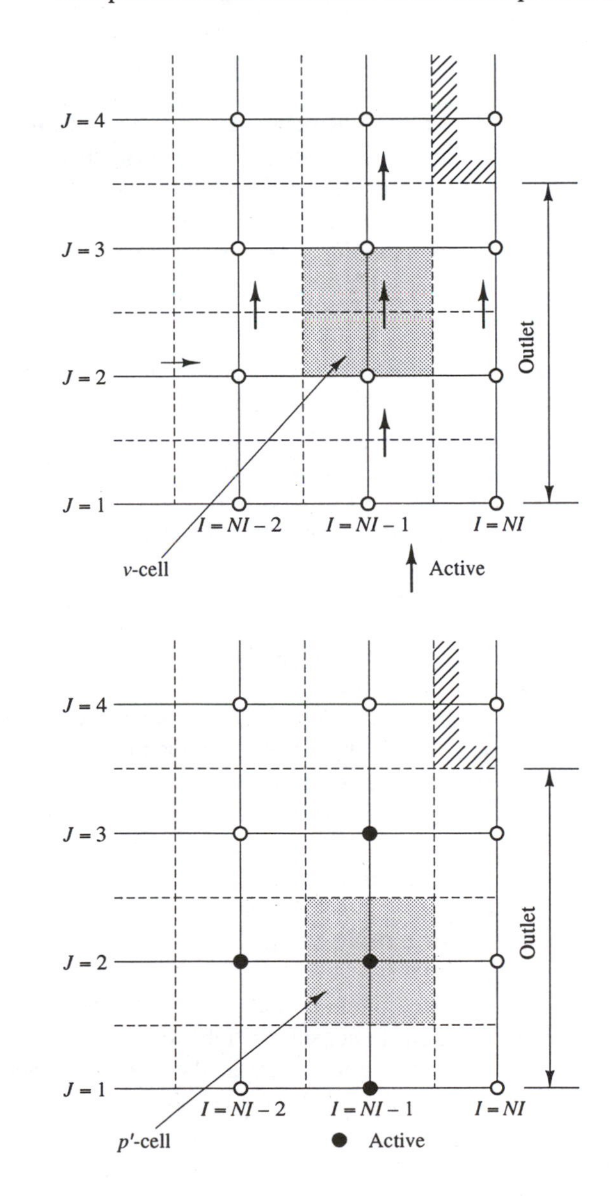

Fig. 9.7 v-control volume at an outlet boundary

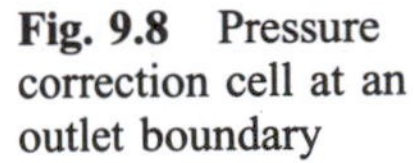

Fig. 9.8 Pressure correction cell at an outlet boundary

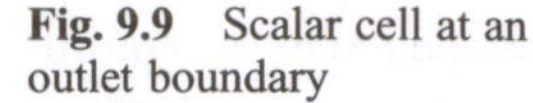
Fig. 9.9 Scalar cell at an outlet boundary

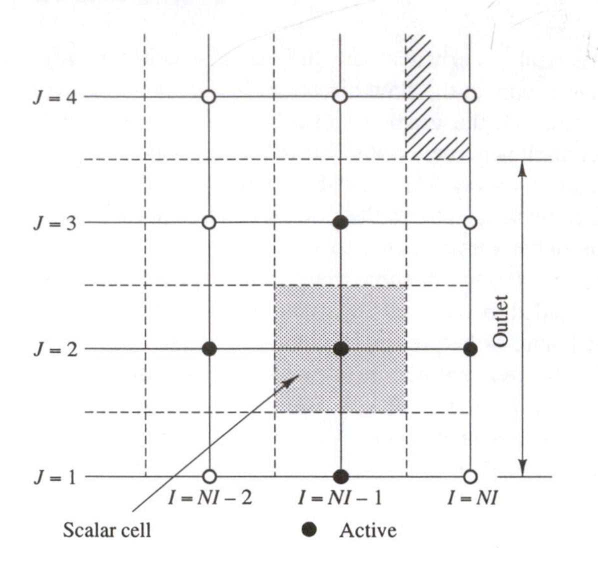

For the v- and scalar equations this implies setting $v_{NI,j} = v_{NI-1,j}$ and $\phi_{NI,J} = \phi_{NI-1,J}$. Figures 9.7 and 9.9 show that all links are active for these variables so their discretised equations can be solved as normal.

Special care should be taken in the case of the u-velocity. Calculation of u at the outlet plane $i = NI$ by assuming a zero gradient gives

$$u_{NI,J} = u_{NI-1,J} \tag{9.5}$$

During the iteration cycles of the SIMPLE algorithm there is no guarantee that these velocities will conserve mass over the computational domain as a whole. To ensure that overall continuity is satisfied the total mass flux going out of the domain (M_{out}) is first computed by summing all the extrapolated outlet velocities (9.5). To make the mass flux out equal to the mass flux M_{in} coming into the domain all the outlet velocity components $u_{NI,J}$ of (9.5) are multiplied by the ratio M_{in}/M_{out}. Thus the outlet plane velocities with the continuity correction are given by

$$u_{NI,J} = u_{NI-1,J} \times \frac{M_{in}}{M_{out}} \tag{9.6}$$

These values are subsequently used as the east neighbour velocities in the discretised momentum equations for $u_{NI-1,J}$.

The velocity at the outlet boundaries is not corrected by means of pressure corrections. Hence in the discretised p'-equation (6.32) the link to the outlet boundary side (east) is suppressed by setting $a_E = 0$. The contribution to the source term in this equation is calculated as normal, noting that $u_E^* = u_E$; no additional modifications are required.

9.4 Wall boundary conditions

The wall is the most common boundary encountered in confined fluid flow problems. In this section we consider a solid wall parallel to the x-direction. Figures 9.10 to 9.12 illustrate the grid details in the near wall regions for the u-velocity component (parallel to the wall), for the v-velocity component (perpendicular to the wall) and for scalar variables.

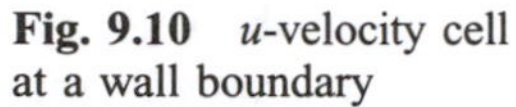
Fig. 9.10 u-velocity cell at a wall boundary

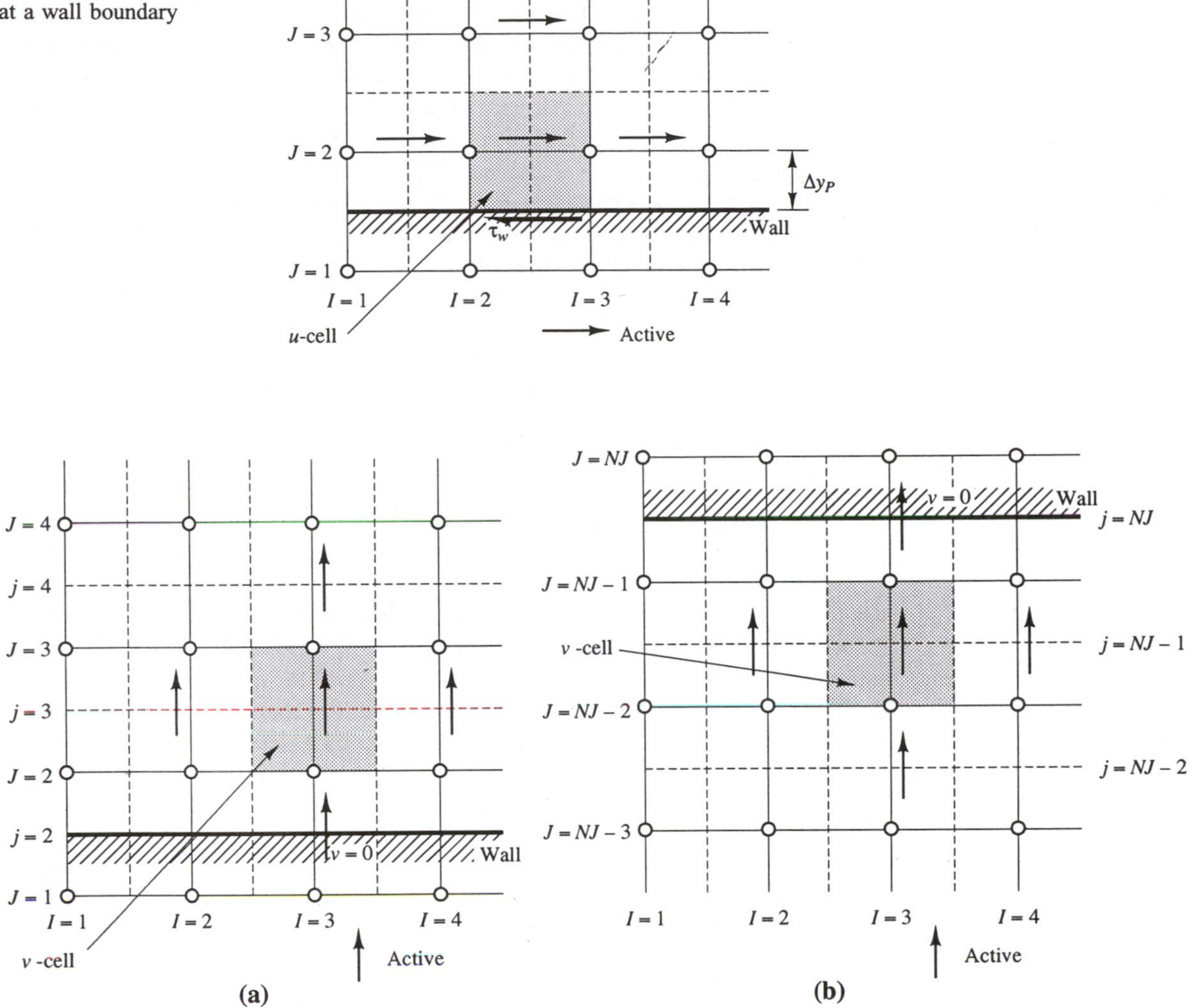

Fig. 9.11 v-cell at a wall boundary (a) $j = 3$ and (b) $j = NJ$

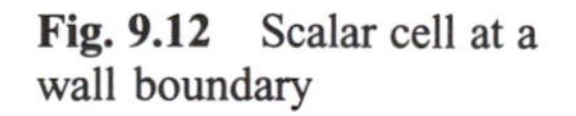
Fig. 9.12 Scalar cell at a wall boundary

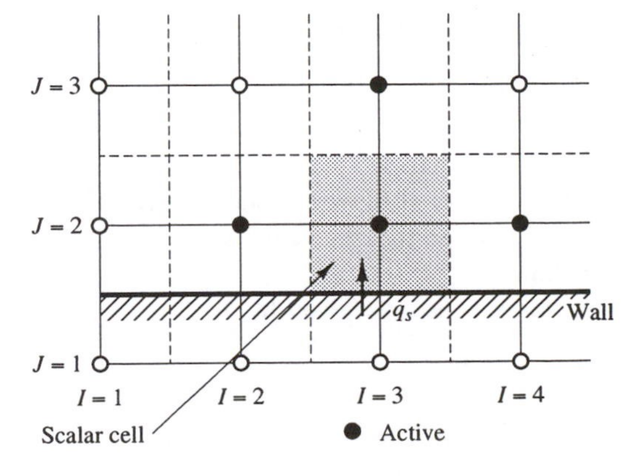

The no-slip condition ($u = v = 0$) is the appropriate condition for the velocity components at solid walls. The normal component of the velocity can simply be set to zero at the boundary ($j = 2$) and the discretised momentum equation at the next v-cell in the flow ($j = 3$) can be evaluated without modification. Since the wall velocity is known it is also unnecessary to perform a pressure correction here. In the discretised p'-equation (6.32) for the cell nearest to the wall the wall link (south) is, therefore, cut by setting $a_S = 0$ and we take $v_S^* = v_S$ in its source term.

For all other variables special sources are constructed, the precise form of which depends on whether the flow is laminar or turbulent. In Chapter 3 we studied the multi-layered structure of the near wall turbulent boundary layer. Immediately adjacent to the wall we have an extremely thin viscous sub-layer followed by the buffer layer and the turbulent core. The number of mesh points required to resolve all the details in a turbulent boundary layer would be prohibitively large and normally we employ the 'wall functions' introduced in Chapter 3 to represent the effect of the wall boundaries.

The implementation of wall boundary conditions in turbulent flows starts with the evaluation of

$$y^+ = \frac{\Delta y_P}{\nu}\sqrt{\frac{\tau_w}{\rho}} \tag{9.7}$$

where Δy_P is the distance of the near wall node P to the solid surface (see Figure 9.10). A near-wall flow is taken to be laminar if $y^+ \leq 11.63$. The wall shear stress is assumed to be entirely viscous in origin. If $y^+ > 11.63$ the flow is turbulent and the wall function approach is used. The criterion places the changeover from laminar to turbulent near wall flow in the buffer layer between the linear and log-law regions of a turbulent wall layer. The exact value of $y^+ = 11.63$ is the intersection of the linear profile and the log-law so it is obtained from the solution of

$$y^+ = \frac{1}{\kappa}\ln(Ey^+) \tag{9.8}$$

In this formula κ is von Karman's constant (0.4187) and E is an integration constant that depends on the roughness of the wall (see section 3.4.2). For smooth walls with constant shear stress E has a value of 9.793.

Laminar Flow/Linear Sub-layer

The wall conditions described under this heading apply in two cases: for solutions of (i) laminar flow equations and (ii) turbulent flow equations when $y^+ \leq 11.63$. In both cases the near wall flow is taken to be laminar. The wall force is entered into the discretised u-momentum equation as a source. The wall shear stress value is obtained from

$$\tau_w = \mu\frac{u_P}{\Delta y_P} \tag{9.9}$$

where u_p is the velocity at the grid node. Figure 9.13 illustrates that this formula is based on the assumption that the velocity varies linearly with distance from the wall in a laminar flow.

The shear force F_s is now given by

$$\begin{aligned} F_s &= -\tau_w A_{Cell} \\ &= -\mu\frac{u_P}{\Delta y_P}A_{Cell} \end{aligned} \tag{9.10}$$

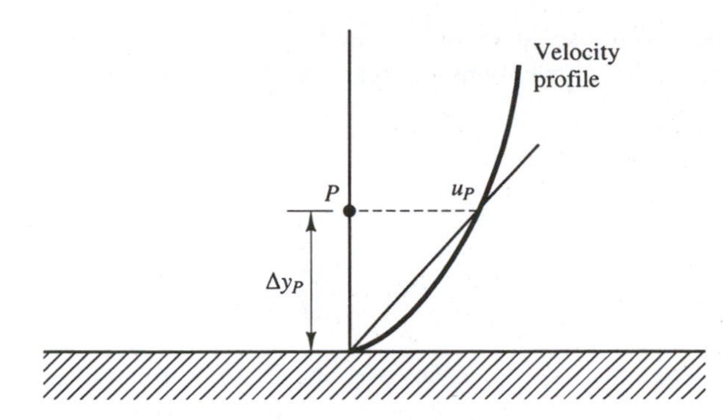

Fig. 9.13 Velocity distribution at a wall

where A_{Cell} is the wall area of the control volume. The appropriate source term in the u-equation is defined by

$$S_p = -\frac{\mu}{\Delta y_P} A_{Cell} \tag{9.11}$$

Heat transfer from a wall at fixed temperature T_w into the near wall cell in laminar flow is calculated from

$$q_s = -\frac{\mu}{\sigma} \frac{C_P(T_P - T_w)}{\Delta y_P} A_{Cell} \tag{9.12}$$

where C_P is the specific heat of the fluid, T_P is the temperature at the node P and σ is the laminar Prandtl number. It is easy to see that the corresponding source terms for the temperature equation are given by

$$S_p = -\frac{\mu}{\sigma} \frac{C_P}{\Delta y_P} A_{Cell} \quad \text{and} \quad S_u = \frac{\mu}{\sigma} \frac{C_P T_w}{\Delta y_P} A_{Cell} \tag{9.13}$$

A fixed heat flux enters the source terms directly by means of the normal source term linearisation:

$$q_s = S_u + S_p T_P \tag{9.14}$$

For an adiabatic wall we have, of course, $S_u = S_p = 0$.

Turbulent Flow

If the value of y^+ is greater than 11.63 node P is considered to be in the log-law region of a turbulent boundary layer. In this region wall function formulae (3.40) and (3.41) associated with the log-law are used to calculate shear stress, heat flux and other variables. The formulae have been applied in many different ways but Table 9.1 gives the optimum near wall relationships from extensive computing trials.

Table 9.1 Near wall relationships for the standard k–ε model

- *Momentum equation tangential to wall*
  - wall shear stress $\tau_w = \rho C_\mu^{1/4} k_P^{1/2} u_P / u^+$ (9.15)
  - wall force $F_s = -\tau_w A_{Cell} = -(\rho C_\mu^{1/4} k_P^{1/2} u_P / u^+) A_{Cell}$ (9.16)
- *Momentum equation normal to wall*
  - normal velocity = 0
- *Turbulent kinetic energy equation*
  - net k-source per unit volume $= (\tau_w u_P - \rho C_\mu^{3/4} k_P^{3/2} u^+) \Delta V / \Delta y_P$ (9.17)
- *Dissipation rate equation*
  - set nodal value $\varepsilon_P = C_\mu^{3/4} k_P^{3/2} / (\kappa \Delta y_P)$ (9.18)
- *Temperature (or energy) equation*
  - wall heat flux $q_w = -\rho C_P C_\mu^{1/4} k_P^{1/2} (T_P - T_w) / T^+$ (9.19)

These relationships should be used in conjunction with the universal velocity and temperature distributions for near wall turbulent flows in (3.40–3.41):

$$u^+ = \frac{1}{\kappa} \ln(Ey^+) \tag{3.40}$$

and

$$T^+ = \sigma_{T,t}\left(u^+ + P\left[\frac{\sigma_{T,l}}{\sigma_{T,t}}\right]\right) \tag{3.41}$$

In these equations the values of κ and E are as given in (9.8), $\sigma_{T,l}$ is the laminar (or molecular) Prandtl number and $\sigma_{T,t}$ is the turbulent Prandtl number ($\approx$0.9) and function $P(\sigma_{T,l}/\sigma_{T,t})$ is called the 'pee-function' that can be evaluated using the following expression derived by Jayatilleke (1969).

$$P\left(\frac{\sigma_{T,l}}{\sigma_{T,t}}\right) = 9.24\left[\left(\frac{\sigma_{T,l}}{\sigma_{T,t}}\right)^{0.75} - 1\right] \times \left\{1 + 0.28 \exp\left[-0.007\left(\frac{\sigma_{T,l}}{\sigma_{T,t}}\right)\right]\right\} \tag{9.20}$$

In order of their appearance in Table 9.1 variables are treated as follows in their discretised equations.

u-velocity component parallel to the wall. The link with the wall (south) is suppressed by setting $a_S = 0$ and the wall force F_s from (9.16) is introduced into the discretised u-equation as a source term, so

$$S_p = -\frac{\rho C_\mu^{1/4} k^{1/2}}{u^+} A_{Cell} \tag{9.21}$$

k-equation. The link at the boundary is suppressed; we set $a_S = 0$. In the volume source (9.17) the second term contains $k^{3/2}$. This is linearised as $k_P^{*1/2}.k_P$, where k* is the k-value at the end of the previous iteration, which yields the following source terms S_p and S_u in the discretised k-equation:

$$S_p = -\frac{\rho C_\mu^{3/4} k_P^{*1/2} u^+}{\Delta y_P} \Delta V \quad \text{and} \quad S_u = \frac{\tau_w u_P}{\Delta y_P} \Delta V \tag{9.22}$$

ε-equation. In the discretised ε-equation the near wall node is fixed to the value given by (9.18) by means of setting the source terms S_p and S_u as follows:

$$S_p = -10^{30} \quad \text{and} \quad S_u = \frac{C_\mu^{3/4} k_P^{3/2}}{\kappa \Delta y_P} \times 10^{30} \tag{9.23}$$

Temperature equation. The link with the wall is suppressed in the T-equation by setting the boundary side coefficient a_S to zero. The wall heat flux is calculated using equation (9.19) and introduced by means of the following source terms:

$$S_p = -\frac{\rho C_\mu^{1/4} k_P^{1/2} C_P}{T^+} A_{Cell} \quad \text{and} \quad S_u = \frac{\rho C_\mu^{1/4} k_P^{1/2} C_P T_{wall}}{T^+} A_{Cell} \tag{9.24}$$

A fixed heat flux enters the source terms directly by means of the normal source term linearisation:

$$q_s = S_u + S_p T_P \tag{9.25}$$

For an adiabatic wall we have $S_u = S_p = 0$, as before.

Rough walls

In the wall function approach described above, the changeover from laminar to turbulent flow as the distance from the wall increases was assumed to occur at $y^+ = 11.63$ which is the solution of equation (9.8) with $E = 9.8$. This criterion applies to smooth walls; if the walls are not smooth E should be adjusted accordingly and a new limiting value of y^+ would result. E may be estimated on the basis of measured absolute roughness values. Schlichting (1979), among others, gives further details.

Moving walls.

Note that it has been tacitly assumed that the wall is stationary. Wall movement in the x-direction is felt by the fluid by a change in the wall shear stress. Its value is adjusted by replacing velocity u_P by the relative velocity $u_P - u_{wall}$. This modifies the laminar wall force formula (9.10) as follows:

$$F_s = -\mu \frac{(u_P - u_{wall})}{\Delta y_P} A_{Cell} \tag{9.26}$$

and the turbulent wall force formula (9.16) as

$$F_s = -\frac{\rho C_\mu^{1/4} k^{1/2} (u_P - u_{wall})}{u^+} A_{Cell} \tag{9.27}$$

The relevant source terms (9.11) and (9.21) are similarly adjusted.

Wall motion also alters the volume source term of the k-equations which becomes

$$\left[\tau_w (u_P - u_{wall}) - \rho C_\mu^{3/4} k_P^{3/2} u^+\right] \Delta V / \Delta y_P \tag{9.28}$$

It should be noted that the wall functions described above have been derived on the basis of the following assumptions:

- the velocity is parallel to the wall and varies only in the direction normal to the wall
- no pressure gradients in the flow direction
- no chemical reactions at the wall
- high Reynolds number

If any one of these assumptions does not hold the accuracy of the predictions using this wall function approach may be reduced or even seriously compromised.

9.5 The constant pressure boundary condition

The constant pressure condition is used in situations where exact details of the flow distribution are unknown but the boundary values of pressure are known. Typical problems where this boundary condition is appropriate include external flows around objects, free surface flows, buoyancy-driven flows such as natural ventilation and fires, and also internal flows with multiple outlets.

In applying the fixed pressure boundary the pressure correction is set to zero at the nodes. The grid arrangement of the p'-cells near a flow inlet and outlet is shown in Figures 9.14 and 9.15.

A convenient way of dealing with a constant pressure boundary condition is to fix pressure at the nodes just inside the physical boundary as indicated in the diagrams by solid squares. The pressure corrections are set to zero by taking $S_u = 0.0$ and $S_p = -10^{30}$ and the nodal pressure is set to the required boundary pressure p_{fix}. The u-momentum equation is solved from $i = 3$ and the v-momentum and other equations from $I = 2$ onwards. The main outstanding problem is the unknown flow direction which is governed by the conditions inside the calculation domain. The u-velocity component across the domain boundary is generated as part of the solution process by ensuring that continuity is satisfied at every cell. For example, in Figure

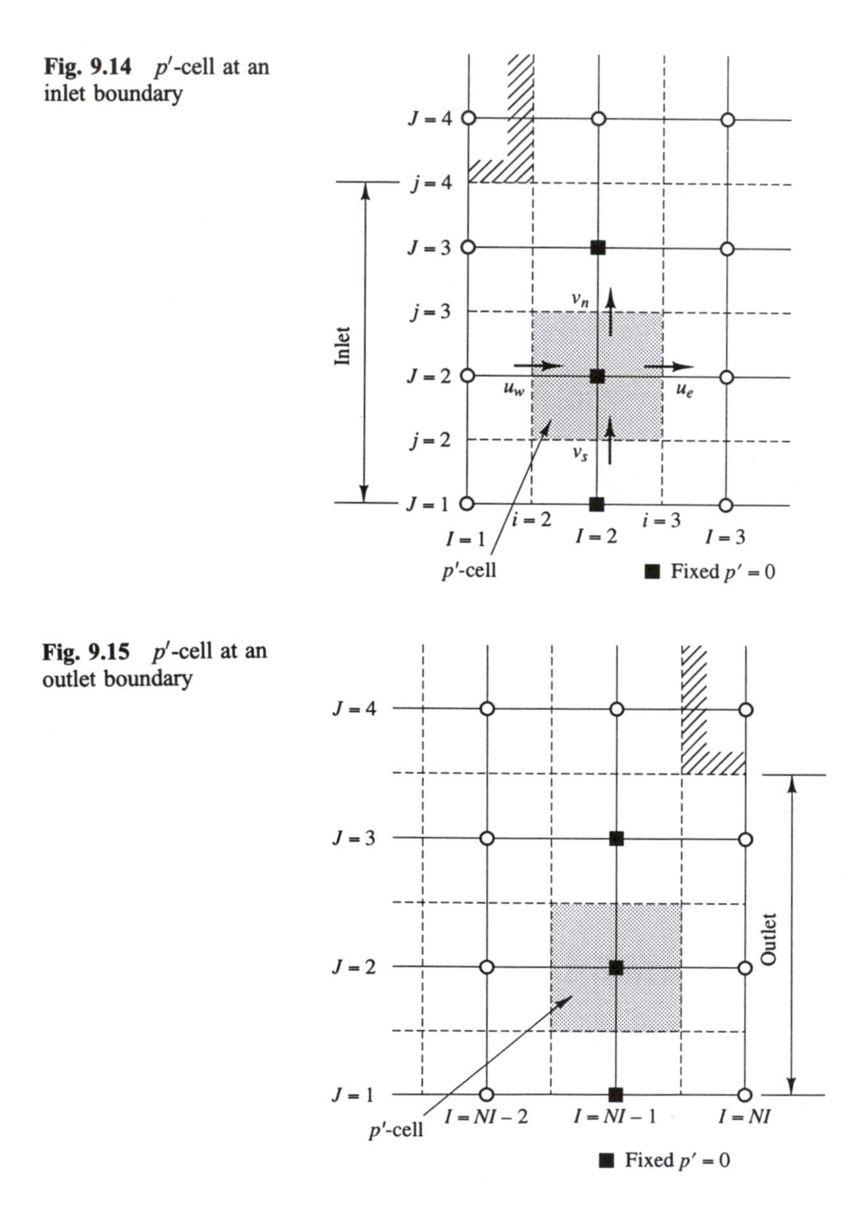

Fig. 9.14 p'-cell at an inlet boundary

Fig. 9.15 p'-cell at an outlet boundary

9.14 the values of u_e and of v_s and v_n emerge from solving the discretised u- and v-momentum equations inside the domain. Given these values we can compute u_w by insisting that mass is conserved for the p'-cell. This yields

$$u_w = \frac{(\rho vA)_n - (\rho vA)_s + (\rho uA)_e}{(\rho A)_w} \tag{9.29}$$

This implementation of the boundary condition causes the p'-cell nearest to the boundaries to act as a source or sink of mass. The process is repeated for each pressure boundary cell. Other variables such as v, T, k and ε must be assigned inflow values where the flow direction is *into* the domain. Where the flow is *outwards* their values just outside the domain may be obtained by means of extrapolation (see section 9.3).

There are several variations that can be useful in practical circumstances. Some codes apply (i) a condition at inlet that fixes the stagnation pressure of the inlet flow just outside the domain (at $i = 2$) instead of the static pressure just inside the domain (at $i = 3$) and/or (ii) the extrapolation procedure at outlets for all variables including u.

9.6 Symmetry boundary condition

The conditions at a symmetry boundary are: (i) no flow across the boundary and (ii) no scalar flux across the boundary. In the implementation, normal velocities are set to zero at a symmetry boundary and the values of all other properties just outside the solution domain (say I or $i = 1$) are equated to their values at the nearest node just inside the domain (I or $i = 2$):

$$\phi_{1,J} = \phi_{2,J} \tag{9.30}$$

In the discretised p'-equations the link with the symmetry boundary side is cut by setting the appropriate coefficient to zero; no further modifications are required.

9.7 Periodic or cyclic boundary condition

Periodic or cyclic boundary conditions arise from a different type of symmetry in a problem. Consider for example swirling flow in a cylindrical furnace shown in Figure 9.16. In the burner arrangement gaseous fuel is introduced through six

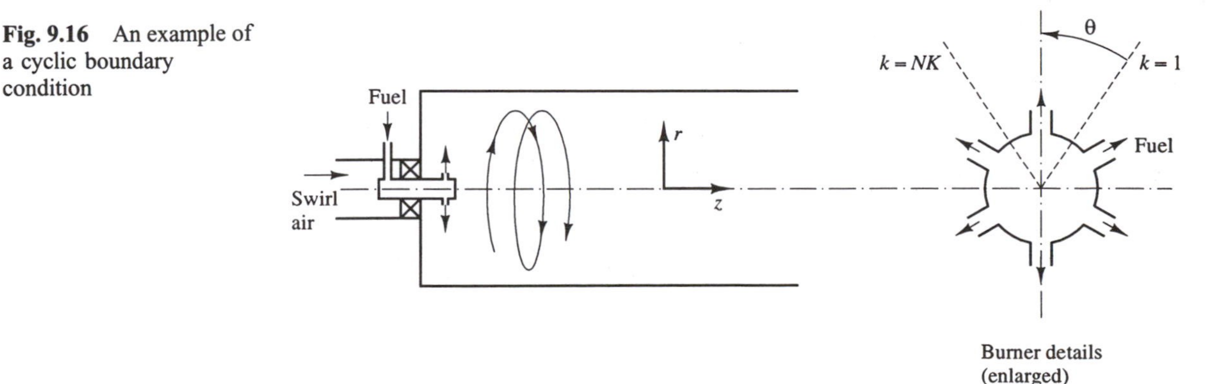

Fig. 9.16 An example of a cyclic boundary condition

symmetrically placed holes and swirl air enters through the outer annulus of the burner.

This problem can be solved in cylindrical polar co-ordinates (z, r, θ) by considering a 60 ° angular sector as shown in the diagram where k refers to $r-z$ planes in the θ-direction. The flow rotates in this direction, and under the given conditions the flow entering the first k-plane of the sector should be exactly the same as that leaving the last k-plane. This is an example of cyclic symmetry. The pair of boundaries $k = 1$ and $k = NK$ are called periodic or cyclic boundaries.

To apply cyclic boundary conditions we need to set the flux of all flow variables leaving the outlet cyclic boundary equal to the flux entering the inlet cyclic boundary. This is achieved by equating the values of each variable at the nodes just upstream and downstream of the inlet plane to the nodal values just upstream and downstream of the outlet plane. For all variables except the velocity component across the inlet and outlet planes (say w) we have

$$\phi_{1,J} = \phi_{NK-1,J} \quad \text{and} \quad \phi_{NK,J} = \phi_{2,J} \tag{9.31}$$

For the velocity component across the boundary we have

$$w_{1,J} = w_{NK-1,J} \quad \text{and} \quad w_{NK+1,J} = w_{3,J} \tag{9.32}$$

9.8 Potential pitfalls and final remarks

Flows inside a CFD solution domain are driven by the boundary conditions. In a sense the process of solving a field problem (e.g. a fluid flow) is nothing more than the extrapolation of a set of data defined on a boundary contour or surface into the domain interior. It is, therefore, of paramount importance that we supply physically realistic, well-posed boundary conditions, otherwise severe difficulties are encountered in obtaining solutions. The single most common cause of rapid divergence of CFD simulations is the inappropriate selection of boundary conditions.

In Chapter 2 we summarised a set of 'best' boundary conditions for viscous fluid flows which included the inlet, outlet and wall condition. Their finite volume method implementation was discussed in sections 9.2 to 9.4 and in sections 9.5 to 9.7 we developed three further conditions, constant pressure, symmetry and periodicity,

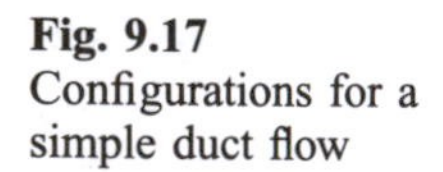
Fig. 9.17 Configurations for a simple duct flow

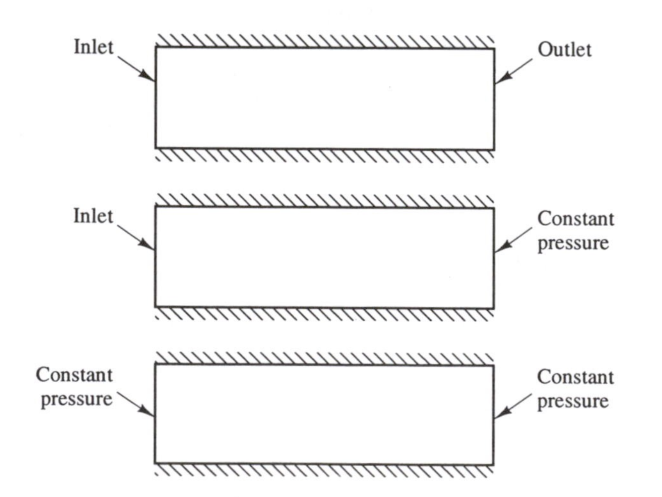

which are physically realistic and very useful in practical calculations. These are by no means the only boundary conditions. Commercial CFD packages may include the time-dependent movement of boundaries, facilities to include rotating and accelerating boundaries and special conditions for transonic and supersonic flows. It would be beyond the scope of this introductory book to discuss the ways of implementing all of them.

A simple illustration of the poor selection of boundary conditions might be an attempt to generate a steady state solution in a domain with wall boundaries and a flow inlet but without an outlet boundary. It is obvious that mass cannot be conserved in the steady state and CFD calculations will 'blow up' swiftly. This almost trivial example also suggests that certain types of boundary conditions must be accompanied by particular other ones. We now briefly state some permissible combinations in subsonic flows:

- walls only
- walls and inlet and at least one outlet
- walls and inlet and at least one constant pressure boundary
- walls and constant pressure boundaries

Figure 9.17 illustrates these configurations for a simple duct flow.

Particular care must be taken in applying the outlet boundary condition. It can only be used if all flows entering the calculation domain are given by means of inlet boundary conditions (i.e. velocity and scalars fixed at inlet) and is only recommended for flow domains with a single exit. Physically the exit pressures govern the flow split between multiple outlets so it is better to specify this quantity at exits than (zero gradient) outlet conditions. It is *not permitted* to combine an outlet condition with one or more constant pressure boundaries, because the zero gradient outlet condition specifies neither the flow rate nor the pressure at the exit, thus leaving the problem under-specified.

We have glossed over a number of very complex problems by only considering subsonic flows. We merely warn the CFD user to tread very carefully when attempting to tackle flows that may have regions of transonic and supersonic flows.

Accuracy limitations of the individual boundary conditions have already been pointed out. Here we note a small selection of the more subtle pitfalls of practical CFD that need to be avoided to ensure that simulation accuracy is optimal.

Positioning of outlet boundaries

If outlet boundaries are placed too close to solid obstacles it is possible that the flow will not have reached a fully developed state (zero gradients in the flow direction) which may lead to sizeable errors. Figure 9.18 gives typical velocity profiles downstream of an obstacle, which illustrate the potential hazards.

If the outlet is placed close to an obstacle it may range across a wake region with recirculation. Not only does the assumed gradient condition not hold, but there is an area of reverse flow where the fluid enters the domain whilst we had assumed an outward flow. Of course, we cannot trust the solution if this condition arises. Somewhat further downstream there may not be reverse flow, but the zero gradient condition does not hold since the velocity profile still changes in the flow direction. It is imperative that the outlet boundary is placed much further downstream than 10 heights downstream of the last obstacle to give accurate results. For high accuracy it is necessary to demonstrate that the interior solution is unaffected by the choice of

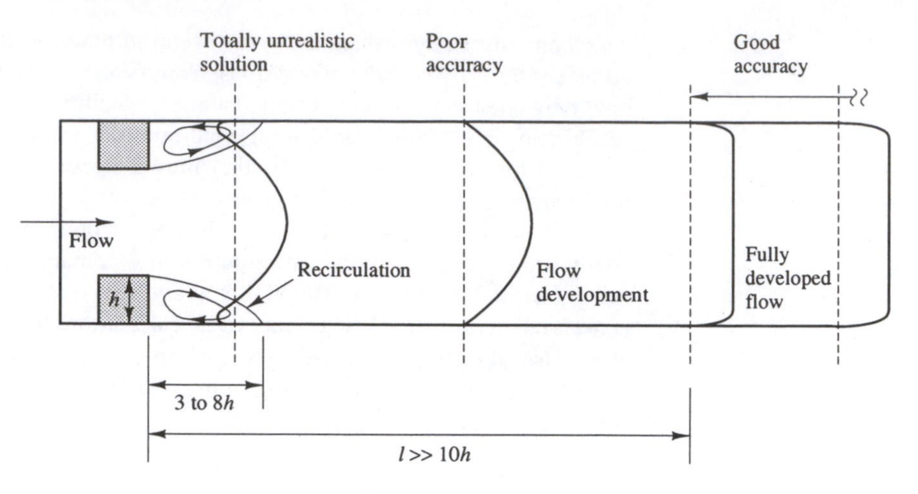

Fig. 9.18 Velocity profiles at different locations downstream of an obstacle

location of the outlet by means of a sensitivity study for the effect of different downstream distances.

Near wall grid

The most accurate way of solving turbulent flows in a general purpose CFD code is to make use of the good empirical fits provided by the wall function approach. To obtain the same accuracy by means of a simulation which includes points inside the (laminar) linear sub-layer the grid spacing must be so fine as to be uneconomical. The criterion that y^+ must be greater than 11.63 sets a **lower** limit to the distance from the wall Δy_P of the nearest grid point. The main mechanism for improving accuracy available to us is grid refinement, but in a turbulent flow simulation we must ensure that, whilst refining the grid, the value of y^+ stays greater than 11.63 and is preferably between 30 and 500.

It is very often impossible to ensure that this is the case everywhere in a general flow; one pertinent example is a flow with recirculation. Near the re-attachment point the velocity component parallel to the wall is zero, so by virtue of the criterion that y^+ must be greater than 11.63 the simulation reverts to the laminar case. There are additional problems associated with the k–ε model in these regions that give rise to further, even more important, inaccuracies. Nevertheless, the point that it is difficult to keep y^+ above its lower limit is well illustrated.

Misapplication of the symmetry condition

It is important to realise that geometric symmetry of the flow domain does not always imply that the flow possesses the same symmetry. An example shown in Figure 9.19 is the flow through a circular pipe with a side jet.

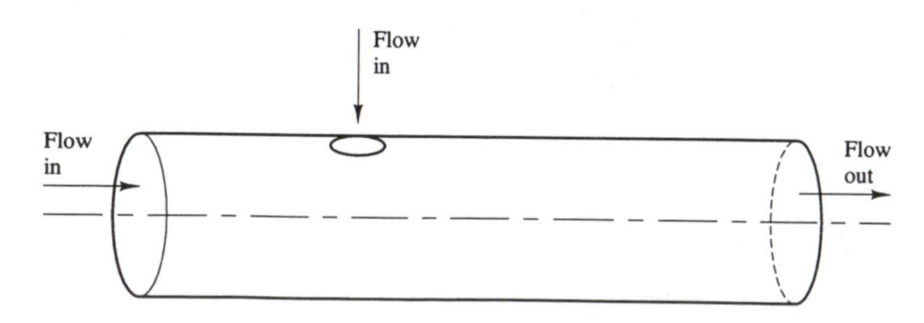

Fig. 9.19 A non-symmetric flow situation in a cylindrical geometry

In spite of the fact that the domain has axisymmetry the occurrence of the cross-flow jet makes the flow non-axisymmetric. Although it is tempting to solve the problem in cylindrical polar co-ordinates the flow solution will be inaccurate because flow may not cross the centreline.

We have discussed the implementation of the most important boundary conditions. Moreover, we have outlined suitable combinations of boundary conditions and highlighted particular problem areas. It is of crucial importance that the CFD user has a good understanding of all the relevant issues as a first step towards accurate flow simulations with the finite volume method.

10

Advanced Topics and Applications

10.1 Introduction

Engineering applications of CFD involve not only fluid flow and heat transfer but also combustion, phase change, multi-phase flow and chemical reactions. Examples of such complex flow systems are furnaces, internal combustion engines, pipelines for the transport of solid–liquid mixtures, heat exchangers with evaporation and/or condensation, and mixers of polymers in moulding processes. Adequate physical models that are appropriate to the problem under consideration must be incorporated in a CFD code to predict the wide variety of processes involved in practical engineering situations. It is impossible to discuss the additional physics and chemistry needed to solve all engineering problems, but we shall demonstrate the method of incorporating extra models into the finite volume framework by looking at some simple combustion modelling concepts. This is particularly appropriate since CFD has been very successful in the prediction of combusting flows. The application of CFD to the modelling of flows in buildings, which has become popular in recent years, is also discussed in this chapter. Attention is paid to the special turbulence modelling required for these buoyancy-driven flows.

The geometrical configurations of many flow problems are complicated. Cartesian/cylindrical grid systems are compared with advanced CFD techniques of modelling irregular geometries by means of body-fitted co-ordinate systems. Finally, we give examples of the use of CFD for the prediction of industrial flows.

10.2 Combustion modelling

During combustion a fuel (e.g. a mixture of hydrocarbons) reacts with an oxidant stream (e.g. air) to form products of combustion. The products are not usually formed in a single chemical reaction; the fuel components and the oxidant undergo a series of reactions. For example, over 40 elementary reactions are involved in the combustion of methane (CH_4), the simplest hydrocarbon fuel. In addition to all the flow equations, the transport equations for the mass fraction m_j of each species j must be solved. The species equations can be written down by using the general transport equation (2.39):

$$\frac{\partial(\rho m_j)}{\partial t} + div(\rho m_j \mathbf{u}) = div(\Gamma_j \, grad \, m_j) + S_j \tag{10.1}$$

The volumetric rate of generation (or destruction) of a species due to chemical reactions appears as the source (or sink) term S_j in each of their transport equations. The total of the mass fractions of fuel, oxidant and inert species is equal to 1, so

$$\sum_{all\ species\ j} m_j = 1 \tag{10.2}$$

Chemical energy is released as heat during combustion and the resulting enthalpy is obtained by solving its transport equation:

$$\frac{\partial(\rho h)}{\partial t} + div(\rho h \mathbf{u}) = div(\Gamma_h \, grad \, h) + S_h \tag{10.3}$$

The source term of the transport equation for enthalpy includes the radiation loss or gain, pressure work as well as the chemical energy. Viscous energy dissipation is normally assumed to be negligible in low Mach combusting number flows. The temperature can be calculated from the enthalpy by means of

$$T = \frac{h - m_{fu} H_{fu}}{\bar{C}_P} \tag{10.4}$$

where H_{fu} is the calorific value of fuel,

$$\bar{C}_P = \frac{1}{(T - T_{ref})} \int_{T_{ref}}^{T} C_P \, dT$$

and $$C_P \equiv \sum_{all\ species\ j} m_j C_j$$

and C_j is the specific heat of species j.

The local density of the mixture is dependent on the reactant and product concentrations and on the mixture temperature. Its value can be calculated from

$$\rho = \frac{P}{RT \sum_{all\ j} \frac{m_j}{M_j}} \tag{10.5}$$

where M_j is the molecular weight of species j.

The flow field is in turn affected by changes in temperature and density, so in addition to the species and enthalpy equations we must solve all the flow equations. The resultant set of PDEs can be very large. Models that consider many intermediate reactions require a vast amount of computing resource, so simple models that incorporate only a few reactions are often preferred in numerical combustion procedures used in CFD. The simplest known procedure is the simple chemical reacting system (SCRS) of Pun and Spalding (1967) described below in some detail. Other approaches of modelling turbulent combustion, such as the eddy break-up model and the laminar flamelet model, are briefly discussed later.

10.2.1 The simple chemical reacting system (SCRS)

If we are concerned with the global nature of the combustion process and with final species concentrations only detailed kinetics are unimportant and a global one-step, infinitely fast, chemical reaction can be assumed where the oxidant combines with the fuel in stoichiometric proportions to form products:

$$1 \text{ kg of fuel} + s \text{ kg of oxidant} \rightarrow (1+s) \text{ kg of products} \tag{10.6}$$

For methane combustion the equation becomes

$$\begin{array}{ccccccc} CH_4 & + & 2O_2 & \rightarrow & CO_2 & + & 2H_2O \\ 1\,\text{mol of } CH_4 & & 2\,\text{mols of } O_2 & & 1\,\text{mol of } CO_2 & & 2\,\text{mols of } H_2O \end{array}$$

i.e.

$$1 \text{ kg of } CH_4 + \tfrac{64}{16} \text{ kg of } O_2 \rightarrow \left(1+\tfrac{64}{16}\right) \text{ kg of products} \tag{10.7}$$

The stoichiometric oxygen/fuel ratio by mass is 4 for the above reaction.

In SCRS infinitely fast chemical reactions are assumed and the intermediate reactions are ignored. The transport equations for the fuel and oxygen mass fraction may be written as

$$\frac{\partial(\rho m_{fu})}{\partial t} + div(\rho m_{fu}\mathbf{u}) = div(\Gamma_{fu}\, grad\, m_{fu}) + S_{fu} \tag{10.8}$$

$$\frac{\partial(\rho m_{ox})}{\partial t} + div(\rho m_{ox}\mathbf{u}) = div(\Gamma_{ox}\, grad\, m_{ox}) + S_{ox} \tag{10.9}$$

Let us consider a variable defined by

$$\phi = s m_{fu} - m_{ox} \tag{10.10}$$

We also assume that all mass exchange coefficients (Γ_j) which appear in the transport equations of species are equal and constant; hence $\Gamma_{fu} = \Gamma_{ox} = \Gamma_\phi$. Equations (10.8) and (10.9) can now be re-written as a transport equation for ϕ:

$$\frac{\partial(\rho\phi)}{\partial t} + div(\rho\phi\mathbf{u}) = div(\Gamma_\phi\, grad\, \phi) + (s.S_{fu} - S_{ox}) \tag{10.11}$$

From the one-step reaction assumption (10.6) we conclude that $(s.S_{fu} - S_{ox}) = 0$, and equation (10.11) reduces to

$$\frac{\partial(\rho\phi)}{\partial t} + div(\rho\phi\mathbf{u}) = div(\Gamma_\phi\, grad\, \phi) \tag{10.12}$$

Here ϕ is a passive scalar: it obeys the scalar transport equation with no source terms. A non-dimensional variable f called the mixture fraction may be defined in terms of ϕ as

$$f = \frac{\phi - \phi_0}{\phi_1 - \phi_0} \tag{10.13}$$

where the suffix *0* denotes the oxidant stream and *1* denote the fuel stream. The local value of f equals 0 if the mixture at a point contains only oxidant and equals 1 if it contains only fuel.

Equation (10.13) may be written in expanded form as

$$f = \frac{[s m_{fu} - m_{ox}] - [s m_{fu} - m_{ox}]_0}{[s m_{fu} - m_{ox}]_1 - [s m_{fu} - m_{ox}]_0} \tag{10.14}$$

If the fuel stream has fuel only we have

$$[m_{fu}]_1 = 1, \quad [m_{ox}]_1 = 0$$

and if the oxidant stream contains no fuel we have

$$[m_{fu}]_0 = 0, \quad [m_{ox}]_0 = 1$$

In such conditions equation (10.14) may be simplified as

$$f = \frac{[sm_{fu} - m_{ox}] - [-m_{ox}]_0}{[sm_{fu}]_1 - [-m_{ox}]_0} = \frac{sm_{fu} - m_{ox} + m_{ox,\,0}}{sm_{fu,\,1} + m_{ox,\,0}} \tag{10.15}$$

In a stoichiometric mixture neither fuel nor oxygen is present in the products and the stoichiometric mixture fraction f_{st} may be defined as

$$f_{st} = \frac{m_{ox,\,0}}{sm_{fu,\,1} + m_{ox,\,0}} \tag{10.16}$$

Fast chemistry implies that if there is an excess of oxidant at a certain point there will be no fuel present in the products; hence $m_{fu} = 0$ if $m_{ox} > 0$ and

$$\text{if } f < f_{st}, \qquad f = \frac{-m_{ox} + m_{ox,\,0}}{sm_{fu,\,1} + m_{ox,\,0}} \tag{10.17}$$

Conversely, if there is a local excess of fuel in the mixture there will be no oxidant in the products, so $m_{ox} = 0$ if $m_{fu} > 0$ and

$$\text{if } f > f_{st}, \qquad f = \frac{sm_{fu} + m_{ox,\,0}}{sm_{fu,\,1} + m_{ox,\,0}} \tag{10.18}$$

The above formulae show that the mass fractions of the fuel m_{fu} and oxygen m_{ox} are linearly related to the mixture fraction f.

By equation (10.13) the mixture fraction f is linearly related to ϕ so it is also a passive scalar and obeys the transport equation

$$\frac{\partial(\rho f)}{\partial t} + div(\rho f \mathbf{u}) = div(\Gamma_f \, grad\, f) \tag{10.19}$$

Equation (10.19) can be solved, subject to suitable boundary conditions – the mixture fractions of fuel and oxidant streams are known and zero normal flux of f across solid walls – to obtain the distribution of f. Given the resulting mixture fraction we can employ equations (10.17–10.18) to give values for the oxygen and fuel mass fractions after combustion:

$$f_{st} \le f < 1 \quad m_{ox} = 0; \quad m_{fu} = \frac{f - f_{st}}{1 - f_{st}} m_{fu,\,1} \tag{10.20}$$

$$0 < f < f_{st} \quad m_{fu} = 0; \quad m_{ox} = \frac{f_{st} - f}{f_{st}} m_{ox,\,0} \tag{10.21}$$

The reactants may be accompanied by inert species, such as N_2, that do not take part in the reaction. The mass of inert species in the mixtures can be obtained by the linear relationship for inert mixing illustrated in Figure 10.1. Simple geometry gives the total mass fraction of the inert species m_{in} after combustion at any value of f as

$$m_{in} = m_{in,\,0}(1 - f) + m_{in,\,1} \cdot f \tag{10.22}$$

The mass fraction of the products of combustion may be obtained from

$$m_{pr} = 1 - (m_{fu} + m_{ox} + m_{in}) \tag{10.23}$$

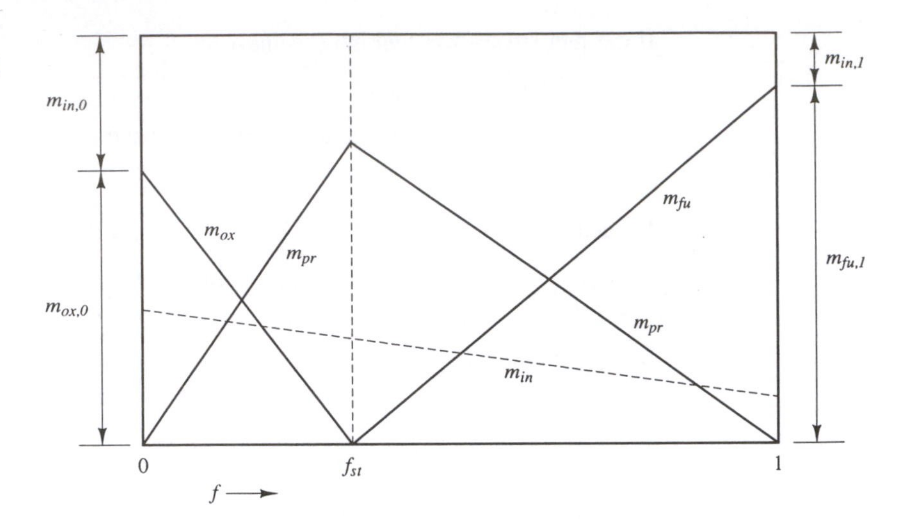

Fig. 10.1 Mixing and fast reaction between fuel and oxidant streams

The above equations represent the SCRS and are shown graphically in Figure 10.1.

When the reaction product contains two or more species, the ratio of the mass fraction of each component to the total product mass fraction is known from the equation for the chemical reaction and can be used to deduce the mass fraction of different product components. For example, consider the burning of methane with O_2:

$$\begin{array}{ccccccc} CH_4 & + & 2O_2 & \rightarrow & CO_2 & + & 2H_2O \\ \text{1 mol of } CH_4 & & \text{2 mols of } O_2 & & \text{1 mol of } CO_2 & & \text{2 mols of } H_2O \\ 16\,\text{kg} & & 64\,\text{kg} & & 44\,\text{kg} & & 36\,\text{kg} \end{array} \quad (10.24)$$

Ratio of CO_2 in products by mass $(r_{CO_2}) = 44/80$
Ratio of H_2O in products by mass $(r_{H_2O}) = 36/80$

If the product mass fraction from equation (10.23) is m_{pr} then the CO_2 mass fraction in the products is $m_{pr}.r_{CO_2}$ and the H_2O mass fraction is $m_{pr}.r_{H_2O}$.

The SCRS model has made the following simplifications: (i) single step reaction between fuel and oxidant and (ii) one reactant which is locally in excess causes *all* the other reactants to be consumed stochiometrically to form reaction products. These assumptions fix algebraic relationships between the mixture fraction f and all the mass fractions m_{fu}, m_{ox}, m_{in} and m_{pr}. As a consequence it is only necessary to solve *one* extra PDE (for f) to calculate combusting flows rather than individual PDEs for each mass fraction. An example which uses this approach for combustion calculations will be presented later in this chapter.

As density varies in combusting flows, the transport equations solved in turbulent combusting flows are those obtained by *Favre* averaging (Favre, 1969; Jones and Whitelaw, 1982). In the SCRS approach, fluctuations of temperature are often taken into account by incorporating a probability density function (pdf) to calculate mean properties. In the pdf method (which originates from turbulence modelling) the average value of a scalar variable ($\overline{m_{fu}}, \overline{m_{ox}}$ etc.) is obtained by weighting the instantaneous value with a probability density function for mixture fraction f. The mean value $\bar{\varphi}$ of a property φ is given by

$$\bar{\varphi} = \int_0^1 \varphi(f)p(f)df$$

where φ is any variable which is a function of f alone, and $p(f)$ is the probability density function. Various different probability distribution functions have been used, but the clipped Gaussian and beta functions give the best results. The interested reader is referred to Bilger and Kent (1974), Lockwood and Naguib (1975), Tamanini (1975), Bilger (1976), Pope (1976), Jones (1979), Lockwood and Monib (1980) and Pope (1985), among others, for further details.

10.2.2 Eddy break-up model of combustion

In the eddy break-up model, due to Spalding (1971), the rate of consumption of fuel is specified as a function of local flow properties. The mixing-controlled rate of reaction is expressed in terms of the turbulence time scale k/ε, where k is the turbulent kinetic energy and ε is the rate of dissipation of k. The model considers the dissipation rates of fuel, oxygen and products, and takes the slowest rate as the reaction rate of fuel. The turbulent dissipation rate of fuel, oxygen and products may be expressed as

$$R_{fu} = -C_R \rho m_{fu} \frac{\varepsilon}{k} \tag{10.25a}$$

$$R_{ox} = -C_R \rho \frac{m_{ox}}{s} \frac{\varepsilon}{k} \tag{10.25b}$$

$$R_{pr} = -C'_R \rho \frac{m_{pr}}{(1+s)} \frac{\varepsilon}{k} \tag{10.25c}$$

A transport equation for the mass fraction of fuel is solved, where the reaction rate of fuel is taken as the smallest of the turbulent dissipation rates of fuel, oxygen and products:

$$S_{fu} = -\rho \frac{\varepsilon}{k} \min\left[C_R m_{fu},\ C_R \frac{m_{ox}}{s},\ C'_R \frac{m_{pr}}{1+s}\right] \tag{10.26}$$

C_R and C'_R are model constants. In addition to the equation for m_{fu} a transport equation for mixture fraction f is also solved to deduce the product and oxygen mass fractions using relationships (10.16–10.18). Figures 10.2a and b show the results of Magnussen and Hjertager (1976) who obtained good predictions of the temperature field in furnace configurations with the eddy break-up model. Figure 10.3 shows a further application of the eddy break-up approach combined with the pdf method to account for scalar fluctuations by Gosman *et al* (1978) and again the prediction compares very well with the experimental data.

The eddy break-up model can also accommodate kinetically controlled reaction terms. When the combustion processes are kinetically controlled, the fuel dissipation rate may be expressed by the Arrhenius kinetic rate expression

$$R_{fu,\ kinetic} = -A_1 \rho^{-a} m_{fu}^{b} m_{ox}^{c} \exp(-E/RT) \tag{10.27}$$

where A_1 is the pre-exponential constant for the Arrhenius reaction rate, a, b and c are model constants, T the temperature in K, E the activation energy and R the gas constant. Now the reaction rate of fuel is given by

$$S_{fu} = -\min\left[\rho \frac{\varepsilon}{k} C_R m_{fu},\ \rho \frac{\varepsilon}{k} C_R \frac{m_{ox}}{s},\ \rho \frac{\varepsilon}{k} C'_R \frac{m_{pr}}{1+s},\ -R_{fu,\ kinetic}\right] \tag{10.28}$$

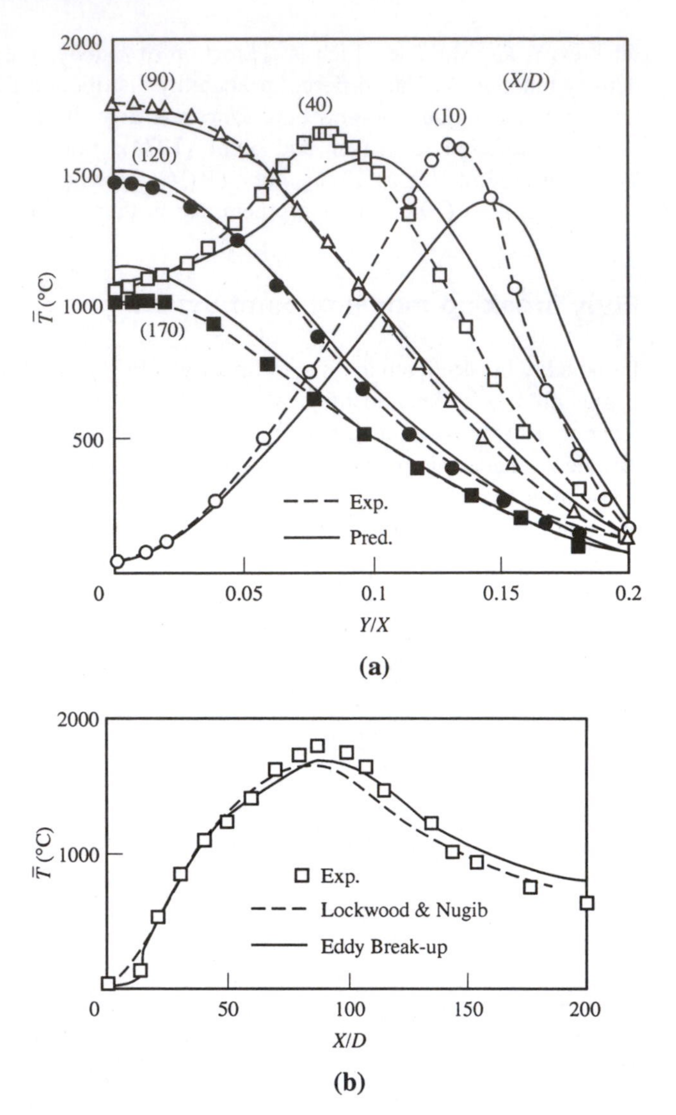

Fig. 10.2 Results of the eddy break-up model (Magnussen and Hjertager, 1976): (a) comparison of experimental (Lockwood and Odidi, 1975) and eddy break-up model predictions of local mean temperatures of a city gas diffusion flame (*Re* 24000); (b) experimental mean temperatures on the axis of the city gas diffusion flame (*Re* 24000) compared with prediction by Lockwood and Naguib (1975) and the predictions of the eddy break-up model.

Figure 10.4 shows the predictions reported by Nikjooy *et al* (1988) who used the above approach in a two-step reaction mechanism for the prediction of combustion in axisymmetric combustor geometries.

The eddy break-up model makes reasonably good predictions and is fairly straightforward to implement in CFD procedures, but the quality of the predictions depends on the performance of the turbulence model. Where a turbulence model fails to make accurate flow predictions the quality of combustion simulations will, of course, also be limited.

10.2.3 Laminar flamelet model

Another popular combustion modelling approach is the laminar flamelet model of combustion. Whereas the SCRS assumes linear relationships (10.16–10.18) between mixture fraction, mass fractions and temperature, the laminar flamelet model allows

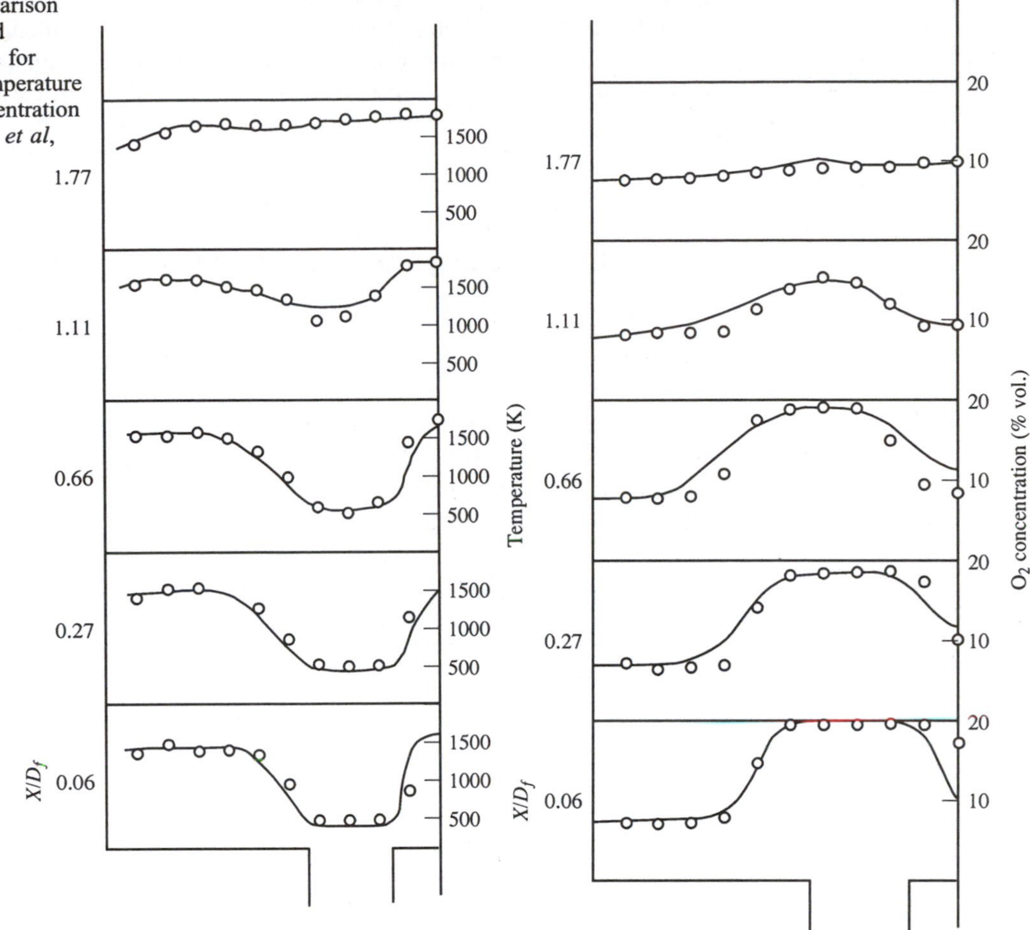

Fig. 10.3 Comparison of predictions and experimental data for Case 6: radial temperature and oxygen concentration profiles (Gosman *et al*, 1978)

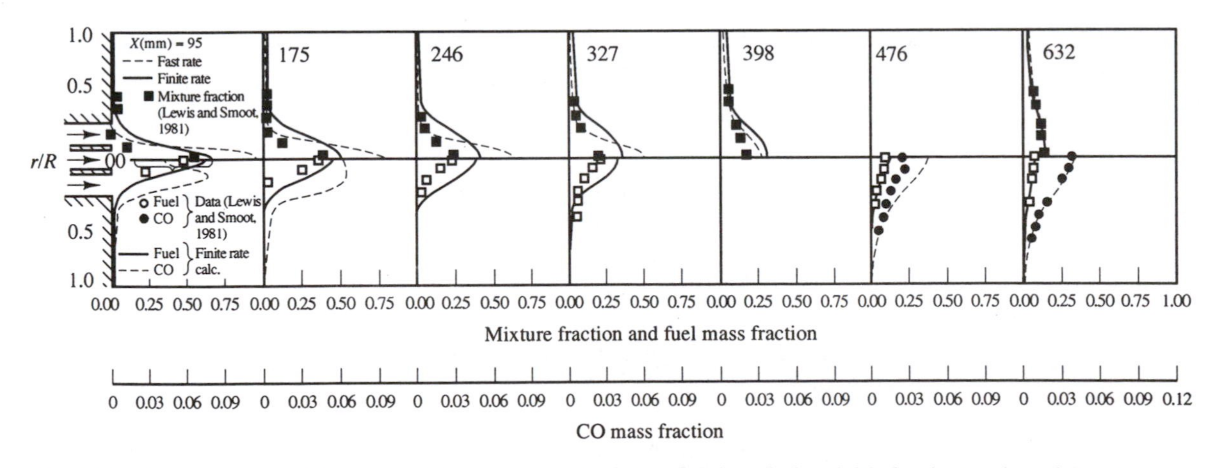

Fig. 10.4 A comparison of the calculated and measured mixture fraction, fuel and CO for the Lewis and Smoot (1981) experiment by Nikjooy *et al* (1988)

the inclusion of experimental information to describe more sophisticated relationships between these variables. As implied by the name of the model the necessary additional data are acquired from measurements in laminar diffusion flames. A transport equation for the mixture fraction is solved and the species mass fractions are deduced from laminar flamelet relationships. Interested readers are referred to Liew *et al* (1984), Bray *et al* (1985), Askari-Sardhai *et al* (1985) and Peters (1986).

10.3 Calculation of buoyant flows and flows inside buildings

The modelling of buoyant flows also requires additional modelling effort. Flows inside buildings fall within the buoyant flow category because they are frequently driven by natural ventilation resulting from temperature differences inside the building. When modelling buoyant flows, the momentum equation in the direction of gravity should include the body force resulting from buoyancy. For example, in two-dimensional flows with buoyancy in the y-direction, the v-momentum equation is given by

$$\frac{\partial}{\partial t}(\rho v) + \frac{\partial}{\partial x}(\rho u v) + \frac{\partial}{\partial y}(\rho v v) = \frac{\partial}{\partial x}\left[\mu \frac{\partial v}{\partial x}\right] + \frac{\partial}{\partial y}\left[\mu \frac{\partial v}{\partial y}\right] - g(\rho - \rho_0) - \frac{\partial p}{\partial y} + S_v \tag{10.29}$$

Here $-g(\rho - \rho_0)$ is the buoyancy term, where ρ_0 is a reference density. The buoyancy term in the discretised form of the above equation can give rise to serious instabilities in the solution processes. Severe under-relaxation is often required in buoyancy-related problems and sometimes a transient approach is recommended for obtaining steady state solutions.

Standard turbulence models need additional modifications when applied to buoyant flows. For example, an additional generation term, recommended by Rodi (1978), in the k-equation of the k–ε turbulence model is used in modelling turbulent buoyant flows. The k-equation takes the form

$$\frac{\partial(\rho k)}{\partial t} + div(\rho k \mathbf{u}) = div(\Gamma_k \, grad \, k) + G + B - \rho\varepsilon \tag{10.30}$$

where G is the usual production or generation term (see section 3.5.2) and B is the generation term relating to buoyancy. The latter is given by

$$B = \beta g_i \frac{\mu}{\sigma_T} \frac{\partial T}{\partial x_i} \tag{10.31}$$

where T is the temperature, and g_i is the gravitational acceleration in the x_i-direction. The volumetric expansion coefficient β is defined by

$$\beta = -\frac{1}{\rho} \frac{\partial \rho}{\partial T} \tag{10.32}$$

The modelled transport equation for the dissipation of turbulence kinetic energy (ε) is given by

$$\frac{\partial(\rho\varepsilon)}{\partial t} + div(\rho\varepsilon\mathbf{u}) = div(\Gamma_\varepsilon \, grad \, \varepsilon) + C_{1\varepsilon}\frac{\varepsilon}{k}(G + B)\left(1 + C_3 R_f\right) - C_{2\varepsilon}\rho\frac{\varepsilon^2}{k} \tag{10.33}$$

where R_f is the flux Richardson number and C_3 is an additional model constant (Rodi, 1978). Hossain and Rodi (1976) defined R_f by the relation $R_f = -B/G$. A single value of C_3 cannot be used in the definition of R_f because C_3 is close to unity in vertical buoyant shear layers and close to zero in horizontal shear layers. Rodi (1978) proposed an alternative definition of the flux Richardson number which allows the use of a single value of $C_3 \approx 0.8$ both for horizontal and vertical layers:

$$R_f = \frac{-G_l}{2(B+G)} \tag{10.34}$$

were G_l is the buoyancy production in the lateral energy component. In a horizontal shear layer where the lateral velocity component is in the direction of gravity, the entire buoyancy production is in the direction of gravity so that

$$G_l = 2B \tag{10.35}$$

In vertical shear layers, the lateral component is normal to the direction of gravity and has no buoyancy contribution so that $G_l = 0$. Accordingly, the flux Richardson number is

$$R_f = -\frac{B}{B+G} \qquad \text{for horizontal layers}$$

$$R_f = 0 \qquad \text{for vertical layers}$$

If the flow in the problem considered is dominated by vertical shear layers, then R_f may be set to zero and C_3 taken as 0.8. The importance of C_3 to CFD prediction of fire problems in buildings was closely studied by Markatos *et al* (1982) to which the reader is referred for more details.

10.4 The use of body-fitted co-ordinate systems in CFD procedures

Computational fluid dynamics methods based on Cartesian or cylindrical co-ordinate systems have certain limitations in irregular geometries. Practical boundary geometries can be complex and often irregular and they can only be approximated in Cartesian and cylindrical co-ordinate systems by treating surfaces in a stepwise manner as illustrated in Figure 10.5. To calculate the flow past the half cylinder of Figure 10.5 using a Cartesian coordinate system, the cylindrical surface may be represented by a step approximation and cells inside the solid part of the cylinder are blocked in the calculation. This has considerable disadvantages since the

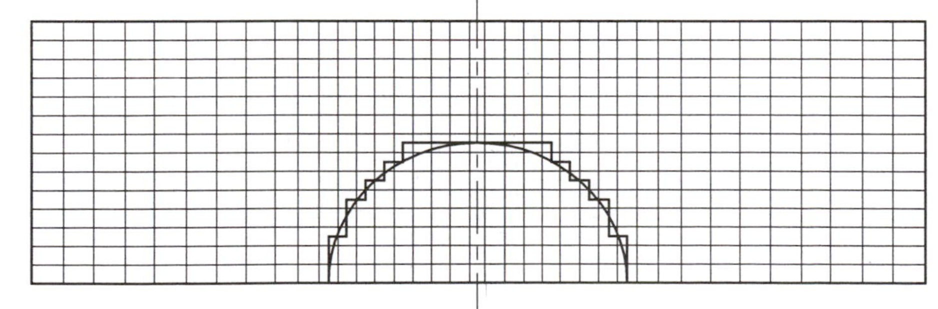

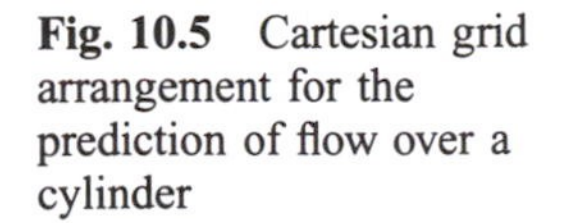

Fig. 10.5 Cartesian grid arrangement for the prediction of flow over a cylinder

approximate boundary description is tedious to set up and introduces errors, especially if the wall shear stresses need to be calculated to good accuracy. Further disadvantages of the Cartesian/cylindrical co-ordinate system include a wastage of computer storage and resources due to (i) blocking of the cells in solid regions and (ii) the introduction of a fine Cartesian mesh in one region of particular interest could imply unnecessary refinement in another region of minimal interest.

Methods based on body-fitted grid or non-orthogonal grid systems have been developed to overcome the limitations referred to above (Rhie and Chow, 1983; Peric, 1985; Demirdzic *et al*, 1987; Shyy *et al*, 1988; Karki and Patankar, 1988) and are used increasingly in present-day CFD procedures. Figure 10.6 shows a body-fitted grid for the cylinder problem. The geometrical flexibility offered by body-fitted grid techniques is useful in the modelling of practical problems involving irregular geometries because (i) all geometrical details can be accurately incorporated and (ii) the grid properties can be controlled to capture useful features in regions of interest. The governing equations with body-fitted grids are, however, much more complex than their Cartesian equivalents. Detailed discussions of the available methods of formulating the governing equations can be found in Demirdzic (1982) and Shyy and Vu (1991). The main difference between the formulations lies in the grid arrangement and in the choice of dependent variables in the momentum equations. In CFD procedures based on body-fitted co-ordinates the use of non-staggered or collocated grid systems for velocities is increasingly preferred to staggered grids which require additional storage allocations. However, non-staggered grids require special procedures to ensure proper velocity and pressure coupling and to avoid the unrealistic pressure fields discussed in section 6.2. Details of some of these special procedures can be found in Hirt *et al* (1974), Rhie and Chow (1983), Peric (1985), Reggio and Camarero (1986) and Rodi *et al* (1986), among others.

Figure 10.7 shows a part of a tube bank where CFD can be used to predict the flow field. Considering symmetry only the shaded part of the geometry needs to be

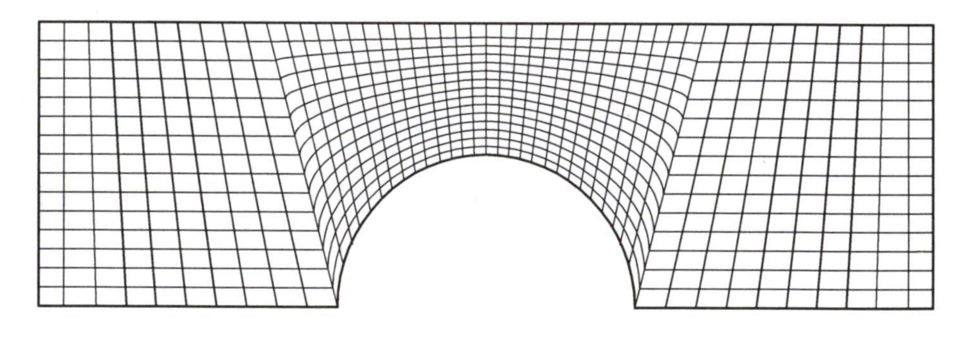

Fig. 10.6 Body-fitted grid arrangement for the prediction of flow over a cylinder

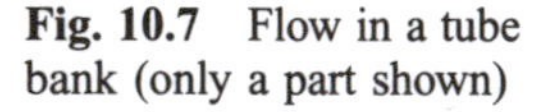

Fig. 10.7 Flow in a tube bank (only a part shown)

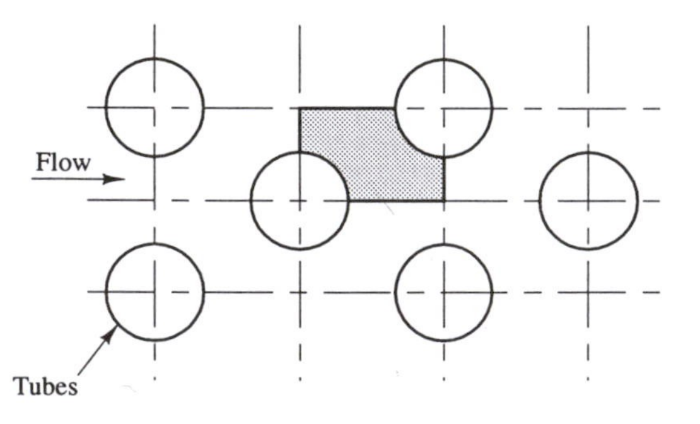

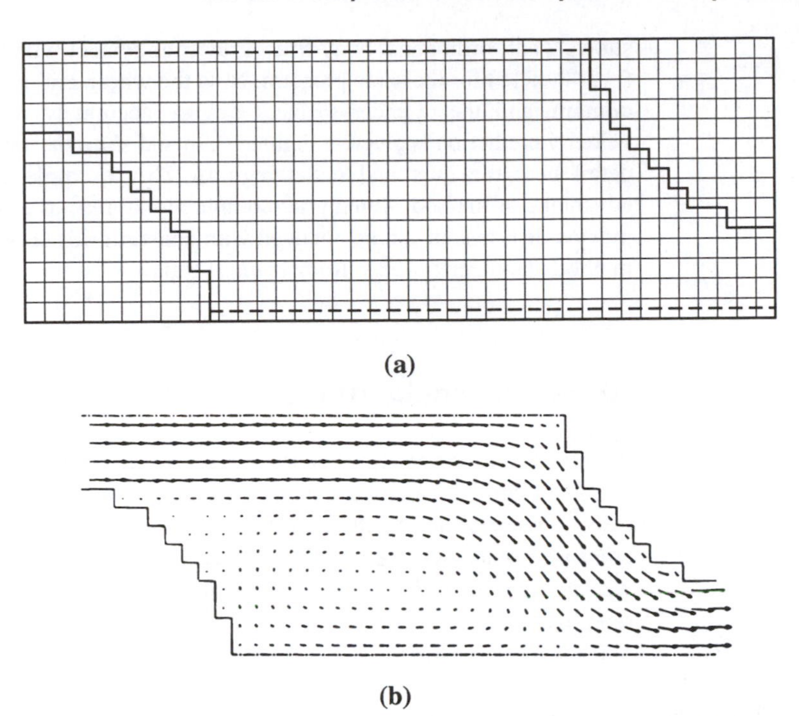

Fig. 10.8 (a) Cartesian grid; (b) predicted flow pattern using the 40 × 15 Cartesian grid

considered. Figure 10.8a shows the way a Cartesian grid arrangement is used to predict the flow. We use a 40 × 15 grid, block the cylinder off with solid wall cells and approximate the surface by means of a step arrangement. Figure 10.8b shows the resulting velocity field. Much of the grid (approximately 25%) is wasted in dealing with the objects so fewer cells are available to represent the flow region. Figure 10.9a

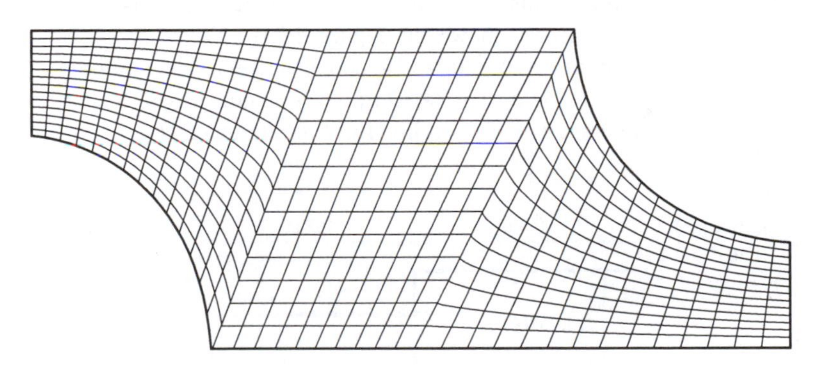

Fig. 10.9 (a) Non-orthogonal body-fitted grid; (b) predicted flow pattern using the 40 × 15 body-fitted grid

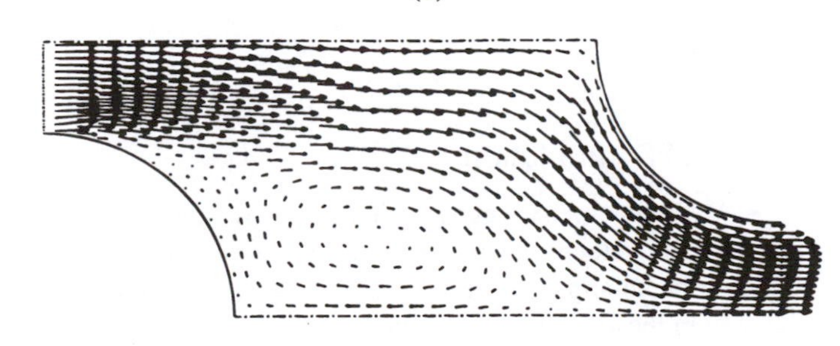

(b)

shows a non-orthogonal body-fitted grid arrangement with the same number of cells (i.e. 40×15) for the same problem. Now the whole grid occupies the computational domain, and the cylinder surfaces can be accurately represented. The resulting velocity prediction is given in Figure 10.9b and shows a considerably greater level of detail near the inlet and outlet regions. This example clearly demonstrates the advantage of the body-fitted grid: computational resources are well utilised, so grid-independent results can be obtained with coarser grids compared to Cartesian-based methods (see Peric, 1985; Rodi *et al*, 1989).

10.5 Advanced applications

In this section we give examples of industrially relevant CFD applications. The boundary conditions and problem specification are briefly described and specimen results are presented to illustrate how CFD and the modelling of combustion and other phenomena can be applied to practical situations. The examples are presented without the finer details of the calculations which can be found in the cited references.

10.5.1 Flow in a sudden pipe contraction

The problem considered

This problem was selected to illustrate the application of CFD to a benchmark problem with a set of well-documented data for the comparison of predictions with experiments. The problem considered here is laminar pipe flow in a sudden contraction shown in Figure 10.10. Durst and Loy (1985) have provided the experimental data for a range of Reynolds numbers. The flow with a Reynolds number ($Re = \rho UD/\mu$) of 372 was considered for CFD modelling, where U is the average velocity in the pipe with diameter D.

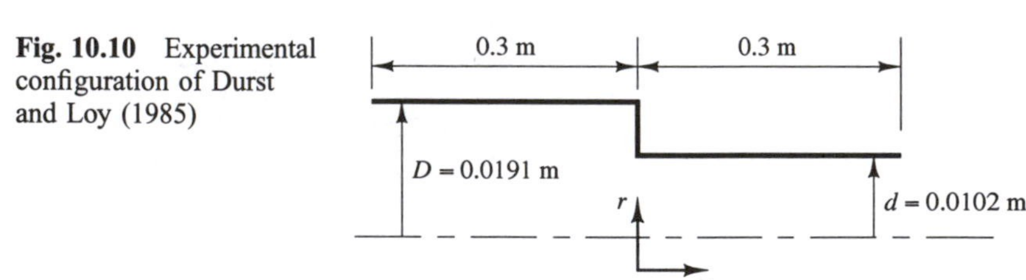

Fig. 10.10 Experimental configuration of Durst and Loy (1985)

CFD simulation

The geometry was modelled with a two-dimensional axisymmetric grid of 100×60. The velocity profile for fully developed laminar flow was imposed at the inlet and the no-slip condition was applied at wall boundaries. At the exit plane, all derivatives in the axial direction were set to zero. The CFD calculation was carried out using the SIMPLER algorithm and the hybrid differencing scheme.

Specimen results

Since the flow is laminar, the governing equations are exact (i.e. no turbulence modelling involved here). The predicted streamlines of the flow are shown in Figure 10.11. The velocity profiles are shown in Figure 10.12 for six different cross-sections of the domain, three before, and three after, the contraction. The experimental data of Durst and Loy (1985) are also included for comparison. It can be seen that the predictions agree well with the experimental measurements. Further grid refinement did not cause significant changes in the predictions and therefore these results can be considered to be grid independent. It should be noted that comparisons for locations other than those shown in the figure and for other Reynolds numbers also agree well with the experimental data. This simple example shows the capability of CFD to predict practical flow situations, with a good degree of accuracy.

Fig. 10.11 Predicted streamline pattern

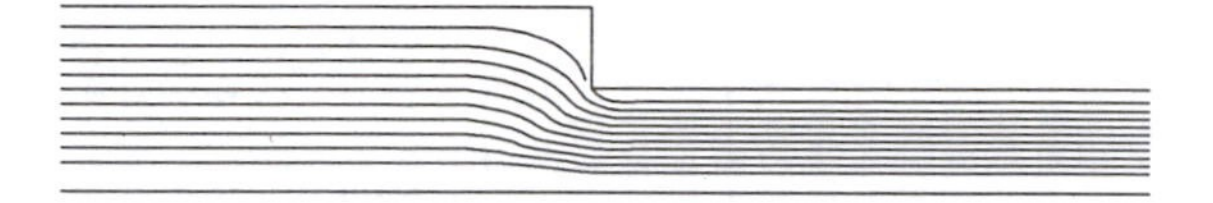

10.5.2 Modelling of a fire in a test room

The problem considered

In contrast to the previous benchmark problem we now study a case at the other end of the spectrum of complexity. We compare CFD calculations with experimental fire tests carried out by the Lawrence Livermore National Laboratory (LLNL) in the test room shown in Figure 10.13. The details of the experiments have been reported in Alvarez *et al* (1984). The fire was at the centre of the floor and clean air was introduced along the floor of the test cell, which is approximated in the model by a 0.12 m high and 2 m long slot for air entry, located 0.1 m above the floor. The fire sources in the experiments were a burner, a spray and a pool of fuel in a tray. The products of combustion were extracted near the top of the cell using an axial flow fan through a rectangular 0.65 m square duct placed 3.6 m above the floor as shown in Figure 10.13. A total of 27 tests were reported by Alvarez *et al* (1984), and the one designated MOD08 has been selected for CFD modelling here. In this test, a spray of isopropyl alcohol from an opposed-jet nozzle located at the centre of the pan was used, and the fuel evaporated quickly to burn in a way similar to a natural pool fire. The fuel injection rate was 13.1 g/s with a total heat release rate of 400 kW. These data were used to specify burner conditions at the fire source. The measured extraction rate, 400 l/s in the steady state, was used to specify the outflow. The mass flow rate of air into the domain and the inlet and outlet velocities are calculated as part of the solution. The walls, the floor and the ceiling of the compartment were of 0.1 m thick refractory. The estimated thermal conductivity, density and specific heat were, respectively, 0.39 W/m/K, 1400 kg/m^3 and 1 kJ/kg/K for the walls and 0.63 W/m/K, 1920 kg/m^3 and 1 kJ/kg/K for the ceiling and the floor. The walls were assumed to be perfectly black for radiation calculations.

CFD simulation

The simulation of the aerodynamics and combustion was carried out using a three-dimensional CFD procedure based on the SIMPLE algorithm and the hybrid

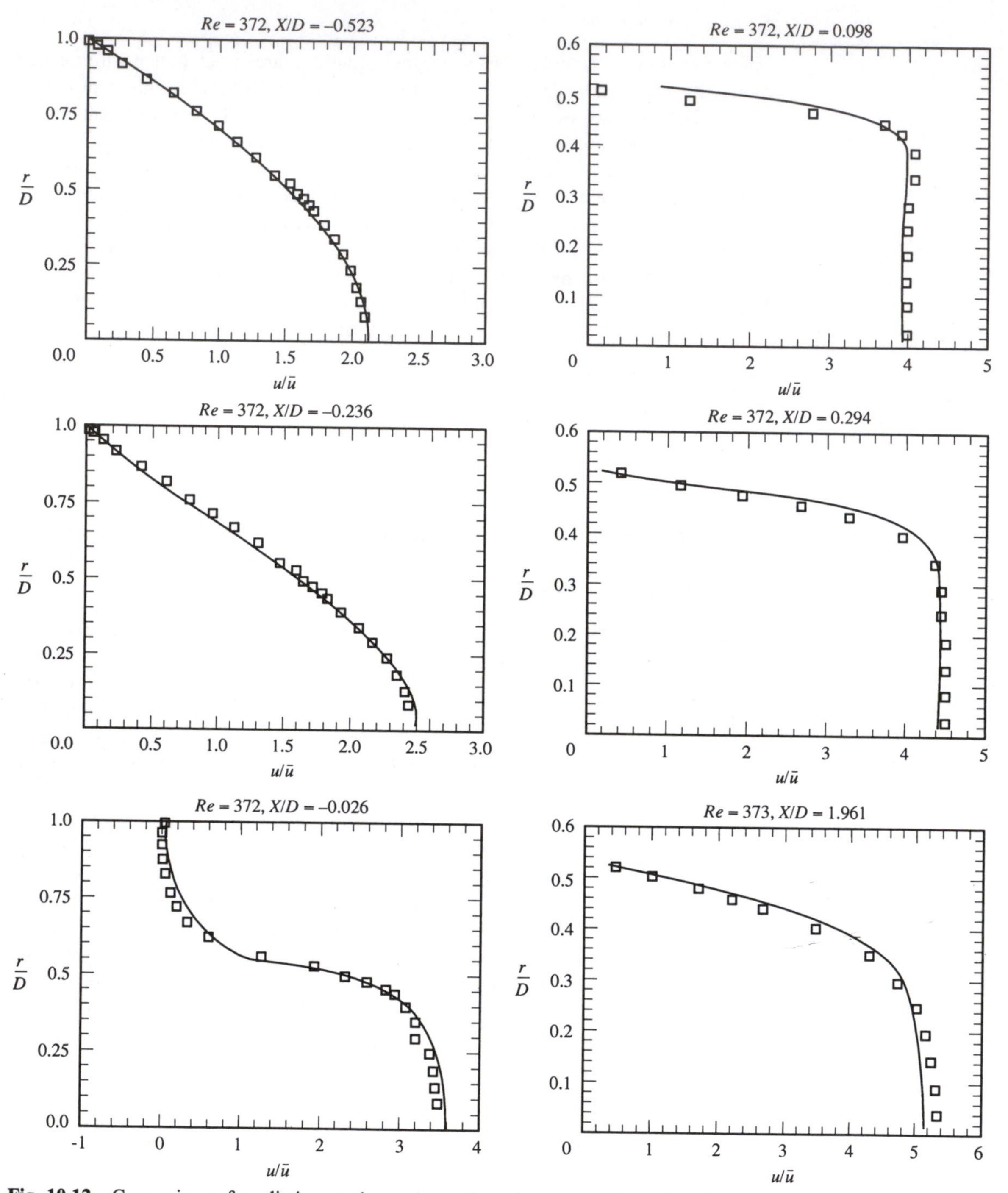

Fig. 10.12 Comparison of predictions and experimental results at six different locations

differencing scheme for discretisation. Turbulence was modelled with the k–ε turbulence model with buoyancy terms and combustion modelling assumed fast chemistry (SCRS). The discrete transfer model of thermal radiation (Lockwood and Shah, 1981) was used to calculate radiative heat transfer. The wall temperatures were obtained from a one-dimensional wall heat transfer model. A numerical grid of $14 \times 13 \times 12$, although not very fine, was considered adequate to predict the overall

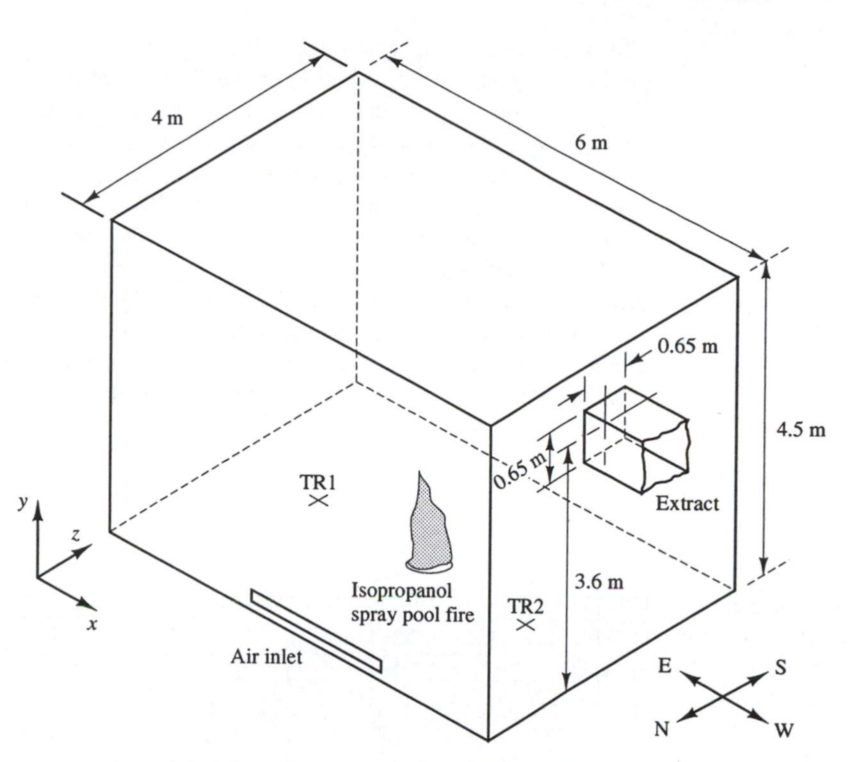

Fig. 10.13 Schematic diagram of the Lawrence Livermore National Laboratory (LLNL) fire test cell

properties of the fire. Further details of the model can be found in Malalasekera (1988) and Lockwood and Malalasekera (1988). Some specimen results are presented below.

Specimen results

Figure 10.14 shows the predicted steady state flow pattern in the $Y - Z$ plane at $X = 3.25$ m. The buoyancy-generated flow is clearly reproduced by the simulation

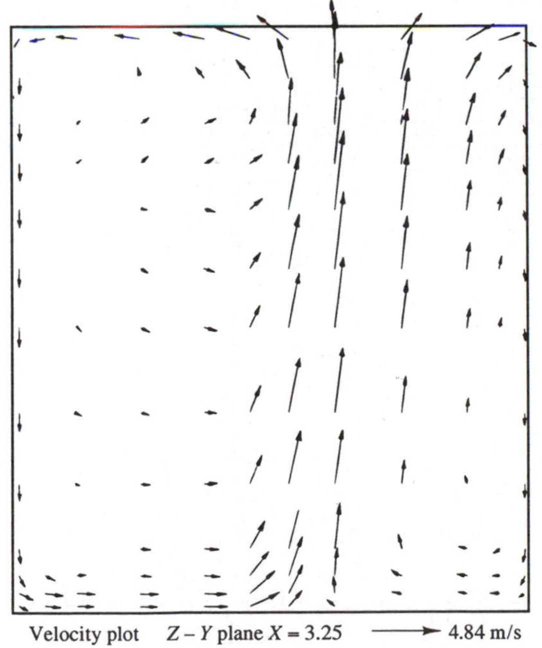

Fig. 10.14 Predicted flow inside the compartment: velocity vector plots in the Z–Y plane at $X = 5.25$ m

which also shows the entrainment induced by the strong buoyancy effects. The predicted temperature distribution in the Y–Z plane at $X = 3.00$ m (Figure 10.15) shows the hot gases around the central flame and the formation of a hot layer at ceiling level. The flame structure and tilt due to induced air flow are also clearly visible. Figure 10.16 compares the room temperature predictions with the experimental data of Alvarez *et al* (1984). The experimental temperatures were recorded using two thermocouple rakes (TR1–east rake – and TR2–west rake) with 15 thermocouples each placed 1.5 m on either side of the fire and located in the central plane as shown in Figure 10.13. The predictions and experiments show good agreement which illustrates the capability of CFD in predicting complex flows. The predictions reproduce the main features of the experiments and, despite the coarse grid, the predictions agree well with the experimental data.

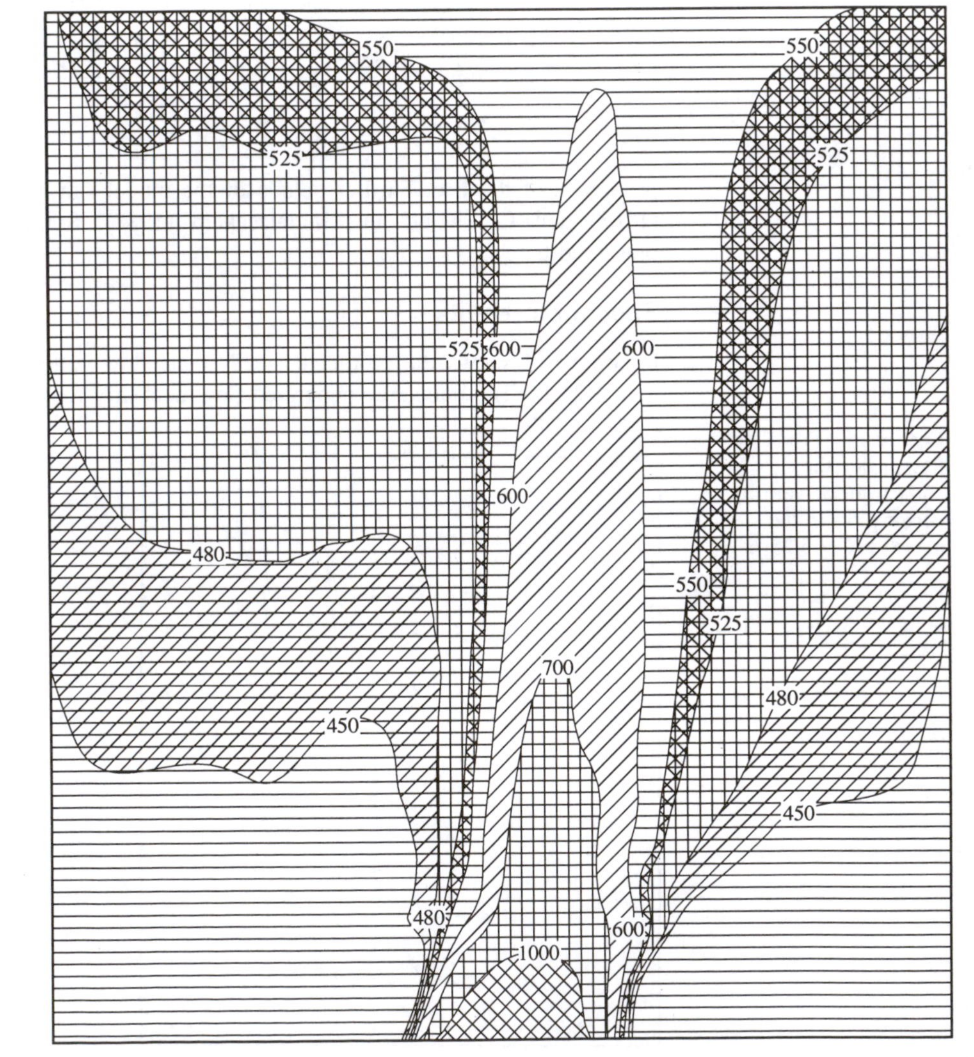

Fig. 10.15 Predicted temperature (K) field in the Y–Z plane at $X = 3.00$ m

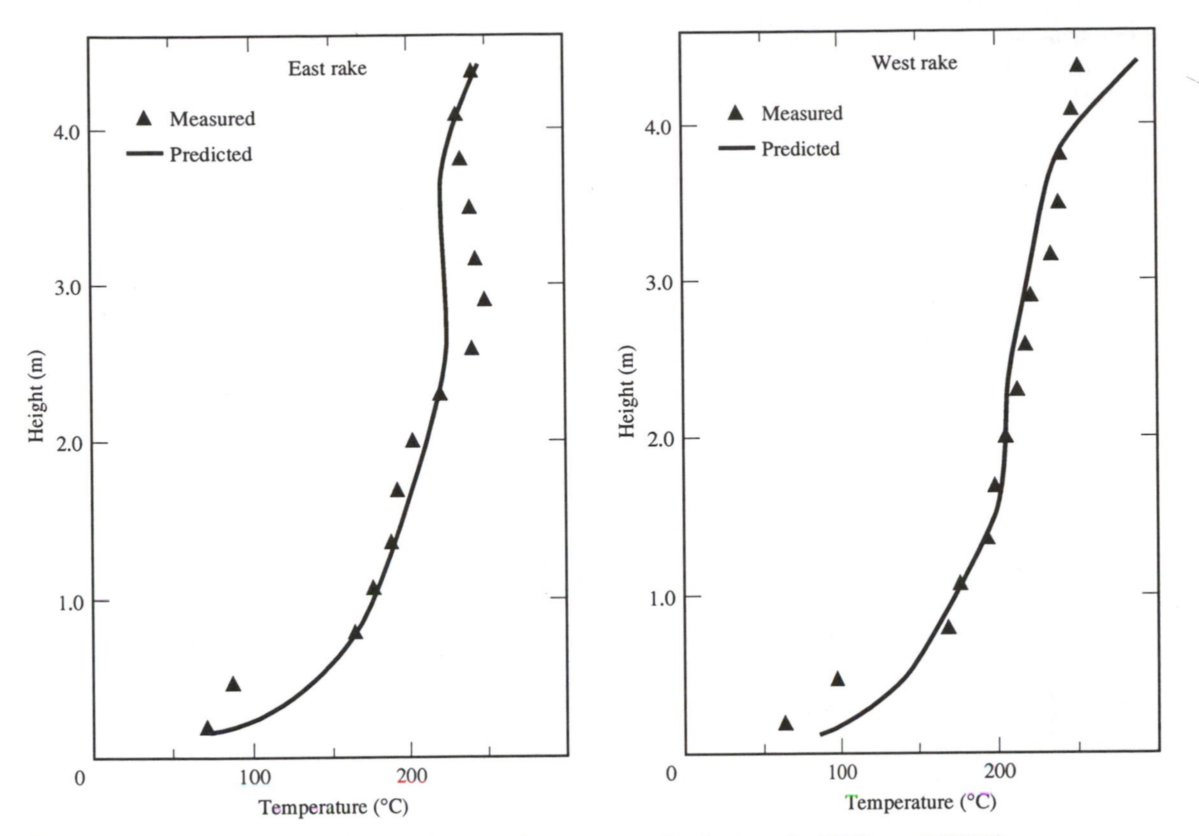

Fig. 10.16 Comparison of predicted and measured temperature distributions for LLNL test MOD08

10.5.3 Prediction of flow and heat transfer in a complex tube matrix

Problem considered

The application of CFD to the prediction of flow and heat transfer in the secondary heat exchanger of a condensing boiler is considered. Figure 10.17 shows a schematic diagram of a part of the heat exchanger which consists of six layers of tubes arranged in a cross-matrix manner. The ends of the tubes are welded to an outer jacket and water enters the jacket at 40 °C before circulating through the tubes. Each layer comprises 32 tubes, each 200 mm in length and with a constant diameter of 4.8 mm. The transverse pitch is 5.8 mm while the longitudinal pitch is 13 mm. The overall dimensions of the tube matrix are 200 mm × 200 mm × 60 mm as shown in Figure 10.18 which also gives the co-ordinate system used in the modelling. In operation, the flue gas initially enters the heat exchanger through the first X–Y plane, flows in the positive direction of the Z-axis and leaves at the last X–Y plane. During flow, the flue gas passes through the six layers of tubes (Figure 10.18) to exchange heat with the tubes carrying water so that its temperature is reduced on exit from the heat exchanger.

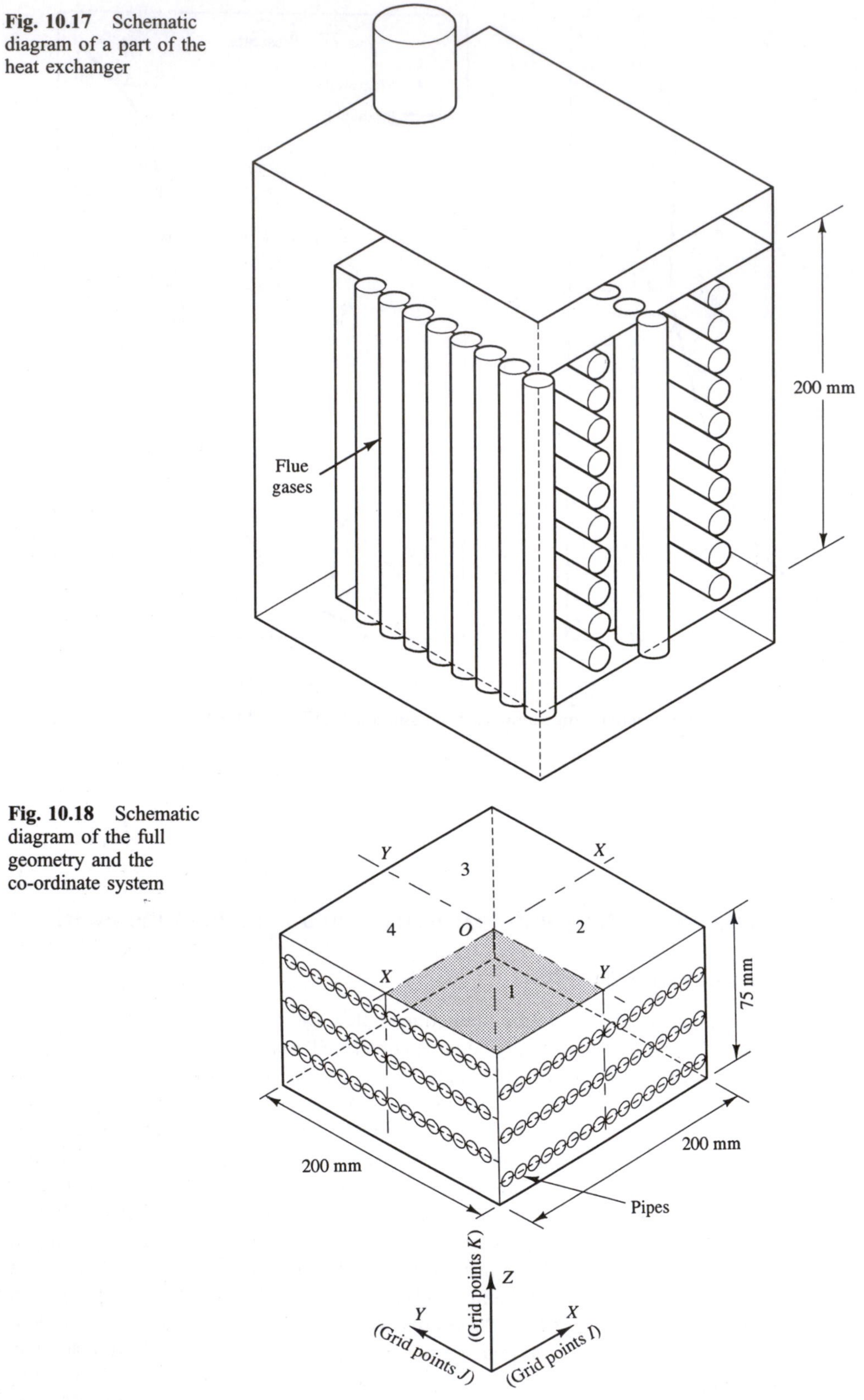

Fig. 10.17 Schematic diagram of a part of the heat exchanger

Fig. 10.18 Schematic diagram of the full geometry and the co-ordinate system

CFD simulation

The geometry shown in Figure 10.18 has four-fold symmetry and only a quarter of the full geometry is required to be considered for modelling. Symmetry boundary conditions are employed on the *O–X* and *O–Y* boundaries. Ideally, a CFD code written in a body-fitted co-ordinate system is necessary to represent the tube surfaces accurately. Here the tube arrangement is very complex and generating a grid for this geometry is a difficult task in Cartesian or body-fitted co-ordinates. It was decided to carry out the simulation in a three-dimensional Cartesian framework. The tube surfaces are approximated by stepwise surfaces. A fine grid of 85 × 85 × 33 cells was used to accommodate all the necessary geometrical details; the grid arrangement adopted is shown in Figures 10.19 and 10.20 which show the calculation grid in the *X–Y* plane and in the *X–Z* plane respectively. Figure 10.20 also compares the approximated tube cross-sections used in the calculations with the actual cross-sections. As can be seen, there is very little difference between them and therefore the approximation of the surfaces does not undermine the accuracy of the simulation.

The simulation of the flow was carried out using the SIMPLE algorithm. Heat transfer from the hot flue gases to the water circulating in the tubes was evaluated by solving the transport equation for enthalpy defined in this case as $h = C_P \Delta T$ with C_P assigned a value of 1092.0 J/kg/°C. This is compatible with the typical composition of flue gas at a temperature of 170 °C at the inlet to the heat exchanger.

Fig. 10.19 Computational grid in the *X–Y* plane

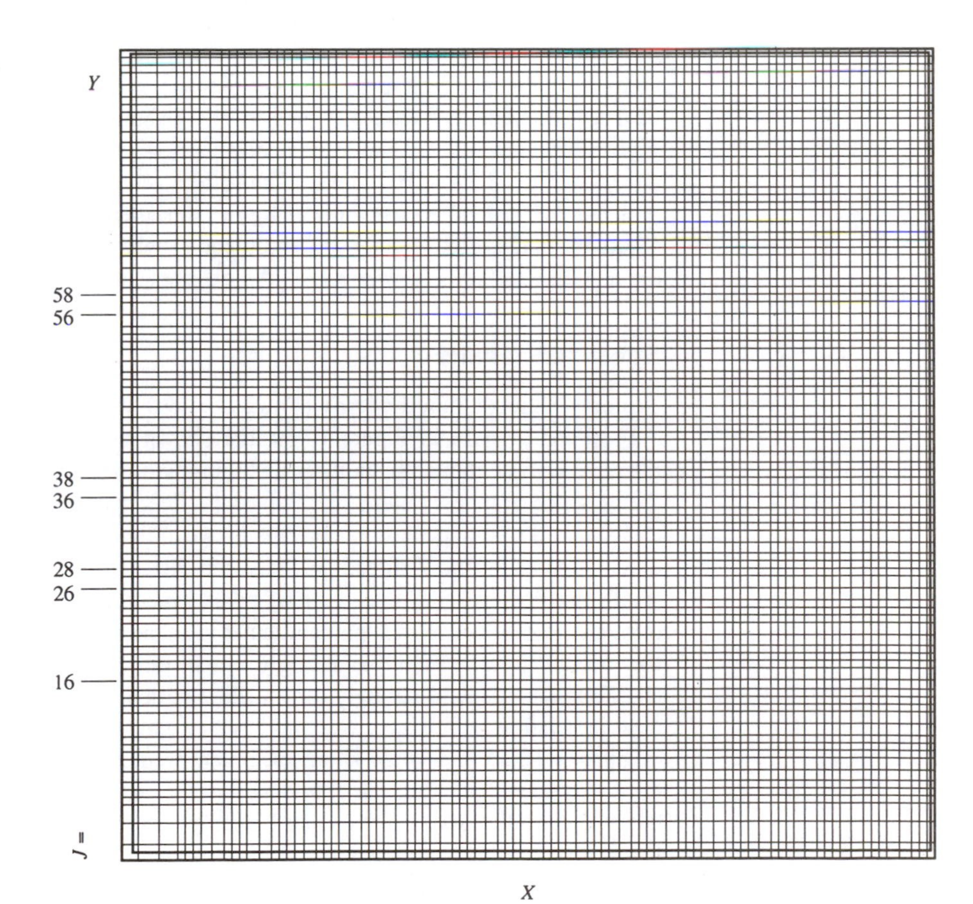

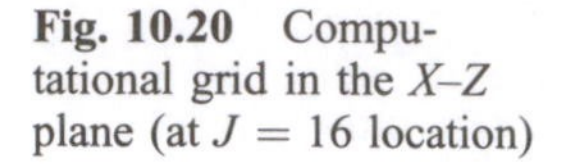

Fig. 10.20 Computational grid in the X–Z plane (at $J = 16$ location)

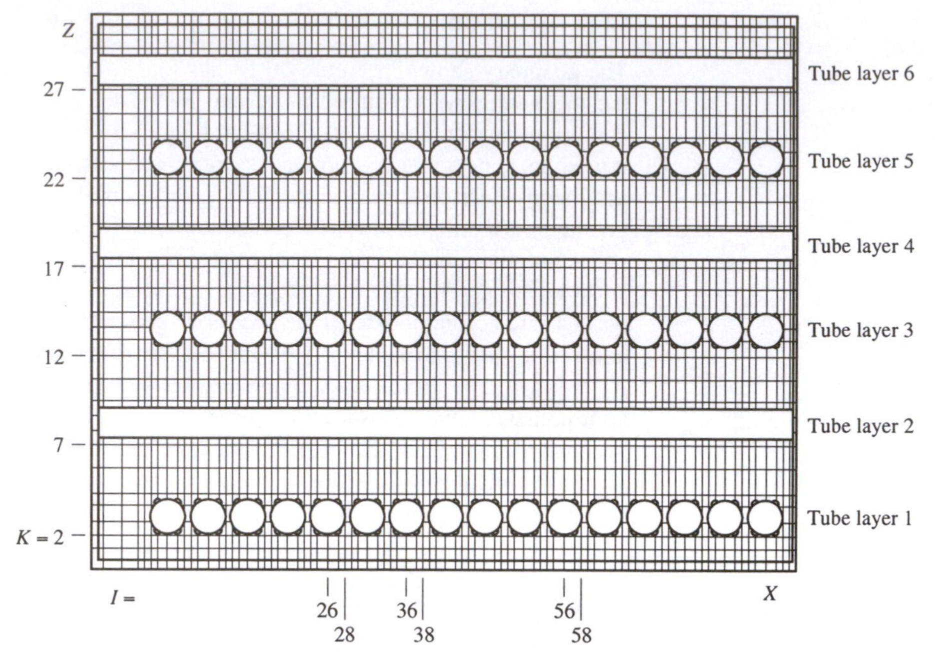

The variation of C_P with temperature was neglected in the simulation. At the inlet to the heat exchanger, the velocity (derived from the specified mass flow rate) and the enthalpy (corresponding to the inlet gas temperature) were specified. The outer walls of the heat exchanger were assumed to be adiabatic. The tube surface temperature was specified as 40 °C. The results shown below correspond to a mass flow rate of gas equal to 8.24×10^{-3} kg/s. The other inlet conditions of the gas were: temperature 170 °C, density 0.797 kg/m^3, turbulent kinetic energy $k_{in} = 0.03w_{in}^2$, where w_{in} is the inlet velocity, and turbulence energy dissipation $\varepsilon_{in} = k^{3/2}/\lambda$, where λ was taken as $0.001D$, where D is the external tube diameter.

Specimen results

Typical predicted flow patterns at two sections in the X–Z planes, identified as $J = 36$ and $J = 38$ in Figure 10.19, are shown in Figures 10.21 and 10.22. The plane $J = 36$ contains all six rows of tubing and the flow is severely obstructed by the presence of tubes. Figure 10.22 gives the flow between the planes $J = 38$ and 39 and shows the flow pattern in the gaps between the planes. Here, the flow pattern is much stronger in the Z direction. Since the distance between the layers of tubes is small, the flow passing one layer of tubes immediately changes direction to pass through the next set of gaps perpendicular to those of the preceding layer. Therefore, the overall flow pattern is strongly three dimensional, and the heat transfer characteristics depend strongly on the flow structure in the tube matrix as will be described below.

Figures 10.23 and 10.24 show specimen predictions of the temperature field in the X–Z plane inside the matrix at locations $J = 36$ and 38. Most of the heat transfer takes place in the first two rows and the temperature field appears to follow the flow features seen in the velocity plots. The gas temperature undergoes a large decrease when it passes the first row of tubes. Near to the walls, the temperature remains

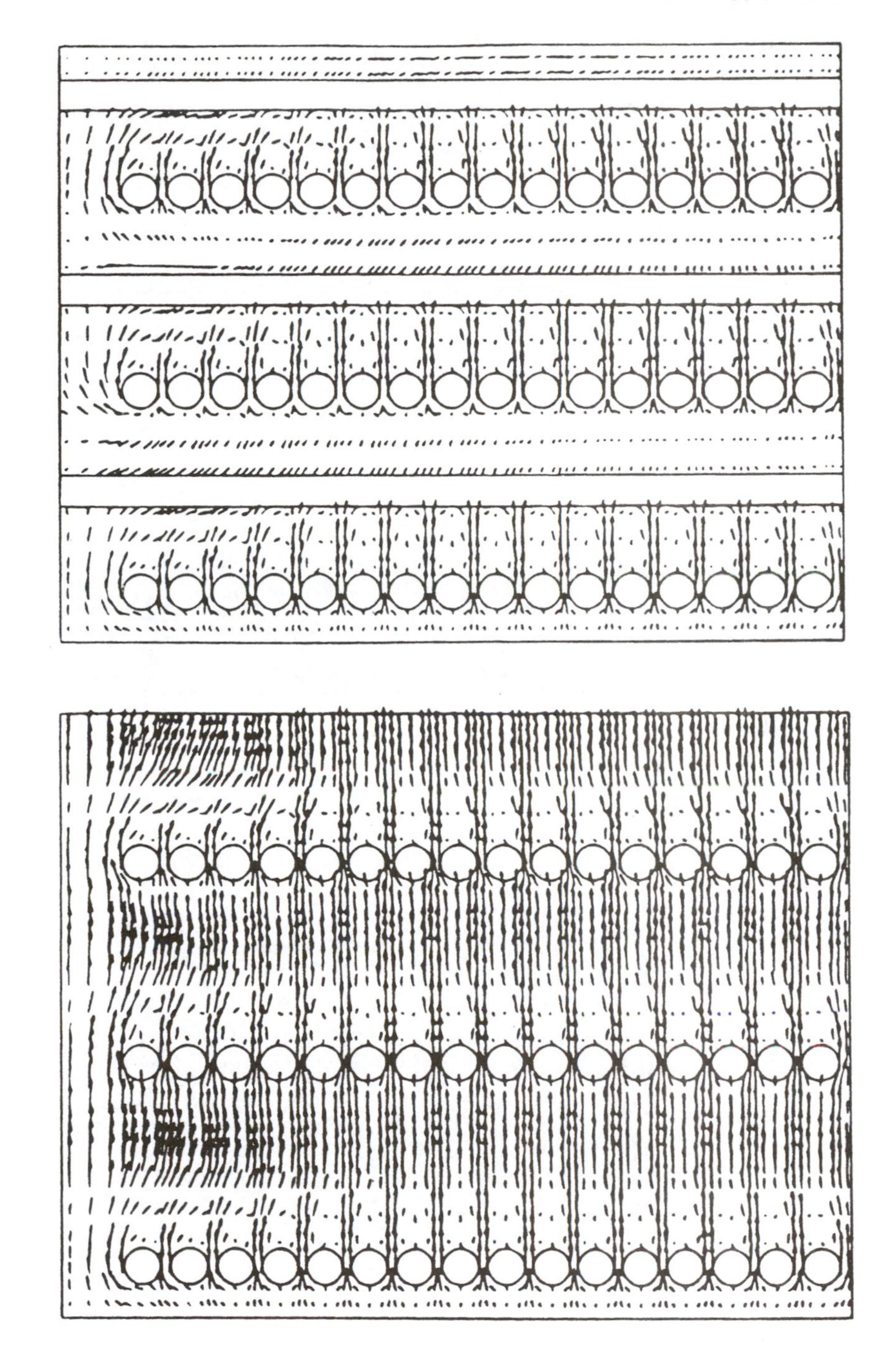

Fig. 10.21 Velocity vector field in the X–Z plane at $J = 36$ ($y = 0.0460$ m)

Fig. 10.22 Velocity vector field in the X–Z plane at $J = 38$ ($y = 0.0484$ m)

higher than at other locations. This is largely due to mixing at these points with the hot gases which escape from the larger gaps available at the wall ends.

Typical heat flux distributions are given for the upstream side of two layers of tubes (in the X–Y plane) in Figures 10.25 and 10.26 by plotting the predicted convective heat fluxes (W/m^2). The heat flux is given on the basis of the projected area of the flow section ignoring the gaps between tubes. Figure 10.25 shows a large convective flux at the first layer. Although the tubes in the second layer are at right

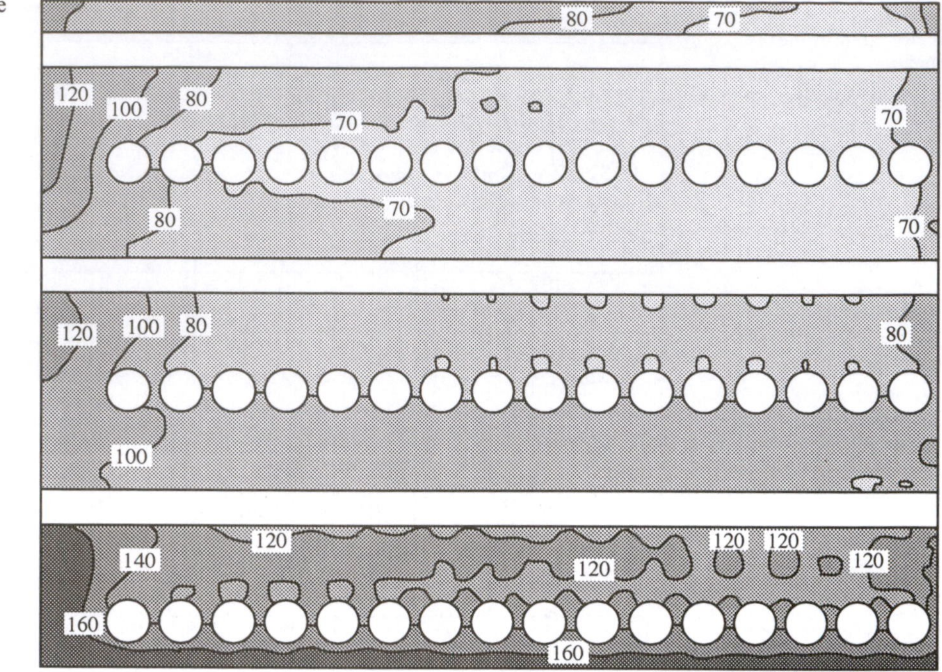

Fig. 10.23 Temperature distribution in the X–Z plane at $J = 36$ ($y = 0.0460$ m)

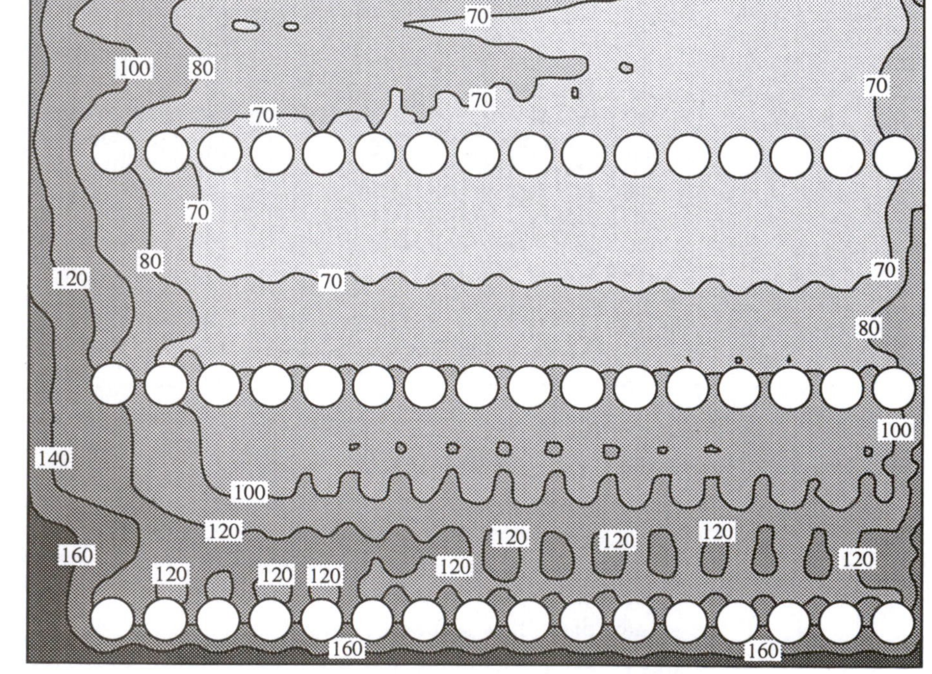

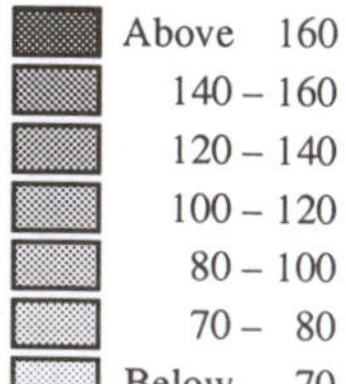

Fig. 10.24 Temperature distribution in the X–Z plane at $J = 38$ ($y = 0.0484$ m)

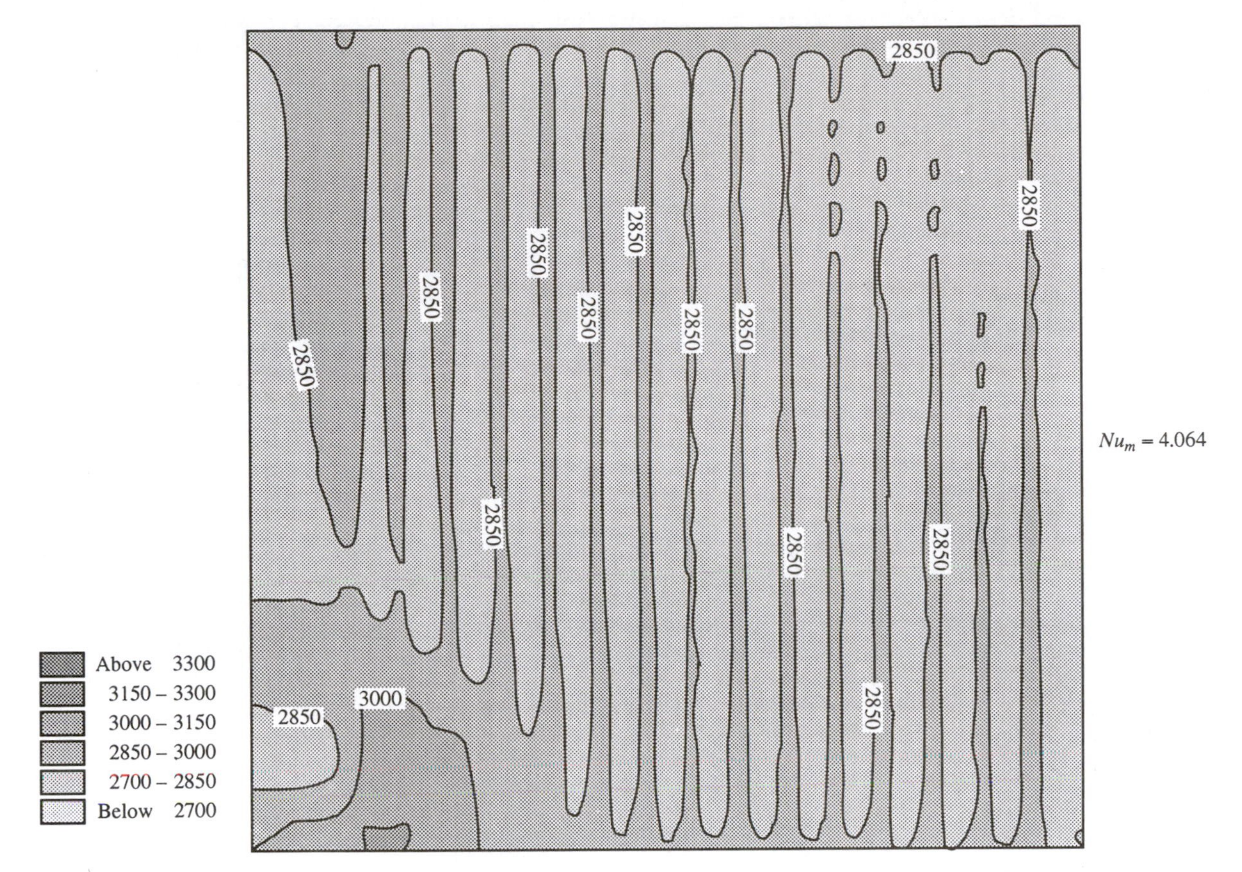

Fig. 10.25 Convective heat flux (W/m^2) at the upstream side of the first tube layer ($K = 2$, $z = 0.0070$)

angles to those of the first, the heat transfer pattern resembles that for the first layer. This is because the flow emerging from the first layer dominates the heat transfer on the upstream side of the second layer. This clearly shows that the flow features have a direct bearing on the resultant heat transfer. It is desirable to maximise the heat transfer rates to achieve a larger temperature difference across the heat exchanger, but this could result in condensation on the tube surfaces at some locations. From the point of view of heat transfer, this may not be desirable since the condensate could act as an impediment to the heat transfer process. The mean Nusselt numbers, calculated on the basis of total convective heat transfer in each tube layer, are given on the right hand side of Figures 10.25 and 10.26. From an engineering point of view these can be used to compare the performance of different heat transfer arrangements. It should be noted that the calculated Nusselt numbers are small because of low flow velocities and the Nusselt number is defined on the basis of the overall difference in temperature ($T_{in} - T_{tube}$).

Figure 10.27 shows the predicted temperature contours at the exit plane of the heat exchanger ($K = 31$, X–Y plane). The high temperature regions can be clearly seen in the corner where the hot gases entering the heat exchanger escape through the larger gaps. There is a marked decrease in temperature as the centre of the heat exchanger is approached. The points marked A, B, C and D are the locations where experimental temperature measurements were made and the bracketed values are the

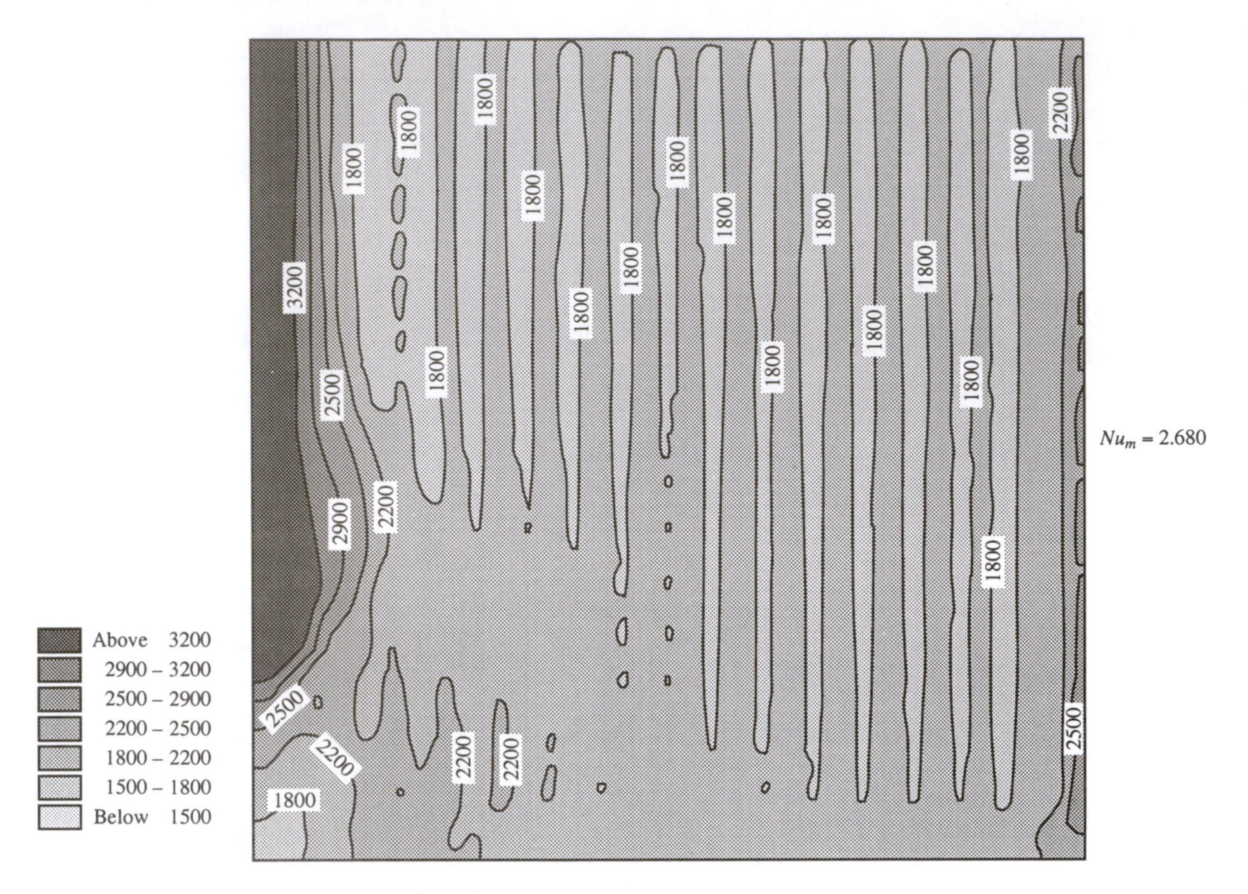

Fig. 10.26 Convective heat flux (W/m^2) at the upstream side of the second tube layer ($K = 7$, $z = 0.0200$)

experimentally measured temperatures. The measured temperatures are in reasonably good agreement with the predicted values. For further details of this simulation see Malalasekera *et al* (1993).

10.5.4 Laminar flow in a circular pipe driven by periodic pressure variations

The problem considered

Many engineering problems involve unsteady behaviour. In some cases, for instance paint mixing, the flow may be steady but the distribution of a transported scalar variable changes with time. In the present example we consider one of the simplest cases of the class of problems with genuinely unsteady flow fields: the periodic oscillations of an incompressible laminar flow in a circular pipe driven by harmonic pressure variations between inlet and outlet. Blood flows in veins and arteries, pressure waves in oil pipelines and air flows in intake manifolds of internal combustion engines can be modelled as periodic duct flows.

The applied pressure difference between the pipe ends is varied according to

$$\Delta P = K \cos nt \qquad (10.36)$$

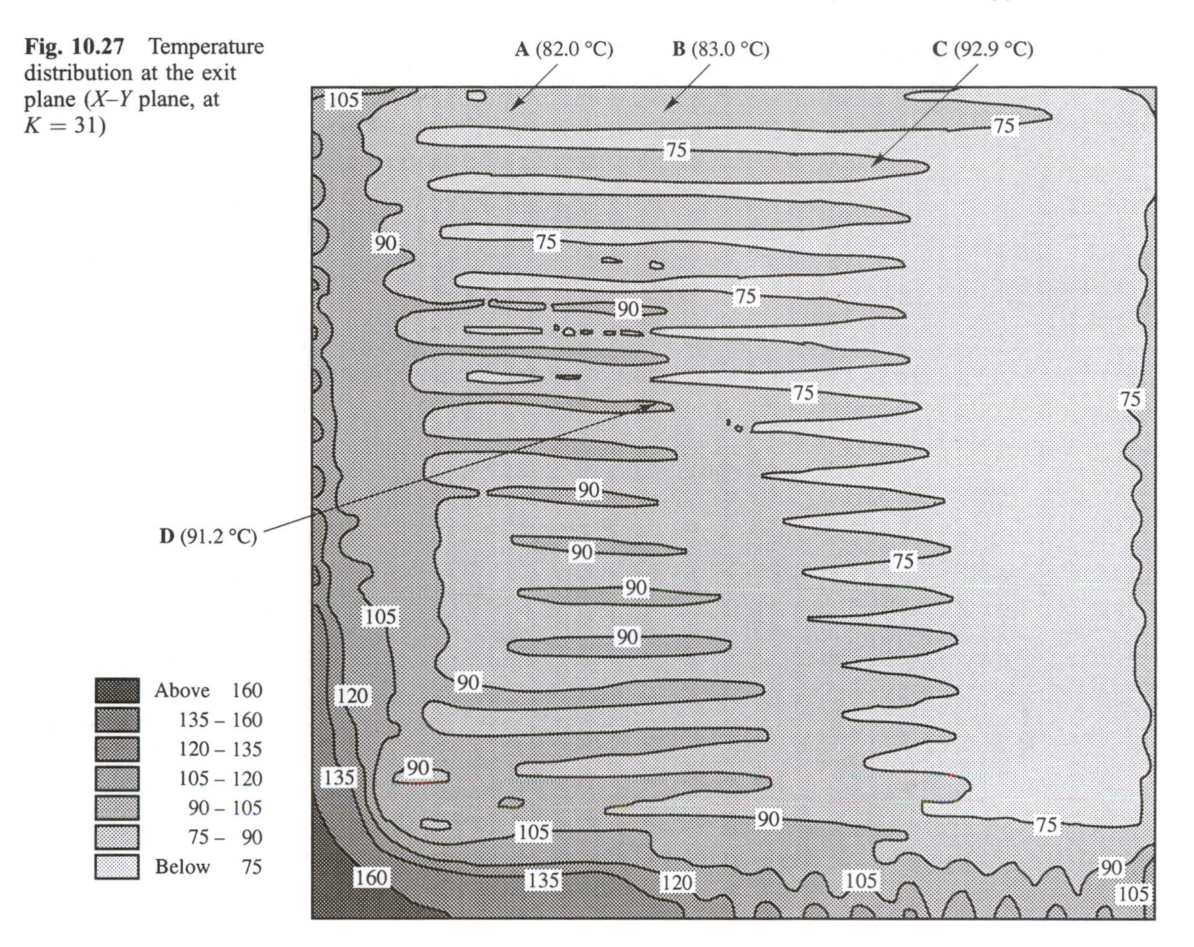

Fig. 10.27 Temperature distribution at the exit plane (X–Y plane, at $K = 31$)

The amplitude K is taken to be 50000 (Pa) and the circular frequency n is equal to 2π Hz giving an oscillation period of 1 s. Schlichting (1979) gives the analytical solution for the axial velocity component $u(r, t)$ as a function of radius r and time t for periodic laminar flow in a very long pipe as the real part of the following expression:

$$u(r,\ t) = -i\left(\frac{K}{n\rho L}\right)e^{int}\left[1 - \frac{J_0\left(r\sqrt{\frac{-in}{v}}\right)}{J_0\left(R\sqrt{\frac{-in}{v}}\right)}\right] \tag{10.37}$$

In this formula ρ, v and L are the fluid density, kinematic viscosity and the length of the pipe respectively, J_0 denotes the Bessel function of the first kind of order 0 and i is $\sqrt{-1}$. The general features of the velocity distributions are dependent on the value of the non-dimensional parameter $\sqrt{(n/v)}R$. The solution behaviour for small and large values of this parameter will be discussed below. Here we calculate the flow for two intermediate values of n by taking the pipe radius R equal to 0.01 m and frequency n as 2π Hz in conjunction with a constant fluid density of 1000 kg/m^3 and dynamic viscosity of 0.4 and 0.1 kg/m/s. This yields values of $\sqrt{(n/v)}R$ of 1.253 and 2.507 respectively.

CFD simulation

In order to set up a valid comparison between the analytical and finite volume solution of this problem we need to consider a pipe of sufficient length. The boundary layer flow near the inlet of a pipe changes in the downstream direction and in a steady flow the velocity distribution becomes fully developed after a distance l_E given by (Schlichting, 1979)

$$\frac{l_E}{R} = 0.25\,\frac{\bar{u}R}{\nu} \tag{10.38}$$

An estimate of the maximum possible mean velocities $\bar{u}$ (approximately 4 m/s here) can be obtained from the Hagen–Poiseuille formula (Schlichting, 1979). This leads to a maximum Reynolds number of 800 and hence a value of l_E of 1 m. Since the flow switches direction in the course of a cycle it is necessary to employ a computational domain of length greater than two times l_E to ensure that there is always a section of fully developed flow halfway along the duct. In this simulation we use a domain with a length L equal to 2.5 m and consider the solution in a cross-sectional plane at a distance of 1.25 m from its ends.

The flow is axisymmetric and we use a grid of 250 axial and 20 radial nodes distributed uniformly in the z and r directions. Figure 10.28 shows a sketch of the solution domain and part of the mesh used. At $r = 0$ a symmetry boundary condition ensures that there is no flow across the axis and that the gradients of all variables in the radial direction are locally zero. At $r = R = 0.01$ m the usual wall boundary condition is maintained. The cosinusoidal driving pressure difference given by Figure 10.29 and equation 10.36 is applied by means of prescribed pressure boundary conditions at $z = 0$ and $z = L = 2.5$ m. The solution procedure is SIMPLER with fully implicit time marching; the time step is 1 ms. The parabolic velocity profile of a steady laminar pipe flow is used as the initial velocity field.

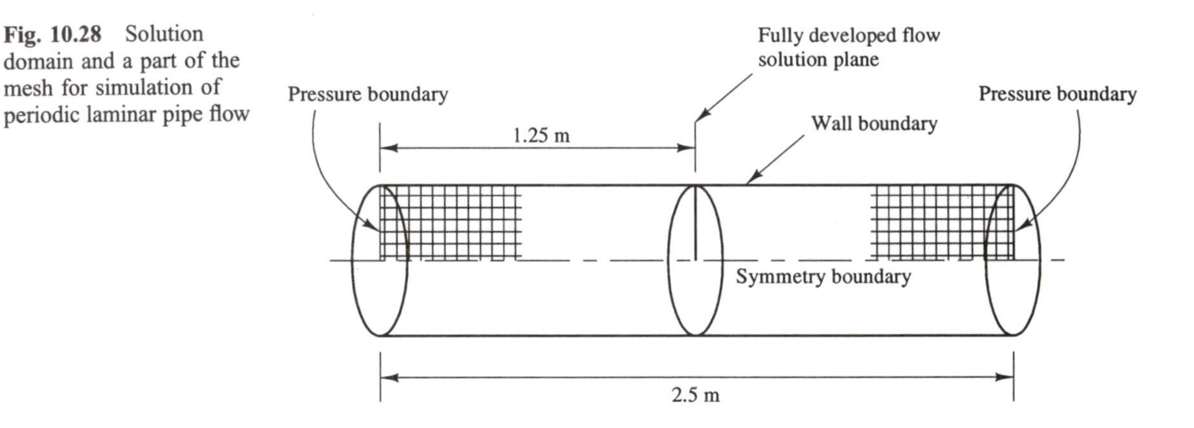

Fig. 10.28 Solution domain and a part of the mesh for simulation of periodic laminar pipe flow

Specimen results

Figures 10.30 and 10.31 compare the numerical and analytical solutions halfway along the pipe at time intervals of 0.125 s. The finite volume solution is studied after three pressure cycles, allowing time for the initial transients to die out. It is clear from the solution that the agreement between numerical and analytical solutions is generally excellent. There are minor discrepancies in the simulation with $\sqrt{(n/\nu)}R = 2.507$ during those parts of the solution cycle where the flow near

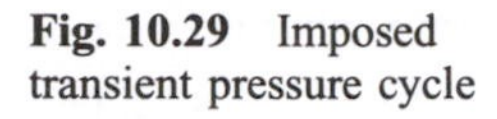

Fig. 10.29 Imposed transient pressure cycle

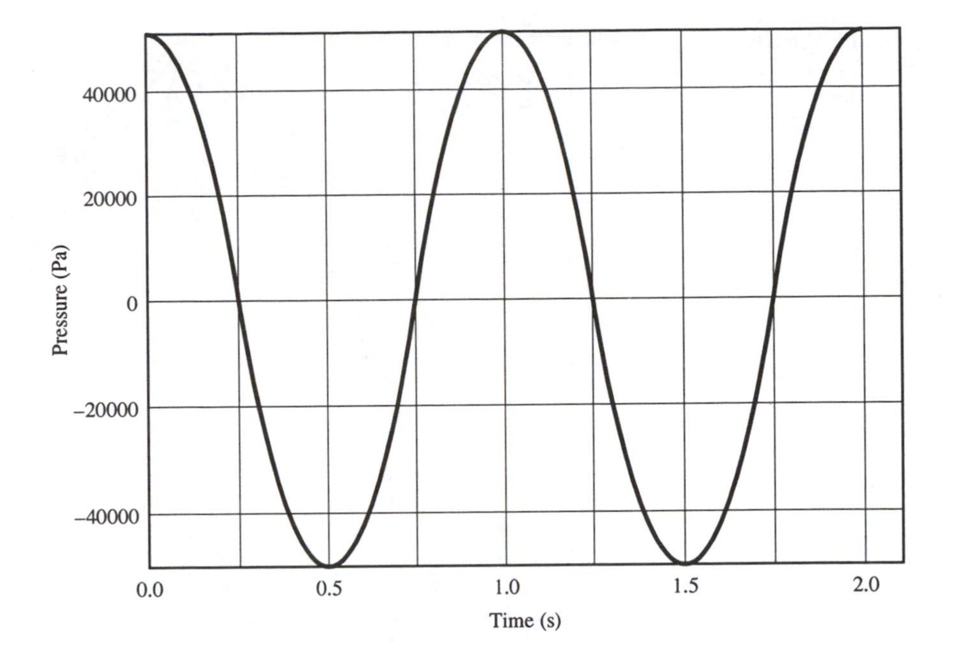

the boundary moves in the opposite direction to that in the core of the pipe. These can be explained by the fact that the local pressure gradient $\partial p/\partial x$ is somewhat different from the overall pressure gradient $\Delta p/L$ owing to energy losses between the inlet and the solution cross-section.

Overall flow behaviour can be explained by considering appropriate expressions (Abramowitz and Stegun, 1964) of the Bessel function J_0 in the analytical solution.

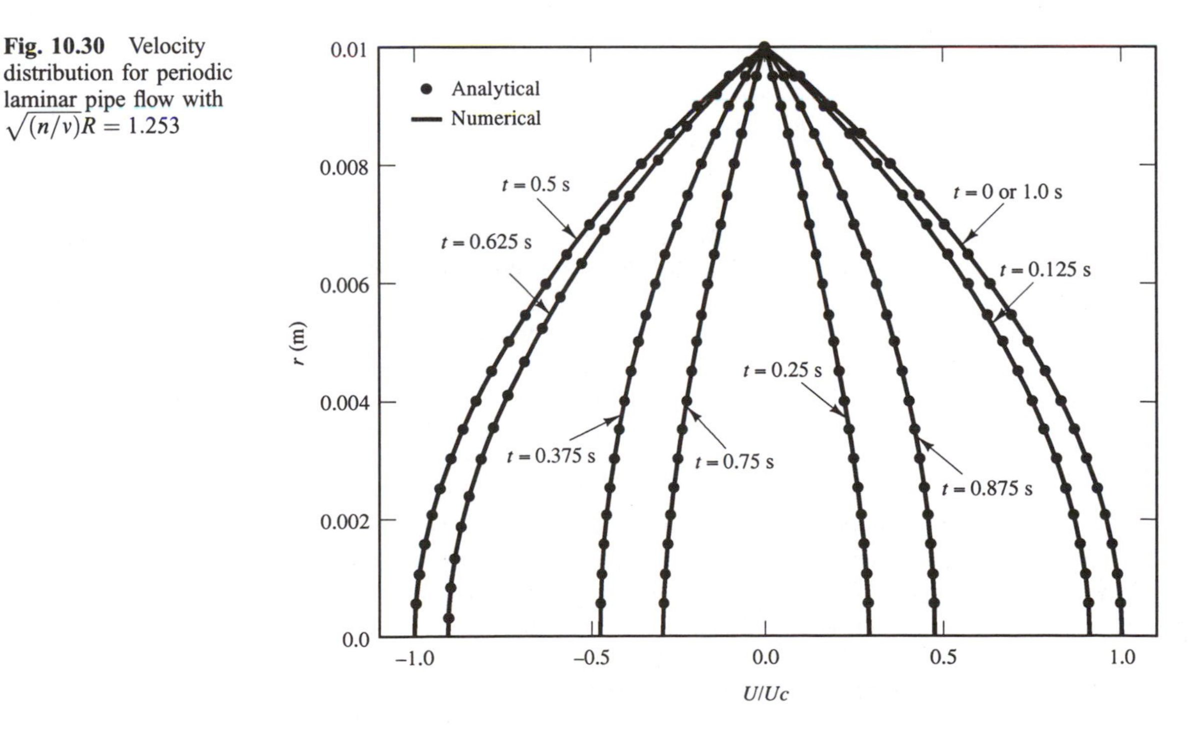

Fig. 10.30 Velocity distribution for periodic laminar pipe flow with $\sqrt{(n/\nu)}R = 1.253$

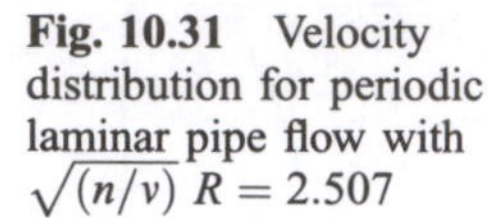

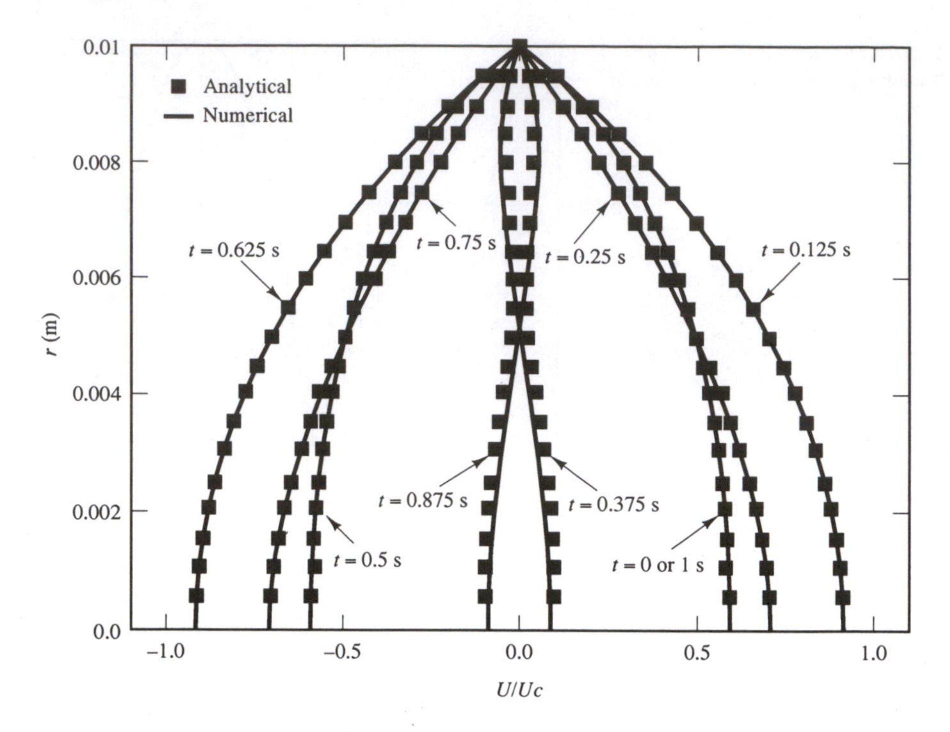

Fig. 10.31 Velocity distribution for periodic laminar pipe flow with $\sqrt{(n/\nu)}\ R = 2.507$

For very slow oscillations $\sqrt{(n/\nu)}R \to 0$ we obtain

$$u(r,\ t) = \frac{K}{4\nu}\left(R^2 - r^2\right)\cos\ nt \tag{10.39}$$

This exhibits the parabolic velocity distribution of a steady, fully developed, laminar pipe flow with a periodic time variation. The amplitude depends on the fluid viscosity and the oscillations are in phase with the driving pressure difference. For fast oscillations $\sqrt{(n/\nu)}R \to \infty$ we have

$$u(r,\ t) = \frac{K}{\nu}\left[\sin\ nt - \sqrt{\frac{R}{r}}\ \exp\left(-\sqrt{\frac{n}{2\nu}}(R-r)\right)\sin\left(nt - \sqrt{\frac{n}{2\nu}}(R-r)\right)\right] \tag{10.40}$$

Expression (10.40) contains two sinusoidal terms, the first of which is independent of viscosity. It describes the flow in the central core of the pipe which has a uniform velocity distribution with an amplitude inversely proportional to the oscillation frequency and a phase lag of $\pi/2$ radians behind the excitation force. The amplitude and the phase of the second term are viscosity dependent. The term decays quickly with distance $(R - r)$ from the pipe wall owing to the exponential factor. It can be shown that this boundary layer flow lags behind the driving pressure difference by $\pi/4$ radians. The phase difference between the core and the boundary layer gives rise to an annular flow pattern during fast oscillations. It is clear that the results of Figures 10.30 and 10.31 exhibit the main characteristics of the slow and fast solution respectively.

The above flow can be comfortably calculated on a workstation or small mainframe computer, but this success should not mislead the prospective user of commercial CFD codes. Other types of unsteady flow problems with complex geometries and/or fluid physics such as turbulent intake manifold flows (Chen,

1994), pulsed combustion (Benelli *et al*, 1992), transient free convective cooling of warm crude oil in storage tanks (Cotter and Charles, 1993) or hydrodynamic instabilities such as vortex shedding require very large computing resources. Often such flow calculations are only practical within reasonable time limits on dedicated large mainframes or supercomputers with advanced architecture and specially adopted algorithm structures.

10.6 Concluding remarks

In this chapter we have discussed ways in which the CFD techniques, which were developed in this book, can be applied to a number of taxing practical problems. It has become clear that the real world usually has a wide range of physical phenomena arrayed alongside fluid flow. In conjunction with fluid flow we have developed energy conservation principles and transport equations for general conserved scalars and our discussion of combustion modelling amply demonstrated how these general conservation equations can be put to good use in the modelling of chemically reacting flows. The application of these principles and buoyancy corrections to the turbulence equation for turbulent kinetic energy k and dissipation rate ε showed how a steady state fire could be modelled adequately.

The advantages of body-fitted co-ordinate systems in complex geometries were highlighted by demonstrating that greater flow detail can be resolved. However, it was shown that even a simplified approximation of a very complex geometry resulted in good predictions of temperatures inside a matrix-type heat exchanger. Standard correlation methods for the design of tube heat exchangers are available only for regularly spaced tubes with effectively two-dimensional flows in between them. CFD results such as those presented in this chapter represent an effective way of gaining an insight into the fluid flow and heat transfer inside the heat exchanger and also provide equipment designers with a tool to perform parametric studies to determine the best combination of number of tubes, tube spacing and inlet conditions. The case studies have highlighted the importance of validation of CFD against experimental measurements. The very sophisticated models used in conjunction with the finite volume computing engine are built on a large number of assumptions. For example, in the case of the matrix heat exchanger:

Which physics/chemistry matters?

- assume constant temperature of water inside the tubes
- neglect effects of possible condensation of flue gas

Physical models and boundary conditions

- turbulence models (e.g. k–ε model and wall functions)
- heat transfer models (e.g. radiative heat transfer)

Domain geometry

- level of detail of geometry representation
- grid design – fitness for purpose

Without these simplifications results cannot be acquired, but owing to the sweeping nature of some of them it is of critical importance that the modelling assumptions that are the foundations of the CFD work are regularly tested against good quality experiments.

Appendix A

Accuracy of a Flow Simulation

In the derivations of Chapter 4 a linear profile assumption was used to calculate the gradients $(\partial\phi/\partial x)$, $(\partial\phi/\partial y)$ etc. at the faces of the control volume. For simple diffusion problems the practice was shown to give reasonably accurate results even for coarse grids. In Example 4.3 we also observed that by refining the grid the accuracy of the solution can be improved. Grid refinement is the main tool at the disposal of the CFD user for the improvement of the accuracy of a simulation. The user would typically perform a simulation on a coarse mesh first to get an impression of the overall features of the solution. Subsequently the grid is refined in stages until no (significant) differences of results occur between successive grid refinement stages. Results are then called 'grid independent'. Here we briefly demonstrate the theoretical basis of this method of accuracy improvement and compare the *order* of discretisation schemes as a measure of their efficacy.

Consider the equally spaced one-dimensional grid (spacing Δx) shown in Figure A.1.

Fig. A.1.

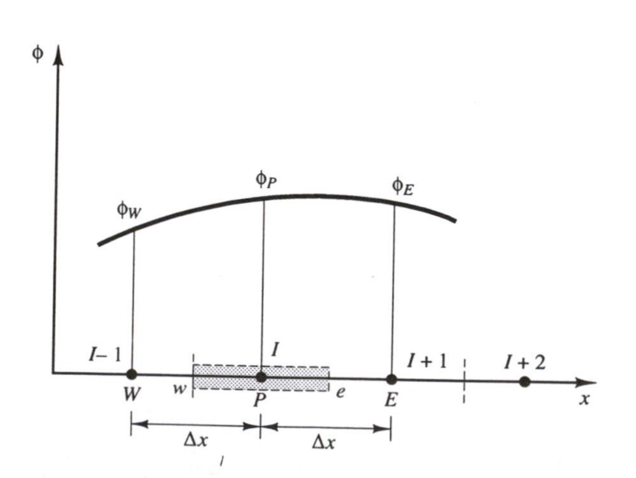

For a function $\phi(x)$ the Taylor series development of $\phi(x+\Delta x)$ around the point i at x is

$$\phi(x+\Delta x) = \phi(x) + \left(\frac{\partial \phi}{\partial x}\right)_x \Delta x + \left(\frac{\partial^2 \phi}{\partial x^2}\right)_x \frac{\Delta x^2}{2} + \cdots \tag{A.1}$$

In our notation we use discrete values ϕ_P and ϕ_E for $\phi(x)$ and $\phi(x+\Delta x)$ respectively so that equation (A.1) can be written as

$$\phi_E = \phi_P + \left(\frac{\partial \phi}{\partial x}\right)_P \Delta x + \left(\frac{\partial^2 \phi}{\partial x^2}\right)_P \frac{\Delta x^2}{2} + \cdots \tag{A.2}$$

This may be re-arranged to give

$$\begin{aligned} \left(\frac{\partial \phi}{\partial x}\right)_P \Delta x &= \phi_E - \phi_P - \left(\frac{\partial^2 \phi}{\partial x^2}\right)_P \frac{\Delta x^2}{2} - \cdots \\ \left(\frac{\partial \phi}{\partial x}\right)_P &= \frac{\phi_E - \phi_P}{\Delta x} - \left(\frac{\partial^2 \phi}{\partial x^2}\right)_P \frac{\Delta x}{2} - \cdots \end{aligned} \tag{A.3}$$

So

$$\left(\frac{\partial \phi}{\partial x}\right)_P = \frac{\phi_E - \phi_P}{\Delta x} + \text{truncated terms} \tag{A.4}$$

By neglecting the truncated terms which involve the multiplying factor Δx we may write

$$\left(\frac{\partial \phi}{\partial x}\right)_P \approx \frac{\phi_E - \phi_P}{\Delta x} \tag{A.5}$$

The error involved in the approximation (A.5) is due to neglecting the truncated terms. Formula (A.3) suggests that the truncation error can be reduced by decreasing Δx. In general the truncated terms of a finite difference scheme contain factors Δx^n. The power n of Δx governs the rate at which the error tends to zero as the grid is refined and is called the *order of the difference approximation*. Therefore equation (A.5) is said to be first order in Δx and we write

$$\boxed{\left(\frac{\partial \phi}{\partial x}\right)_P = \frac{\phi_E - \phi_P}{\Delta x} + O(\Delta x)} \tag{A.6}$$

Since it uses values at point E and P (where $x_E > x_P$) to evaluate the gradient $(\partial \phi / \partial x)$ at P, formula (A.6) is called a forward difference formula with respect to point P.

Similarly we may derive a backward difference formula for $(\partial \phi / \partial x)$ at P from

$$\phi(x-\Delta x) = \phi(x) - \left(\frac{\partial \phi}{\partial x}\right)_x \Delta x + \left(\frac{\partial^2 \phi}{\partial x^2}\right)_x \frac{\Delta x^2}{2} + \cdots \tag{A.7}$$

After some algebra we find the backward difference formula for $(\partial \phi / \partial x)$ at P:

$$\boxed{\left(\frac{\partial \phi}{\partial x}\right)_P = \frac{\phi_P - \phi_W}{\Delta x} + O(\Delta x)} \tag{A.8}$$

Equations (A.7) and (A.8) are both first-order accurate. The backward and forward difference formulae described here involve values of ϕ at two points only.

By subtracting Equation (A.7) from (A.8) we get

$$\phi(x+\Delta x) - \phi(x-\Delta x) = 2\left(\frac{\partial \phi}{\partial x}\right)_P \Delta x + \left(\frac{\partial^3 \phi}{\partial x^3}\right)_P \frac{\Delta x^3}{3!} + \cdots \tag{A.9}$$

A third formula for $(\partial\phi/\partial x)_P$ can be obtained by re-arranging the equation (A.9) as

$$\boxed{\left(\frac{\partial \phi}{\partial x}\right)_P = \frac{\phi_E - \phi_W}{2\Delta x} + O(\Delta x^2)} \tag{A.10}$$

Equation (A.10) uses values at E and W to evaluate the gradient at the mid-point P, and is called a central difference formula. The central differencing formula is second-order accurate. The quadratic dependence of the error on grid spacing means that after grid refinement the error reduces more quickly in a second-order accurate differencing scheme than in a first-order accurate scheme.

In the finite volume discretisation procedure developed in section 4.2 the gradient at a cell face, for example, at 'e' , was evaluated using

$$\left(\frac{\partial \phi}{\partial x}\right)_e = \frac{\phi_E - \phi_P}{\Delta x} = \frac{\phi_E - \phi_P}{2(\Delta x/2)} \tag{A.11}$$

By comparing formulae (A.10) and (A.11) it can be easily recognised that equation (A.11) evaluates the gradient at the mid-point between P and E through a central difference formula at point 'e'; therefore its accuracy is second order for uniform grids.

It is relatively straightforward to demonstrate the third-order accuracy of the QUICK differencing scheme for the convective flux at a mid-point cell face in a uniform grid of spacing Δx. The QUICK scheme calculates the value ϕ_e at the east cell face of a general node as

$$\phi_e = \frac{3}{8}\phi_E + \frac{6}{8}\phi_P - \frac{1}{8}\phi_W \tag{A.12}$$

Taylor series expansion about the east face gives

$$\phi_E = \phi_e + \tfrac{1}{2}\Delta x\left(\frac{\partial \phi}{\partial x}\right)_e + \frac{1}{2}\left(\tfrac{1}{2}\Delta x\right)^2\left(\frac{\partial^2 \phi}{\partial x^2}\right)_e + O(\Delta x^3) \tag{A.13}$$

$$\phi_P = \phi_e - \tfrac{1}{2}\Delta x\left(\frac{\partial \phi}{\partial x}\right)_e + \frac{1}{2}\left(-\tfrac{1}{2}\Delta x\right)^2\left(\frac{\partial^2 \phi}{\partial x^2}\right)_e + O(\Delta x^3) \tag{A.14}$$

$$\phi_W = \phi_e - \tfrac{3}{2}\Delta x\left(\frac{\partial \phi}{\partial x}\right)_e + \frac{1}{2}\left(-\tfrac{3}{2}\Delta x\right)^2\left(\frac{\partial^2 \phi}{\partial x^2}\right)_e + O(\Delta x^3) \tag{A.15}$$

If we add together $3/8 \times$ (A.13) $+ 6/8 \times$ (A.14) $- 1/8 \times$ (A.15) we obtain:

$$\frac{3}{8}\phi_E + \frac{6}{8}\phi_P - \frac{1}{8}\phi_W = \phi_e + O(\Delta x^3) \tag{A.16}$$

The terms involving orders Δx and Δx^2 cancel out in this uniform grid and the QUICK scheme is a third-order accurate approximation.

If necessary formulae involving more points (in backward or forward directions) can be derived which have higher order accuracy; see texts such as Abbott and Basco (1989) or Fletcher (1991) for further details.

Appendix B

Non-uniform Grids

For the sake of simplicity the worked examples have focused on uniform grids of nodal points. However, the derivation of discretised equations in Chapters 4 and 5 used general geometrical dimensions such as $\delta x_{PE}, \delta x_{WP}$ etc. and is also valid for non-uniform grids. In a non-uniform grid the faces e and w of a general node may not be at the mid-points between nodes E and P, and nodes W and P, respectively. In this case the interface values of diffusion coefficients Γ are calculated as follows:

$$\Gamma_w = (1 - f_W)\Gamma_W + f_W \Gamma_P \tag{B.1a}$$

where the interpolation factor f_W is given by

$$f_W = \frac{\delta x_{Ww}}{\delta x_{Ww} + \delta x_{wP}} \tag{B.1b}$$

and

$$\Gamma_e = (1 - f_P)\Gamma_P + f_P \Gamma_E \tag{B.1c}$$

where

$$f_P = \frac{\delta x_{Pe}}{\delta x_{Pe} + \delta x_{eE}} \tag{B.1d}$$

For a uniform grid these expressions simplify to equations (4.5a–b): since $f_W = 0.5$, $f_P = 0.5$, we have $\Gamma_w = (\Gamma_W + \Gamma_P)/2$ and $\Gamma_e = (\Gamma_P + \Gamma_E)/2$.

Basically there are two practices used to locate control volume faces in non-uniform grids (Patankar, 1980).

Practice A

Nodal points are defined first and the control volume faces located mid-way between the grid points. This is illustrated in Figure B.1

Fig. B.1.

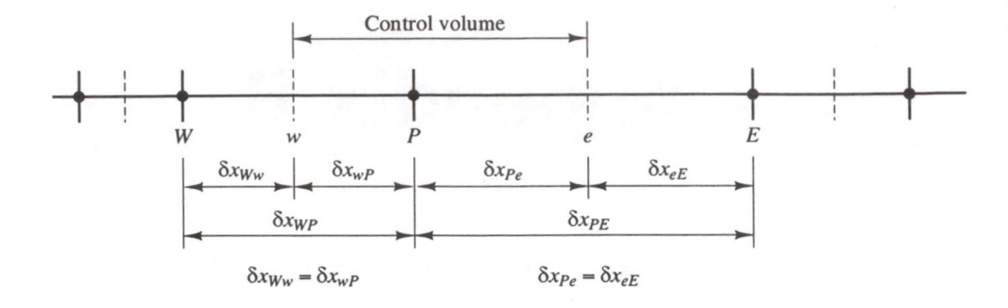

Practice B

Locations of the control volume faces are defined first and the nodal points are placed at the centres of the control volumes. This is illustrated in Figure B.2.

Here the faces of a control volume are not at the mid-point between the nodes. The evaluation of gradients obtained through a linear approximation is unaffected because the gradient remains the same at any point between the nodes in question but the values of diffusion coefficient Γ need to be evaluated using interpolation functions (B.1).

Fig. B.2.

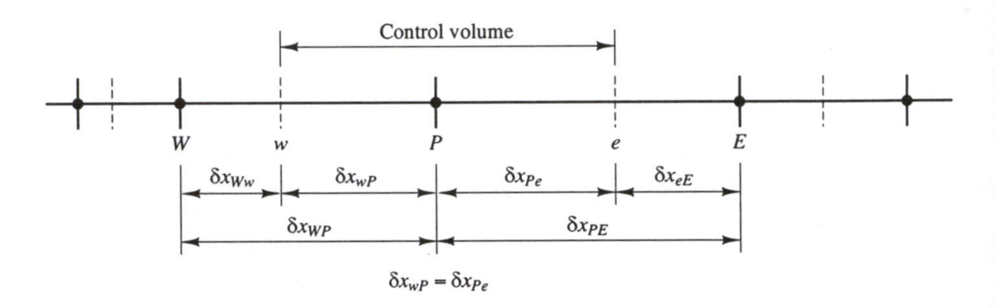

It is very important to note that central difference formulae for the calculation of gradients at cell faces and the QUICK scheme for convective fluxes are only second- and third-order accurate respectively when the control volume face is mid-way between the nodes. In practice A a control volume face, *e* for example, is mid-way between nodes *P* and *E*, so the differencing formula used to evaluate the gradient $(\partial\phi/\partial x)_e$ is second-order accurate. A further advantage of practice A is that property values Γ_e, Γ_w etc. can be easily evaluated by taking the average values. The disadvantage of practice A is that the value of the variable ϕ at *P* may not necessarily be the most representative value for the entire control volume as point *P* is not at the centre of the control volume. In practice B the value of ϕ at *P* is a good representative value for the control volume as *P* lies at the centre of the control volume, but the discretisation schemes lose accuracy. A thorough discussion on these two practices is found in Patankar (1980) to which the reader is referred for further details.

Appendix C

Calculation of Source Terms

Source terms of the discretised equations are evaluated using prevailing values of the variables. Since a staggered grid is employed in the calculation procedures, interpolation is required in the calculation of velocity gradient terms which often appear in source terms. For example, consider the two-dimensional u-velocity equation

$$\frac{\partial}{\partial x}(\rho uu) + \frac{\partial}{\partial y}(\rho vu) = \frac{\partial}{\partial x}\left(\mu \frac{\partial u}{\partial x}\right) + \frac{\partial}{\partial y}\left(\mu \frac{\partial u}{\partial x}\right) - \frac{\partial p}{\partial x} + S_u \qquad \text{(C.1)}$$

where

$$S_u = \frac{\partial}{\partial x}\left(\mu \frac{\partial u}{\partial x}\right) + \frac{\partial}{\partial y}\left(\mu \frac{\partial v}{\partial x}\right)$$

A small part of the solution grid is shown in Figure C.1; we use the standard notation for the backward staggered velocity components introduced in Chapter 6.

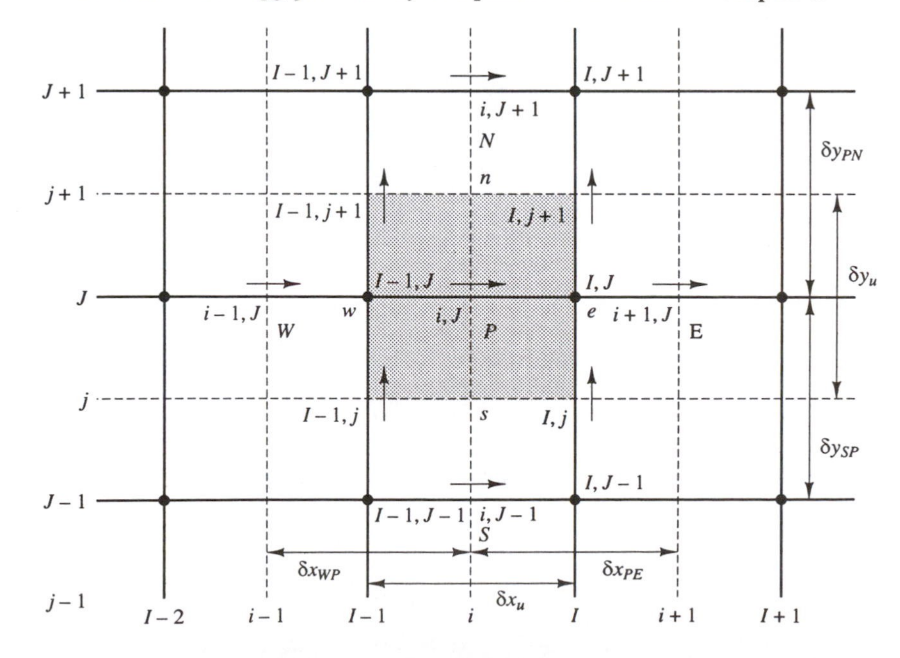

Fig. C.1.

The discretised form of equation (C.1) for the u-control volume centred at (i, J) is

$$a_{i,J}u_{i,J} = \sum a_{nb}u_{nb} - \frac{(P_{I,J} - P_{I-1,J})}{\delta x_u}\Delta V_u + \bar{S}_u\Delta V_u \tag{C.2}$$

where δx_u is the width of the u-control volume and ΔV_u is the volume of the u-control volume.

The source term for the u-velocity equation at the cell shown is evaluated as

$$\begin{aligned}\bar{S}_u\Delta V &= \left[\frac{\partial}{\partial x}\left(\mu\frac{\partial u}{\partial x}\right) + \frac{\partial}{\partial y}\left(\mu\frac{\partial v}{\partial x}\right)\right]_{Cell} . \Delta V \\ &= \left[\frac{\left(\mu\frac{\partial u}{\partial x}\right)_{east\ face} - \left(\mu\frac{\partial u}{\partial x}\right)_{west\ face}}{\delta x_u} + \frac{\left(\mu\frac{\partial v}{\partial x}\right)_{north\ face} - \left(\mu\frac{\partial v}{\partial x}\right)_{south\ face}}{\delta y_u}\right] . \Delta V \\ &= \left[\frac{\left(\mu\frac{u_{i+1,J} - u_{i,J}}{\delta x_{PE}}\right) - \left(\mu\frac{u_{i,J} - u_{i-1,J}}{\delta x_{WP}}\right)}{\delta x_u} + \frac{\left(\mu\frac{v_{I,j+1} - v_{I-1,j+1}}{\delta x_u}\right) - \left(\mu\frac{v_{I,j} - v_{I-1,j}}{\delta x_u}\right)}{\delta y_u}\right] . \delta x_u\delta y_u\end{aligned} \tag{C.3}$$

The source terms of other equations are also calculated in a similar manner.

References

Chapter 1

Gottlieb, D. and Orszag, S. A. (1977). *Numerical Analysis of Spectral Methods: Theory and Applications*, SIAM, Philadelphia.

Hastings, C. (1985). *Approximations for Digital Computers.* Princeton University Press, Princeton, NJ.

Patankar, S. V. (1980). *Numerical Heat Transfer and Fluid Flow*, Hemisphere Publishing Corporation, Taylor & Francis Group, New York.

Smith, G. D. (1985). *Numerical Solution of Partial Differential Equations: Finite Difference Methods*, 3rd edn, Clarendon Press, Oxford

Zienkiewicz, O. C. and Taylor, R. L. (1991). *The Finite Element Method – Vol. 2: Solid and Fluid Mechanics*, McGraw-Hill, New York.

Chapter 2

Bland, D. R. (1988). *Wave theory and applications*, Clarendon Press, Oxford.

Fletcher, C. A. J. (1991). *Computational Techniques for Fluid Dynamics*, Volumes I and II, Springer-Verlag, Berlin.

Gresho, P. M. (1991). Incompressible Fluid Dynamics: Some Fundamental Formulation Issues, *Annu. Rev. Fluid Mech.*, Vol. 23, pp. 413–453.

Issa, R. I. and Lockwood, F. C. (1977). On the Prediction of Two-dimensional Supersonic Viscous Interactions near Walls, *AIAA J.*, Vol. 15, No. 2, pp. 182–188.

McGuirk, J. J. and Page, G. J. (1990). Shock Capturing Using a Pressure-Correction Method, *AIAA J.*, Vol. 28, No. 10, pp. 1751–1757.

Schlichting, H. (1979). *Boundary-layer theory*, 7th ed, McGraw-Hill, New York.

Shapiro, A. H. (1953). *Compressible fluid flow*, Vol. 1, John Wiley & Sons, New York.

The Open University, (1984). *Mathematical methods and fluid mechanics*, Course MST322, The Open University Press, Milton Keynes, England.

Chapter 3

Abbott, M. B. and Basco, D. R. (1989). *Computational Fluid Dynamics – An Introduction for Engineers*, Longman Scientific & Technical, Harlow, England.

Amano, R. S. and Goel, P. (1987). Investigation of Third-order Closure Model of Turbulence for the Computation of Incompressible Flows in a Channel with a Backward-facing Step, *Trans. ASME, J. Fluids Eng.*, Vol. 109, pp. 424–428.

Anderson, D. A., Tannehill, J. C. and Pletcher, R. H. (1984). *Computational Fluid Mechanics and Heat Transfer*, Hemisphere Publishing Corporation, Taylor & Francis Group, New York.

Baldwin, B. S. and Lomax, H. (1978). Thin Layer Approximation and Algebraic Model for Separated Turbulent Flow, AIAA Paper 78–257.

Bradshaw, P., Cebeci, T. and Whitelaw, J. H. (1981). *Engineering Calculation Methods for Turbulent Flow*, Academic Press, London.

Buchhave, P., George, Jr, W. K. and Lumley, J. L. (1979). The Measurement of Turbulence with the Laser-Doppler Anemometer, *Annu. Rev. Fluid Mech.*, Vol. 11, pp. 443–503.

Cebeci, T. (1989). Essential Ingredients of a Method for Low Reynolds-number Airfoils, *AIAA J.*, Vol. 27, No. 12, pp. 1680–1688.

Cebeci, T. and Smith, AMO (1974). Analysis of Turbulent Boundary Layers, *Applied Mathematics and Mechanics*, Vol. 15, Academic Press, New York.

Champagne, F. H., Pao, Y. H. and Wygnanski, I. J. (1976). On the Two-dimensional Mixing Region, *J. Fluid Mech.*, Vol. 74, Pt 2, pp. 209–250.

Comte-Bellot, G. (1976). Hot-wire Anemometry, *Annu. Rev. Fluid Mech.*, Vol. 8, pp. 209–231.

Demuren, A. O. and Rodi, W. (1984). Calculation of Turbulence-driven Secondary Motion in Non-circular Ducts, *J. Fluid Mech.*, Vol. 140, pp. 189–222.

Gatski, T. B. and Speziale, C. G. (1993). On Explicit Algebraic Stress Models for Complex Turbulent Flows, *J. Fluid Mech.*, Vol. 254, pp. 59–78.

Gutmark, E. and Wygnanski, I. (1976). The Planar Turbulent Jet, *J. Fluid Mech.*, Vol. 73, Pt 3, pp. 465–495.

Horiuti, K. (1990). Higher-order Terms in the Anisotropic Representation of Reynolds Stresses, *Phys. Fluids A*, Vol. 2, No. 10, pp. 1708–1710.

Klebanoff, P. S. (1955). Characteristics of Turbulence in a Boundary Layer with Zero Pressure Gradient, NACA Report 1247, National Bureau of Standards, Washington, D.C.

Kleiser, L. and Zang, T. A. (1991). Numerical Simulation of Transition in Wall-bounded Shear Flows, *Annu. Rev. Fluid Mech.*, Vol. 23, pp. 495–537.

Lam, C. K. G. and Bremhorst, K. A. (1981). Modified Form of the k–ε Model for Predicting Wall Turbulence, *Trans. ASME, J. Fluids Eng.*, Vol. 103, pp. 456–460.

Laufer, J. (1952). The Structure of Turbulence in Fully Developed Pipe Flow, NACA Report 1174, National Bureau of Standards, Washington, DC.

Launder, B. E. (1989). Second-moment Closures: Present and Future? *Int. J. Heat Fluid Flow*, Vol. 10, pp. 282–300.

Launder, B. E. and Spalding, D. B. (1974). The Numerical Computation of Turbulent Flows, *Comput. Methods Appl. Mech. Eng.*, Vol. 3, pp. 269–289.

Launder, B. E., Reece, G. J. and Rodi, W. (1975). Progress in the Development of a Reynolds-stress Turbulence Closure, *J. Fluid. Mech.*, Vol. 68, Pt 3, pp. 537–566.

Lumley, J. L. (1978). Computational Modelling of Turbulent Flows, *Adv. Appl. Mech.*, Vol. 18, pp. 123–176.

Monin, A. S. and Yaglom, A. M. (1971). *Statistical fluid mechanics: mechanics of turbulence*, Vol. 1, MIT Press, Cambridge, MA.

Nakayama, Y. (ed.) (1988). *Visualised flow*, Pergamon Press, Oxford.

Naot, D. and Rodi, W. (1982). Numerical Simulation of Secondary Currents in Channel Flow, *J. Hydraul. Div., ASCE*, Vol. 108 (HY8), pp. 948–968.

Patel, V. C., Rodi, W. and Scheuerer, G. (1985). Turbulence Models for Near-Wall and Low Reynolds Number Flows: A Review *AIAA J.*, Vol. 23, No. 9, pp. 1308–1319.

Rodi, W. (1980). Turbulence Models and Their Application in Hydraulics – A State of the Art Review, IAHR, Delft, The Netherlands.

Schlichting, H. (1979). *Boundary-layer theory*, 7th ed., McGraw-Hill, New York.

So, R. M. C., Lai, Y. G., Zhang, H. S. and Hwang, B. C. (1991). Second-order Near-wall Turbulence Closures: A Review, *AIAA J.*, Vol. 29, No. 11, pp. 1819–1835.

Speziale, C. G. (1987). On Non-linear k-l and k–ε Models of Turbulence, *J. Fluid Mech.*, Vol. 178, pp. 459–475.

Speziale, C. G. (1991). Analytical Methods for the Development of Reynolds-stress Closures in Turbulence, *Annu. Rev. Fluid Mech.*, Vol. 23, pp. 107–157.

Tennekes, H. and Lumley, J. L. (1972). *A first course in turbulence*, MIT Press, Cambridge, MA.

Tritton, D. J. (1977). *Physical fluid dynamics*, Van Nostrand Reinhold, Wokingham, England.

Van Dyke, M. (1982). *An Album of Fluid Motion*, Parabolic Press, Stanford, CA, USA.

White, F. M. (1991). *Viscous Fluid Flow*, 2nd edn, McGraw-Hill, New York.

Wygnanski, I., Champagne, F. and Marasli, B. (1986). On the Large-scale Structures in Two-dimensional, Small-Deficit, Turbulent Wakes, *J. Fluid Mech.*, Vol. 168, pp. 31–71.

Yakhot, V., Orszag, S. A., Thangam, S., Gatski, T. B. and Speziale, C. G. (1992). Development of Turbulence Models for Shear Flows by a Double Expansion Technique, *Phys. Fluids A*, Vol. 4, No. 7, pp. 1510–1520.

Chapter 4

MATLAB, (1992). *The Student Edition of MATLAB*, The Math Works Inc., Prentice Hall, Englewood Cliffs, NJ.

Chapter 5

Alvarez, J., Jones, W. P. and Seoud, R. (1993). Prediction of Momentum and Scalar Fields in a Jet Cross-Flow using First and Second Order Turbulence Closures, *AGARD Symposium on Computational and Experimental Assessment of Jets in Cross Flow, Winchester, England, April 1993*.

Borris, J. P. and Brook, D. L. (1973). Flux Corrected Transport I: SHASTA, A Fluid Transport Algorithm that Works, *J. Comput. Phys.*, Vol. 11, pp. 38–69.

Borris, J. P. and Brook, D. L. (1976). Solution of the Continuity Equation by the Method of Flux Corrected Transport, *J. Comput. Phys.*, Vol. 16, pp. 85–129.

FLUENT Users' Manual Version 4.22 (1992), Fluent Europe Ltd, Sheffield, UK.

Han, T., Humphrey, J. A. C. and Launder, B. E. (1981). A Comparison of Hybrid and Quadratic-Upstream Differencing in High Reynolds Number Elliptic Flows, *Comput. Methods Appl. Mech. Eng.*, Vol. 29, pp. 81–95.

Huang, P. G., Launder, B. E. and Leschziner, M. A. (1985). Discretization of Non-linear Convection Processes: A Broad-range Comparison of Four Schemes, *Comput. Methods Appl. Mech. Eng.*, Vol. 48, pp. 1–24.

Hayase, T., Humphrey, J. A. C. and Greif, R. (1992) A Consistently Formulated QUICK Scheme for Fast and Stable Convergence Using Finite-volume Iterative Calculation Procedures, *J. Comput. Phys.*, Vol. 98, pp. 108–118.

Hirsch, C. (1990). *Numerical Computation of Internal and External Flows*, Vol. 2, p. 493. John Wiley & Sons, Chichester, England.

Leonard, B. P. (1979). A Stable and Accurate Convective Modelling Procedure Based on Quadratic Upstream Interpolation, *Comput. Methods Appl. Mech. Eng.*, Vol. 19, pp. 59–98.

Leschziner, M. A. (1980). Practical Evaluation of Three Finite Difference Schemes for the Computation of Steady-State Recirculating Flows, *Comput. Methods Appl. Mech. Eng.*, Vol. 23, pp. 293–312.

Osher, S. (1984). Riemann Solvers, the Entropy Condition and Difference Approximations, *SIAM J. Numer. Anals.*, Vol. 21, pp. 217–235.

Osher, S. and Chakravarthy, S. (1984). High Resolution Schemes and the Entropy Condition, *SIAM J. Numer. Anal.*, Vol. 21, pp. 955–984.

Patankar, S. V. (1980). *Numerical Heat Transfer and Fluid Flow*, Hemisphere Publishing Corporation, Taylor & Francis Group, New York.

Pollard, A. and Siu, A. L. W. (1982). The Calculation of Some Laminar Flows Using Various Discretization Schemes, *Comput. Methods Appl. Mech. Eng.*, Vol. 35, pp. 293–313.

Roache, P. J. (1976). *Computational Fluid Dynamics*, Hermosa, Albuquerque, NM.

Scarborough, J. B. (1958). *Numerical Mathematical Analysis*, 4th edn, Johns Hopkins University Press, Baltimore, MD.

Spalding, D. B. (1972). A Novel Finite-difference Formulation for Differential Expressions Involving Both First and Second Derivatives, *Int. J. Numer. Methods Eng.*, Vol. 4, p. 551.

Van Leer, B. (1973). *Towards the Ultimate Conservative Difference Scheme I. The Quest of Monotonicity*, Lecture Notes in Physics, Vol. 18, pp. 163–168. Springer-Verlag, Berlin.

Van Leer, B. (1974). Towards the Ultimate Conservative Difference Scheme II. Monotonicity and Conservation Combined in a Second-order Scheme, *J. Comput. Phys.*, Vol. 14, pp. 361–370.

Van Leer, B. (1979). Towards the Ultimate Conservative Difference Scheme V. A Second Order Sequel to Gadunov's Method, *J. Comput. Phys.*, Vol. 32, pp. 101–136.

Zhu, J. (1991). A Low-diffusive and Oscillation-free Convective Scheme, *Commun. Appl. Numer. Methods*, Vol. 7, p. 225.

Chapter 6

Anderson, D. A., Tannehill, J. C. and Pletcher, R. H. (1984). *Computational Fluid Mechanics and Heat Transfer*, Hemisphere Publishing Corporation, Taylor & Francis Group, New York.

Harlow, F. H. and Welch, J. E. (1965). Numerical Calculation of Time-dependent Viscous Incompressible Flow of Fluid with Free Surface, *Phys. Fluids*, Vol. 8, pp. 2182–2189.

Issa, R. I. (1986). Solution of the Implicitly Discretised Fluid Flow Equations by Operator-Splitting, *J. Comput. Phys.*, Vol. 62, pp. 40–65.

Issa, R. I., Gosman, A. D. and Watkins, A. P. (1986). The Computation of Compressible and Incompressible Recirculating Flows, *J. Comput. Phys.*, Vol. 62, pp. 66–82.

Jang, D. S., Jetli, R. and Acharya, S. (1986). Comparison of the PISO, SIMPLER, and SIMPLEC Algorithms for the Treatment of the Pressure-Velocity Coupling in Steady Flow Problems, *Numer. Heat Transfer*, Vol. 19, pp. 209–228.

Patankar, S. V. (1980). *Numerical Heat Transfer and Fluid Flow*, Hemisphere Publising Corporation, Taylor & Francis Group, New York.

Patankar, S. V. and Spalding, D. B. (1972). A Calculation Procedure for Heat, Mass and Momentum Transfer in Three-dimensional Parabolic Flows, *Int. J. Heat Mass Transfer*, Vol. 15, p. 1787.

Van Doormal, J. P. and Raithby, G. D. (1984). Enhancements of the SIMPLE Method for Predicting Incompressible Fluid Flows, *Numer. Heat Transfer*, Vol. 7, pp. 147–163.

Chapter 7

Anderson, D. A., Tannehill, J. C. and Pletcher, R. H. (1984). *Computational Fluid Mechanics and Heat Transfer*, Hemisphere Publishing Corporation, Taylor & Francis Group, New York.

Concus, P., Golub, G. H. and O'Leary, D. P. (1976), A Generalised Conjugate Gradient Method for the Numerical Solution of Elliptic Partial Equations, in J. R. Bunch and D. J. Rose (eds) *Sparse Matrix Computations*, Academic Press, New York.

Fletcher, C. A. J. (1991). *Computational Techniques for Fluid Dynamics*, Volumes I and II, Springer-Verlag, Berlin.

Hestenes, M. R. and Stiefel, E. L. (1952). *NBS J. Res.*, Vol. 49, pp. 409–436.

Kershaw, D. S. (1978). The Incomplete Cholesky Conjugate Gradient Method for the Iterative Solution of Linear Equations, *J. Comput. Phys.*, Vol. 26, pp. 43–65.

Press, W. H., Flannery, B. P., Teukolsky, S. A. and Velterling, W. T. (1992). *Numerical Recipes*

– *The Art of Scientific Computing (FORTRAN Version)*, Cambridge University Press, Cambridge.
Reid, J. K. (1971). On the Method of Conjugate Gradients for the Solution of Large Sparse Systems of Linear Equations, in J. K. Reid (ed.) *Large Sparse Sets of Linear Equations*, Academic Press, New York.
Schneider, G. E. and Zedan, M. (1981). A Modified Strongly Implicit Procedure for the Numerical Solution of Field Problems, *Numer. Heat Transfer*, Vol. 4, pp. 1–9.
Stone, H. L. (1968), Iterative Solution of Implicit Approximations of Multidimensional Partial Differential Equations, *SIAM, J. Numer. Anal.*, Vol. 5, No. 3, pp. 530–558.
Thomas, L. H. (1949). Elliptic Problems in Linear Difference Equations over a Network, Watson Sci. Comput. Lab. Report, Columbia University, New York.

Chapter 8

Abbott, M. B. and Basco, B. R. (1989). *Computational Fluid Dynamics – An Introduction for Engineers*, Longman Scientific & Technical, Harlow, UK.
Ahmadi-Befrui, Gosman A. D., Issa, R. I. and Watkins, A. P. (1990). EPISO – An Implicit Non-iterative Solution Procedure for the Calculation of Flows in Reciprocating Engine Chambers, *Comput. Methods Appl. Mech. Eng.*, Vol. 79, pp. 249–279.
Amsden, A. A. and Harlow, F. H. (1970). The SMAC Method: A Numerical Technique for Calculating Incompressible Fluid Flows, Las Alamos Scientific Laboratory Report LA-4370, Los Alamos, New Mexico.
Amsden, A. A., Butler, T. D., O'Rourke, P. J. and Ramshaw, J. D. (1985). KIVA – A Comprehensive Model of 2D and 3D Engine Simulations, SAE Paper No. 850554.
Amsden, A. A., O'Rourke, P. J. and Butler, T. D. (1989). KIVA-II – A Computer Program for Chemically Reactive Flows with Sprays, Los Alamos National Laboratory Report LA-11560-MS.
Anderson, D. A., Tannehill, J. C. and Pletcher, R. H. (1984). *Computational Fluid Mechanics and Heat Transfer*, Hemisphere Publishing Corporation, Taylor & Francis Group, New York.
Blunsdon, C. A., Malalasekera, W. M. G. and Dent, J. C. (1992). Application of the Discrete Transfer Model of Thermal Radiation in CFD Simulation of Diesel Engine Combustion and Heat Transfer, SAE Paper No. 922305.
Blunsdon, C. A., Malalasekera, W. M. G. and Dent, J. C. (1993). Modelling Infrared Radiation from the Combustion Products in a SI Engine, SAE Paper No. 932699.
Crank, J. and Nicolson, P. (1947). A Practical Method for Numerical Evaluation of Solutions of Partial Differential Equations of the Heat-conduction Type, *Proc. Cambridge Phil. Soc.*, Vol. 43, pp. 50–67.
Fletcher, C. A. J. (1991). *Computational Techniques for Fluid Dynamics*, Volumes I and II, Springer-Verlag, Berlin.
Harlow, F. H. and Amsden, A. A. (1971). A Numerical Fluid Dynamics Calculation Method for All Flow Speeds, *J. Comput. Phys.*, Vol. 8, pp. 197–213.
Harlow, F. H. and Welch, J. E. (1965). Numerical Calculation of Time-dependent Viscous Incompressible Flow of Fluid with Free Surface, *Phys. Fluids*, Vol. 8, pp. 2182–2189.
Hirt, C. W., Amsden, A. A. and Cook, J. L. (1974). An Arbitrary Lagrangian–Eulerian Computing Method for All Flow Speeds, *J. Comput. Phys.*, Vol. 14, pp. 227–253.
Issa, R. I. (1986). Solution of Implicitly Discretised Fluid Flow Equations by Operator-Splitting, *J. Comput. Phys.*, Vol. 62, pp. 40–65.
Issa, R. I., Gosman, A. D. and Watkins, A. P. (1986). The Computation of Compressible and Incompressible Recirculating Flows, *J. Comput. Phys.*, Vol. 62, pp. 66–82.
Kim, S. W. and Benson, T. J. (1992). Comparison of the SMAC, PISO and Iterative Time-advancing Schemes for Unsteady Flows, *Comput. Fluids*, Vol. 21, No. 3, pp. 435–454.
Özisik, M. N. (1985). *Heat Transfer – A Basic Approach*, McGraw-Hill, New York.

Zellat, M., Rolland, Th. and Poplow, F. (1990). Three Dimensional Modelling of Combustion and Soot Formation in an Indirect Injection Diesel Engine, SAE Paper No. 900254.

Chapter 9

Jayatilleke, C. L. V. (1969). The Influence of Prandtl Number and Surface Roughness on the Resistance of the Laminar Sublayer to Momentum and Heat Transfer, *Prog. Heat Mass Transfer*, Vol. 1, p. 193.

Patankar, S. V. (1980) *Numerical Heat Transfer and Fluid Flow*, Hemisphere Publishing Corporation, Taylor & Francis Group, New York.

Schlichting, H. (1979). *Boundary-layer theory*, 7th edn, McGraw-Hill, New York.

Chapter 10

Abramowitz, M. and Stegun, A. (eds) (1964). *Handbook of Mathematical Functions*, Dover Publications, New York.

Alvarez, N. J., Foote, K. L. and Pagni, P. J. (1984). Forced Ventilated Enclosure Fires, *Combust. Sci. Technol.*, Vol. 39, p. 55.

Askari-Sardhai, A. Liew, S. K. and Moss, J. B. (1985). Flamelet Modelling of Propane Air Chemistry in Turbulent Non-premixed Combustion, *Combust. Sci. Technol.*, Vol. 44, pp. 89–95.

Benelli, G., Michele, G. D. E., Cossalter, V., Lio, M. D. A. and Rossi, G. (1992). Simulation of Large Non-linear Thermo-acoustic Vibrations in a Pulsating Combustor, *Twenty-Fourth Symposium (International) on Combustion*, The Combustion Institute, pp. 1307–1313.

Bilger, R. W. (1976). Turbulent Jet Diffusion Flames, *Prog. Energ. Combust. Sci.*, Vol. 1, pp. 87–109.

Bilger, R. W. and Kent, J. H. (1974). Concentration Fluctuations in Turbulent Jet Diffusion Flames, *Combust. Sci. Technol.*, Vol. 9, p. 25.

Bray, K. N. C., Libby, P. A. and Moss, J. B. (1985). Unified Modelling Approach for Premixed Turbulent Combustion – Part I: General Formulation, *Combust. Flame*, Vol. 61, pp. 87–102.

Chen, A. (1994). Application of Computational Fluid Dynamics to the Analysis of Inlet Port Design in Internal Combustion Engines, PhD. Thesis, Loughborough University.

Cotter, M. A. and Charles, M. E. (1993). Transient Cooling of Petroleum by Natural Convection in Cylindrical Storage Tanks – I. Development and Testing of a Numerical Simulator, *Int. J. Heat Mass Transfer*, Vol. 36, No. 8, pp. 2165–2174.

Demirdzic, I. (1982). A Finite Volume Method for Computation of Fluid Flow in Complex Geometries, Ph.D. Thesis, Imperial College, London.

Demirdzic, I., Gosman, A. D., Issa, R. I. and Peric, M. (1987). A Calculation Procedure for Turbulent Flow in Complex Geometries, *Comput. Fluids*, Vol. 15, No. 3, pp. 251–273.

Durst, F. and Loy, T. (1985). Investigations of Laminar Flow in a Pipe with Sudden Contraction of Cross Sectional Area, *Comput. Fluids*, Vol. 13, No. 1, pp. 15–36.

Favre, A. (1969). Statistical Equations of Turbulent Gases, in *Problems of Hydrodynamics and Continuum Mechanics*, p. 231. SIAM, Philadelphia.

Gosman, A. D., Lockwood, F. C. and Salooja, A. P. (1978). The Prediction of Cylindrical Furnaces Gaseous Fuelled with Premixed and Diffusion Burners, *Seventeenth Symposium (International) on Combustion*, The Combustion Institute, pp. 747–760.

Harlow, F. H. and Welch, J. E. (1965). Numerical Calculation of Time-dependent Viscous Incompressible Flow of Fluid with Free Surface, *Phys. Fluids*, Vol. 8, pp. 2182–2189.

Hirt, C. W., Amsden, A. A. and Cook, J. L. (1974). An Arbitrary Lagrangian–Eulerian Computing Method for All Flow Speeds, *J. Comput. Phys.*, Vol. 14, pp. 227–253.

Jones, W. P. (1979). Models for Turbulent Flows with Variable Density, in W. Kollmann (ed.) *Prediction Methods for Turbulent Flows*, VKI Lecture Series, pp. 378–421. Hemisphere Publishing Corporation, New York.

Jones, W. P. and Whitelaw, J. H. (1982). Calculation Methods for Reacting Turbulent Flows: A Review, *Combust, Flame*, Vol. 48, No. 1, pp. 1–26.

Karki, K. C. and Patankar, S. V. (1988). Calculation Procedure for Viscous Incompressible Flows in Complex Geometries, *Numer, Heat Transfer*, Vol. 14, pp. 295–307.

Lewis, M. H. and Smoot, L. D. (1981) Turbulent Gaseous Combustion. Part I: Local Species Concentration Measurements, *Comb. Flame*, Vol. 42, p. 183.

Liew, S. K., Bray, K. N. C. and Moss, J. B. (1984). A Stretched Laminar Flamelet Model of Turbulent Nonpremixed Combustion, *Combust. Flame*, Vol. 56, pp. 199–213.

Lockwood, F. C. and Malalasekera, W. M. G. (1988). Fire Computation: The Flashover Phenomenon, *Twenty-second Symposium (International) on Combustion*, The Combustion Institute, pp. 1319–1328.

Lockwood, F. C. and Monib, H. A. (1980). Fluctuating Temperature Measurements in a Heated Round Free Jet, *Combust. Sci. Technol.*, Vol. 22, p. 63.

Lockwood, F. C. and Naguib, A. S. (1975). The Prediction of the Fluctuations in the Properties of Free, Round Jet Turbulent Diffusion Flames, *Combust. Flame*, Vol. 24, pp. 109–124.

Lockwood, F. C. and Odidi, A. O. (1975). Measurement of Mean and Fluctuating Temperature and Ion Concentration in Round Jet Turbulent Diffusion and Premixed Flames, Fifteenth Symposium (International) on Combustion, The Combustion Institute, p. 561.

Lockwood, F. C. and Shah, N. G. (1981). A New Radiation Solution Method for Incorporation in General Combustion Prediction Procedures, *Eighteenth Symposium (International) on Combustion*, The Combustion Institute, pp. 1405–1414.

Magnussen, B. F. and Hjertager, B. H. (1976). On Mathematical Modelling of Turbulent Combustion with Special Emphasis on Soot Formation and Combustion, *Sixteenth Symposium (International) on Combustion*, The Combustion Institute, pp. 719–729.

Malalasekera, W. M. G. (1988). Mathematical Modelling of Fires and Related Processes, Ph.D. Thesis, Imperial College, London

Malalasekera, W. M. G., James, E. H., Tayali, N. E. and Lee, N. E. K. (1993). Flow and Heat Transfer in a Secondary Heat Exchanger of a Condensing Boiler, *29th US National Heat Transfer Conference, Atlanta, GA*, HTD Vol. 237, pp. 83–92.

Markatos, N. C., Malin, M. R. and Cox, G. (1982). Mathematical Modelling of Buoyancy Induced Smoke Flow in Enclosures, *Int. J. Heat Mass Transfer*, Vol. 25, No. 1, pp. 63–75.

Nikjooy, M., So, R. M. C. and Peck, R. E. (1988). Modelling of Jet-Swirl-Stabilised Reacting Flows in Axisymmetric Combustors, *Combust. Sci. Technol.*, Vol. 58, pp. 135–153.

Peric, M. (1985). Finite Volume Method for the Prediction of Three-Dimensional Fluid Flow in Complex Ducts, Ph.D. Thesis, Imperial College, London.

Peters, N. (1986). Laminar Flamelet Concepts in Turbulent Combustion, *Twenty-first Symposium (International) on Combustion*, The Combustion Institute, pp. 1321–1250.

Pope, S. B. (1976). The Probability Approach to the Modelling of Turbulent Reacting Flows, *Combust. Flame*, Vol. 27, p. 299.

Pope, S. B. (1985). PDF Methods for Turbulent Reacting Flows, *Prog. Energ. Combust. Sci.*, Vol. 11, pp. 119–192.

Pun, W. M. and Spalding, D. B. (1967). A Procedure for Predicting the Velocity and Temperature Distributions in Confined, Steady, Turbulent, Gaseous Diffusion Flames, *Proc. Int. Astronautical Federation Meeting, Belgrade.*

Reggio, M. and Camarero, R. (1986). Numerical Solution Procedure for Viscous Incompressible Flows, *Numer. Heat Transfer*, Vol. 10, pp. 131–146.

Rhie, C. M. and Chow, W. L. (1983). Numerical Study of the Turbulent Flow Past an Aerofoil with Trailing Edge Separation, *AIAA J.*, Vol. 21, No. 11, pp. 1525–1532.

Rodi, W. (1978). Turbulence Models and Their Applications in Hydraulics – A State of the Art Review, University of Karlsruhe SFB 80/T/127.

Rodi, W. and Hossain, M. S. (1976). Influence of Buoyancy on the Turbulent Intensities in Horizontal and Vertical Jets, in *Heat Transfer and Turbulent Buoyant Convection Studies*, D. B. Spalding and Afgan N. (eds.), Hemisphere Publishing Corporation, Washington, D.C.

Rodi, W., Majumdar, S. and Schonung, B. (1989). Finite Volume Methods for Two-dimensional Incompressible Flows with Complex Boundaries, *Comput. Methods Appl. Mech. Eng.*, Vol. 75, pp. 369–392.

Schlichting, H. (1979). *Boundary-layer theory*, 7th edn, McGraw-Hill, New York.

Shyy, W. and Vu, T. G. (1991). On the Adaptation of Velocity Variable and Grid Systems for Fluid Flow in Curvilinear Co-ordinates, *J. Comput. Phys.*, Vol. 92, pp. 82–105.

Shyy, W., Correa, S. M. and Braaten, M. E. (1988). Computation of Flow in a Gas Turbine Combustor, *Combust. Sci. Technol.*, Vol. 58, No. 1–3, pp. 97–117.

Spalding, D. B. (1971). Mixing and Chemical Reaction in Steady Confined Turbulent Flames, *Thirteenth Symposium (International) on Combustion*, The Combustion Institute, pp. 649–657.

Tamanini, F. (1975). Numerical Model for the Prediction of Buoyancy Controlled Turbulent Diffusion Flames, Factory Mutual Research, Norwood, MA, Report No. FMRC 22360-1.

Appendix A

Abbott, M. B. and Basco, D. R., (1989). Computational Fluid Dynamics – An Introduction for Engineers', Longman Scientific & Technical, Harlow, England.

Fletcher, C. A. J. (1991). Computational Techniques for Fluid Dynamics, Volumes I and II, Springer-Verlag, Berlin.

Appendix B

Patankar, S. V. (1980). Numerical Heat Transfer and Fluid Flow, Hemisphere Publishing Corporation, Taylor & Francis Group, New York.

Index